Small Mines Development in Precious Metals

Editors
Ta M. Li
Thomas M. Plouf

Published by
Society of Mining Engineers, Inc.
Littleton, Colorado • 1987

Printed in the United States of America
by Cushing-Malloy, Inc., Ann Arbor, Michigan

Library of Congress Catalog Card Number 87-62034
ISBN 0-87335-069-3

Organizing Committee

CHAIRMAN

Ta M. Li
Behre Dolbear - Riverside, Inc.
Denver, CO

CO-CHAIRMEN

Ernest L. Bohnet
Pincock, Allen & Holt, Inc.
Lakewood, CO

Ronald J. Roman
L-Bar Products, Inc.
Ravensdale, WA

COMMITTEE

Roshan B. Bhappu
Mountain States Research
& Development
Vail, AZ

Thomas M. Plouf
MinProc USA, Ltd.
Englewood, CO

Jerry D. Lewis
Boise Cascade Corp.
Boise, ID

George M.L. Robinson
EnecoTech Inc.
Denver, CO

W. Joseph Schlitt
Brown & Root USA, Inc.
Houston, TX

Preface

The Society of Mining Engineers recently changed its meetings policy to provide greater flexibility and member access. The Annual Meeting, now with the Technological Information Exchange Exhibit, will continue to be held early each year. Throughout the rest of the year there will be a number of specialty, topical, and regional meetings that will be thematic and focused on topics of current interest and held near appropriate membership centers.

"Small Mines Development in Precious Metals," being held in Reno, Nevada, on August 31-September 2, 1987, and of which this book is the Proceedings, is the first topical conference held under this new policy. The Society expresses its gratitude to all those involved in putting this successful meeting together and looks forward to many other regional/topical meetings in the future.

Bruce A. Kennedy
1987 President
Society of Mining Engineers

Foreword

Small Mines Development in Precious Metals, sponsored by the Society of Mining Engineers, has evolved into a forum with focus on the advancement of mineral development for small mine operations by both large and small mining firms. Meeting the economic and technical challenges for these operations has been characterized by phrases such as *minimal design, used equipment,* and *contract mining.* Spurred on by improved gold prices, today's small precious metal mines represent cost conscious and highly efficient operations that can economically recover gold from what was considered waste only five years ago.

SME's first topical meeting appropriately documents elements of the mining production cycle from exploration to metals marketing and how operators have met the challenge in successful mining of low grade gold ores. The technical papers appearing in this Proceedings have been presented to provide a first reference source for further discussions in mine planning, operations, processing optimization, and management/marketing of small mines and their products.

The organizing committee is grateful to the numerous authors, session chairmen, and program committeemen whose efforts and diligence have made the Proceedings a valuable reference tool for future small mine operators. Special thanks go to the SME staff, Darline Daley, Ruth Orologio, and Marianne Snedeker, whose commitment and endless patience have made this volume possible. The technical program at this first SME topical meeting will not only serve as an excellent beginning point for addressing the problems facing small mine development, but will also serve as a high standard on which future meetings can be based.

Ta M. Li
Thomas M. Plouf

Contents

1

Introduction: Problems Facing Small Mines

T. M. Li
Behre Dolbear - Riverside, Inc.
Denver, CO

Chapter 1

THE CASE FOR SMALL MINES DEVELOPMENT
PROBLEMS AND PROSPECTS

ALFRED G. HOYL

PRESIDENT, SILVER VENTURES CORPORATION AND COAL FUELS CORPORATION

Abstract. The small miner and small mine have an important role to play in the finding, developing and production of our mineral resources.

"Help, not hinder" must be our byword. We must push in every way we can devise to change the adversary position of government and environmental movements.

We need the small mines and they must be encouraged, not hindered.

Technically excellent and adequate equipment is available, and needed information is also available to the small operation.

Environmental questions and problems must be fully considered regardless of the size of a mining operation.

Text.

As the mineral industry developed from flint flaking through copper, bronze, iron, and now the full industrial and beneficial use of all the elements, i.e. throughout most of historical past, it would have been considered small mining.

Indeed, most mineral deposits are discovered and started by independent individuals who may or may not continue to hold an interest in the deposit as it develops into a large mine.

We would hope, however, that the independent discoverer would be well compensated for his efforts. Many times, unfortunately, this does not happen although hope springs eternal.

Just what is a small mine? Mineral deposits vary from billions of tons of reserves to a few thousand tons and mining operations from thousands of employees to a single person.

The small mine fits somewhere in the lower half of the scale. An individual or small group with limited capital may find and initiate a mining project. A few will grow and develop into substantial companies. Most will be purchased or absorbed into larger companies. The small miner will have served his purpose, however, and can be justly proud of his role.

In addition to discovery and initial development of a mining project which may be small and stay small, no miner need feel inferior as he can always look forward to growing and expanding as far as his imagination carries him.

Small mines often develop into large ones but those that stay small must be operated as small mines. Their production of mineral products is nonetheless as valuable as their big brothers on a unit basis.

The small and independent mining people out there provide a continuing pool of prospects and projects for the larger companies to investigate, as well as a significant production of needed minerals.

In short, the small miner is essential and his future is secure, however, it has become much more difficult for him to proceed and operate.

For the good of the world, America, and the mineral industry, we must have independent people - "small miners" - out there anticipating needs, finding, and developing our minerals. Every person who goes into the mineral industry needs to know that he is a needed and essential element in the maintenance of civilization, and the strength of his country. He must also know that he can start his own effort or business and succeed or be a part of a larger company or endeavor.

Assuming that the small miner is essential, how will he operate under present conditions?

Ta Li and his committee have put together an excellent program here which will address this question.

Until recently there were few restrictions on the independent miner. Today we have many new restrictions and problems. They have grown out of the environmental movement, world economic conditions, the ignorance of the public as to mineral affairs, and the failure of the insurance industry to insure, (if you don't need insurance, it is readily available - if you do need it, you can't get it).

Hinder, not help is the byword. Our governments are adversaries for the most part. Adversaries almost ruthless in actions are the environmentalists, the Environmental Protection Agency, the Mine Safety and Health Administration, Mined Land Reclamation bureaus, land developers, and local citizens groups.

A bright shaft of light among government agencies is the U.S. Bureau of Mines and the U.S. Geological Survey. They continue to be very helpful. Perhaps with consultations with small miners groups, they could become even more helpful.

Permitting is time consuming and needlessly expensive. Bonding is almost impossible. MSHA fines and fines for inconsequential details, hands out a sheaf of violations and wishes you, "Have a nice day".

On the other side, as big reserves are used up and land disappears into the insatiable maw of wilderness with buffer zones protecting buffer zones protecting wilderness areas, the small mine is more in demand and becoming more essential.

Excellent equipment is available and knowledge is in plentiful supply if one seeks it out. No one can beat a determined, independent entrepreneur.

I have often thought that a small operation is harder to manage than a large one since the large operation can command many specialists such as mining engineers, metallurgists, PR men, geologists, business managers, marketing specialists, environmental managers, etc. The small operator has to do all of these things himself.

The answer to this could be consultants. Consultants are expensive, however. As a thought, consultants, labs, research facilities, etc., could cut their fees drastically to the small operation and take bonds, debentures or promissory notes to make up the difference, these to be paid back from production only. Surely a way could be found to accomplish this without compromising the objectivity of the consultant.

As to permitting, through our mining organizations, we must develop computer programs which can turn out a permit application almost overnight and which will be accepted with only essential modifications by the state reclamation divisions, these modifications to be provided by the state reclamation divisions anddone in the "help, not hinder" mode.

Again, we must, through our mining organizations, develop self insurance and self bonding.

I hesitate to mention this, but smelters must provide very reasonable rates and deductions and must pay promptly upon weighing and sampling, or somehow we must develop end product capability to serve a small group or single mining operation.

Most refineries do give prompt reasonable service. Smelters could do much better than they do.

There is much to do to convert governments to the "help, not hinder" outlook instead of their present adversary position. It will be a difficult task, but must be done.

Looking to the future, the small miner must look to a larger future and the big miner must look more and more to modest but profitable mines. This change in policy will add additional technical and financial input and will be of mutual benefit to all sizes of mining projects.

The small miner basically must contend with the same market factors as the big operator. It is therefore as essential to him as to the big operator to develop as much diversification as possible. For instance, a small mining company could have one or more precious metal operations, a coal property, a sand and gravel operation, or a crushed rock operation.

Financing will always be difficult for the small mine. Since timing is such an important factor and so hard to control as to prices and when profitable production will commence, bank loans can be very risky. Banks are not patient. Equity financing is the best route to follow.

Let us not be faint of heart but push on, not only to our personal benefit, but to the benefit of our country, our upcoming youth, and to the world we live in.

As Dr. Charles F. Park, Jr. has so wisely said, "We are EARTHBOUND". Let us all do a thoughtful and excellent job, produce the minerals we need, operate at a profit, and protect and restore the environment as we proceed.

Chapter 2

WHAT FINANCIAL INSTITUTIONS LOOK FOR IN SMALL MINES DEVELOPMENT

by

STANLEY DEMPSEY

CHAIRMAN, DENVER MINING FINANCE COMPANY

ABSTRACT

Mine developers commonly undertake projects that require more funds than they can invest themselves. From that point until they are able to complete a "bankable" feasibility study, they require access to venture capital.

In the United States, most venture capital has been provided to the small mining segment of the industry through tax advantaged limited partnerships. Recent changes in Federal tax law makes other forms of financing competitive, and the use of corporations and private and public offerings of shares may soon become more popular ways of securing venture capital.

A number of new financial services firms have entered the mining finance business in recent years. They provide mine developers with access to venture capital.

Financial institutions, including the specialized mining finance firms, expect to receive from a mine developer seeking funds information about the mining project, including land ownership, ore reserves, mining methods, metallurgy, and marketing of the products of the mine. They also want to see projections of revenues, costs, and profitability, how funds invested will be used, and how an investor will "exit," or get his money back out of the project. They are particularly interested in the background, experience, and general reputation of management personnel.

Mine developers are most effective in making proposals to financial institutions when they provide the right kinds of information in a format that is easy for the institution to use. This paper describes the kind of data that is required, and includes examples and checklists that may assist developers who are preparing proposals.

Chapter 3

HOW BEST TO USE CONSULTING FIRMS TO ACCELERATE THE DEVELOPMENT PROCESS

Hans W. Schreiber
President

Rudi P. Fronk
Vice President

BEHRE DOLBEAR RIVERSIDE, INC.
NEW YORK DENVER

INTRODUCTION

Widespread activities in precious metal projects in the western U.S., Australia, Canada and other parts of the world have opened a flood-gate of consulting expertise to aid and accelerate the development process.

Utilization of outside expertise has taken on many shapes and forms that have in some instances actually "decelerated" or "retarded" project progress. Small mines are especially vulnerable to apparent false-starts and misdirections, because of their generally lower capital commitments, shorter lead times and limited mineral in-house expertise. Mine development for small mines requires fundamental approaches by knowledgeable consultants that are timely and cost-efficient and in line with accelerated development, if they are to be successful.

The development process is obviously that process which brings a given deposit into production. Or, put another way, the process comprises all activities, decisions and commitments which take place from the moment an authorizing entity (usually a board of directors) commits to development and production (usually on the basis of a final feasibility study) until the first ton of ore is processed to a saleable product. The activities, decisions and commitments over a development period--such a period can range from as short as a year or less to as long as 5 years or more--are many, highly varied, and often interdependent. The development period is also particularly risky, not only because of the varied and interdependent decisions and commitments, but also because of capital demand amounts. Delays in the development process are thus particularly costly - sometimes so costly that small and often undercapitalized mining companies, failing to reach production and a cash flow within a planned period, face, if not bankruptcy, at least severe equity dilution or particularly onerous and restrictive covenants on additional loan financing. It therefor behooves especially small mining companies to examine ways to accelerate development. By doing so, the development process may actually meet planned scheduling.

The development process generally comprises environmental impact studies completion, final engineering, permitting, financing, equipment and machinery procurement, contractor selection and construction -- such activities largely being overlapping. The paper addresses what the small mines operator or the small mining company can do with consulting firms to accelerate development. By "small mines" is meant open pit operations of 2000 tpd or less and underground operations of 500 tpd or less. By "consulting firms" is meant entities who do not possess an engineering and construction capability, such as Derry Michener Booth & Wahl, Pincock Allen & Holt, Behre Dolbear-Riverside, etc. The paper concentrates on gold projects.

PRIORITY CONSIDERATIONS

Of the items listed in the previous paragraph which comprise the development process, the two which most consistently appear to cause delays are permitting and financing. The former's delays can be due to political overtones and thus the delays may not be reasonably foreseeable. The latter's delays are most often due to inadequate preparation, and thus are avoidable. For the reason stated above, and considering that financing is fundamental to all of the activities which constitute the development process, the paper focuses on financing as the priority item which can most assist in accelerating or retarding the development process.

FINANCING TERMS

Equity financing, while clearly still very strong - witness activity on the Vancouver and Toronto stock exchanges - may lessen in relation to loan financing and/or inter-related loan and equity financing. This trend may well strengthen as owners become more chary regarding dilution and as banks become more flexible, not only in terms of internal policies, but in broadened areas of activity as permitted by recent regulatory changes. Clearly, banks are making inroads on the investment banking houses. Interrelated loan-equity financing mechanisms may increasingly involve loan principal with warrants or convertible debentures or loans "oiled" by accompanying equity options.

Whether financing is of a straight loan nature or an interrelated loan-equity nature, lenders will look to reserves for security, especially since small mining companies generally do not have balance sheets strong enough to collateralize loans. Reliability of reserve estimates remains - in conjunction with direct cash operating cost estimates - the bulwark of financing. And it is with the use of consultants that such reliability can likely best be established.

ORE RESERVE ESTIMATE RELIABILITY

Delays in financing are very often due to inadequacies in the work undertaken to support ore reserve estimates. The delays are usually significant, because either or both repeated work or additional work are needed. Conversely, financing can be significantly accelerated (thus accelerating the development process) by the undertaking of work appropriate to establishing "comfort" in the reserve estimate and by the adequate presentation of such work in a feasibility report. While this seems obvious, the fact is that practice and experience show that many entities--often small mining companies--do not initially pass the lending entities' "comfort" requirements or guidelines, and thus must seek other lending entities or go back "to the drawing board."

Small mines equity owners and operating staff generally have not had the exposure to either the variety of deposits or the various types of lending institutions, as is the case with large mining companies. Furthermore, during the ore body delineation and feasibility determination phases, work undertaken is often heavily influenced and limited by the availability of funds, and not vice-versa -- that is, the availability of funds should theoretically be that amount which will support the work necessary to meet the likely comfort levels of those institutions who are seen as potential lenders. Thus, the money seen as "saved" during the pre-development phases (granted that the cash may not have been available) is lost and more by the delays in the ensuing development phase. It is the consulting firms, because of their exposure to a variety of ore deposit types and experience with various lending institutions, who can best give the guidelines, commensurate guidance, and eventual quality control to attain the somewhat varying comfort levels of various lending institutions. Consulting firms are best used in the specific areas of:

1. Reserve accounting and/or auditing;
2. Setting the specifications under which bulk ore sample(s) are collected and tested for processing characteristics;

The use of consulting firms in the capacities indicated above has an additional benefit--the inherent eagerness of management to bring deposits into production with as much rapidity as the capabilities of their own staff will allow, engenders propensities towards shortcuts and mistakes. Consulting firms tend to serve as a counter to such propensities.

RESERVE ACCOUNTING AND AUDITING

It is not the scope of the paper to set forth the steps inherent in establishing a reserve or in determining the adequacy of the standards of parameters which support the reserve estimate. It is the scope of the paper, however, to suggest how consulting firms can best be utilized in undertaking the tasks inherent with establishing reserves and the standard of supporting parameters.

The determination at what spacing a drilling pattern or grid has attained near-optimal standards of reliability as regards continuity of mineralization is fundamental to reserve accounting. Such a determination is well within the capabilities of consulting firms, which have the benefit of experiencing the adequacies of various patterns and grids for deposits, ranging from massive sulphide to hot spring, and of applied calculation methodology, ranging from simple polygonal to advanced kriging. Clearly, the consulting firm must take into account cost, time and achieved reliability standards. But in so doing, the consulting firm will be less influenced than the small mines management by the constraints of cost and time as an impingement on attainment of a standard of reliability. Furthermore, the consulting firm is a more objective judge of the reliability standard sought by a lending (financing) entity.

The design of an integrated and complete "check assay" program appropriate to the type of deposit and initial indication of grade variability is likewise fundamental to reserve accounting. An integrated check assay program includes (1) procedures to permit estimation of drill product recovery; (2) determination to what size all recovered drill product is ground before splitting; (3) establishing the splitting sequence and a check on comparability of splits; and (4) determining reproducibility, thus laboratory bias in fire and cyanide soluble assay work. The design of such a program falls well into the capabilities of consulting firms, which frequently audit the reserves of operating mines (usually as part of "due diligence" for additional equity to be offered to the public). The auditing of the reserves of operating mines permits comparing estimated grades with experienced grades, and in turn permits assessing the adequacy or shortfalls of the check assay program (if any) instituted and applied at the time the operating mine was still in a pre-development phase.

An appropriately-designed check assay program is critical to a reliable determination of "down-side" grade risk. The quantification of such risk is usually necessary to achieve an adequate comfort level with many lending entities. Small mines management and key personnel often are not in a position to think through and apply an appropriately designed check assay program. This is not only because of lack of exposure, but also because of conceptions regarding the occurrence of gold in the company's particular deposit which are set early-on and founded on an inadequate data base. Management simply becomes too subjective to accept that its conceptions are prematurely formed and that a more complex and costly program might well be worthwhile in terms of setting a higher standard of reliability.

REPRESENTATIVITY OF PROCESS TESTING

Bulk sampling concurrent with subsequent testing is, with the possible exception of delineation drilling, likely the singlemost expensive step in the series of steps which comprise determination of feasibility. Delineation drilling (if successful) is exciting, and accompanying announcements of results support and promote the company's equity. Process testing by bulk sampling in turn is dull, since results are (subjectively) expected or anticipated to mirror bench scale testing results. Process bulk sample testing is also onerous, not only because of the cost per se, but because the effort often competes with funds for more glamorous activities such as delineation and/or deposit extension (on strike or to depth) drilling. Thus there is a tendency to "shortcut" or become careless about the various considerations which comprise a bulk-sample-process testing program. Examples of projects in the western U.S. are numerous as regards actual recovery and/or experienced recovery rates which were less at full-scale operations than was the case for the testing program. Almost without exception, the reason(s) became clear upon objective reinvestigation. The reasons also showed that the less-than-indicated design performance was also avoidable, and this without the subjectivity which accompanies the benefit of hindsight.

Consulting firms are often used either by the entity experiencing the less-than-expected results or by an entity considering acquisition to "case the problem" and set specifications for a new testing program. Thus consulting firms gain significant exposure not only to original errors but to the solution for correcting such errors. Clearly, such exposure and experience permit consulting firms to judge (and design into the testing program accordingly) the following critical considerations which constitute a bulk sample-process testing program which should yield an adequate standard of reliability:

1. Timing of the collection of the sample in relation to the status of delineation drilling;
2. Make-up of the bulk sample(s) in relation to the status of definition of different ore types and the status of indicated relative abundance of each ore type;
3. The comparability of the sample's size classification with the likely size classification of material resulting from planned blasthole density at full production;
4. Comparability of temperature and humidity over the contemplated period of testing with average annual temperatures and humidity (applicable only in the case of heap leaching).

ECONOMIC FEASIBILITY STUDY REPORT CONTENT

Jerry M. Whiting and Landy A. Stinnett, at the Northwest Mining Association 1984 annual meeting, delivered a paper entitled "Preparation and Use of Venture Feasibility Reports To Help Mine Financing". The paper describes three types of feasibility reports, namely:

1. Technical;
2. Cost estimates;
3. Venture.

The authors indicated that the venture type is relatively unusual, but is an "exceedingly important type of feasibility report." This type of report is indeed becoming increasingly more important. The venture type report interrelates technical and economic findings with the probability of the venture (gold deposit) developing into a "new business" and showing that the economic performance (profitability) will be as projected. The overall probability of the venture meeting profitability projections is based on the individual probabilities of meeting capital cost estimates, capital cost period-of-infusion (development process) estimates, cash operating cost estimates and revenue projections. Probabilities are directly related to a standard of reliability engendered by the work which supports or develops the parameter(s) used for the estimate(s). In essence, the venture type of feasibility report is a risk analysis. It provides the potential lending entity with the means to assess to what extent he can risk decreases in price levels. Put another way, a commercial bank lender may prefer to consider a situation of relatively low operating margins at a given gold price, but a high probability of meeting direct cash operating cost estimates. Alternatively, an investment bank lender may prefer to consider a situation of relatively high operating margins at the same given gold price, but with a low probability of meeting direct cash operating cost estimates.

Risk analysis is a common activity of commercial bank and investment banking houses. Small mines managements and key technical personnel are not commonly involved in risk analysis and, in any event, would have difficulty in objective risk quantification, since the project is "their baby". Consulting firms tend to be exposed to risk analysis through their retention by lending entities to conduct "audits" of reserves, capital and operating cost estimates. Thus, consulting firms should be used to guide the manner of presenting risk analysis in the venture type report, and to provide judgment as to the relative importance of various types of risk to the potential lending entity, whether a commercial bank, an investment banking house, an insurance company, a venture capital syndicate, etc.

CONCLUSION

Consulting firms are usually not more expert than the small mining company in the specific geologic, mining and processing characteristics which constitute the particular deposit. But consulting firms, in contrast to the small mines company management and its key technical personnel, by virtue of their exposure to a greater variety of deposits and their association with a larger number of attempts to gain financing, are more objective interpreters of when the particular deposit has attained an adequate level of reliability in the parameters which most greatly affect the deposit's projected economic performance. Consulting firms, then, are best used to

1. set the guidelines to define a scope of work;
2. assess when the project has met adequate reliability standards; and
3. formulate the presentation of work results.

By using consulting firms in the above-described manner during pre-development phases, the development process is almost certain to be accelerated. Acceleration of the development process will yield highly attractive returns on funds expended for during pre-development which, without the use of consulting firms, would otherwise not have been realized.

Chapter 4

SMALL MINING - A PERSPECTIVE FOR PROFITABILITY

by

Mr. David J. Starbuck

Starbuck Minerals Management
President
Mayer, Arizona

INTRODUCTION

For small precious metals mining to show a comeback a different perspective must be taken than that of larger mining companies. This perspective must be much the same as that of past successful smaller mining enterprises such as when the US had a need for Uranium. Abandoned mines and environmental concerns in older mining districts have discouraged many from entering with the attitudes of why try there if so many unsuccessful mines exist. A small mining venture can be successful there but must have a "think smart" attitude. There must be a thorough understanding of deposits, utilization of advanced technology and a properly managed mining operation.

"THINK SMART" MINING

The small mine operator is usually looking for small high grade deposits, consequently, long term success depends on a series of small successful mining operations. The small miner must succeed right-off or credibility and future project funding will suffer. No veteran of even a few years in the mining industry needs to be told that a dollar spent on production today fails to yield as much as it once did. More efficient equipment and techniques have helped soften the problem, but there is little likelihood that a mining company can do much to control inflation, wages, regulations, or grade. That leaves the small miner to "think smart" as one area where operating dollars can still be stretched to produce more than they presently do (Gardner, 1981).

In the planning stage of a mining venture, when most options are still open, the main controllable elements are market timing, deposit selection, cutoff grade, mining method, mine design, mining sequence, rate of production build up, process design, project financing structure, project management, and project ownership structure. During planning, optimum criteria can be established for each of these. Once a mine is in production, these options are foreclosed or at least severely constrained. The more thought given to a mining program using all information available the less the risk of encountering disabling surprises.

Thus the intent of this paper is to suggest the "think smart" attitude most small mine operations use for long term success. If one looks at a number of successful mining ventures, whether large or small, there is a general logic most have followed to be successful. They have used proven exploration, mining and processing techniques along with economic evaluation and experienced personnel to improve their odds.

LOGIC IN DEVELOPING A FEASIBLE VENTURE

Most successful small mine ventures achieve positive cash flow as quickly as possible by seeking out those deposits which require the least time and expense on activities leading up to production. To do this they look for deposits that are the most accessible, easily investigated, mined and processed with the cheapest fundamental methods. The general sequence of events starting with district selection are deposit selection, preliminary feasibility, exploration, detailed economic feasibility, mine planning, development and production. Each successive step becomes more costly requiring intermediate decisions along the way.

District Selection

Small operations should look for known deposits which favor the occurrence of the type deposit sought rather than efforts in unproductive areas. Known deposits in developed areas have some advantages over remote areas:

1. More alternative methods of transportation are available.
2. A larger stable work force is available.
3. The area is usually more pro-mining.
4. Power, water, housing, technical services, skilled services, schools, recreation and shopping are available.

Some disadvantages of older districts are many have become historic or recreational and local preservation or environmental groups may protest mining. These problems and the fact that there are so many abandoned mines in many districts has not discouraged a number of successful small operations. It requires establishing good rapport in the area and developing a good feasible program to overcome investor doubts.

Many older gold and silver districts resulted from the discoveries of individual prospectors or syndicates. These districts typically developed into a multitude of small to medium sized mines. Many of these small mines were profitable till they encountered water, excessive haulage, complex geology or had low metal recovery and possibly high concentrating costs. Many had decent ore grades but were closed down for these reasons and mismanagement, falling market, war or governmental restrictions. Careful scrutiny, mapping and sampling may disclose the very reason for mine closure or disclose potential the "old timers" were not aware of.

Other opportunities are smaller higher grade orebodies resulting from a larger companies regional exploration which did not fit their larger tonnage programs. Previous operations with secondary base and precious metals production may still have precious metals reserves that may now be economic. Exploring a small deposit sometimes results in a larger one possibly resulting in the adage:"most larger mines are a result of a smaller mine indicating potential for a larger deposit and thus larger mine."

Deposit Selection

Many deposits that were previously uneconomic might now have potential because of changed market conditions, improved technology, or more productive mobile equipment. Advances in milling, such as higher recoveries, compact mobile plants and leaching methods have given new life to some previously uneconomic deposits. Opportunities exist for a variety of ventures such as reprocessing mill tails, heap and in-situ leaching of dumps, mine backfill and unworked deposits, reworking reconcentrated placers, working fine gold placers or continuing where uneconomic operations left off.

Venture funds and resources are more apt to result in timely development of a deposit if the best looking deposits are pursued concurrently rather than sequentially. This may require taking several deposits all the way into preliminary exploration, before making a go-no go decision on the best one. Small mining firms that have little experience in precious metals should be very cautious about being too optimistic. Most successful small operators tend to be slightly conservative. Usually the first to investigate and follow through on prospects is a geologist. The individual should have the background appropriate for the business at hand, preferably a precious metals mine geologist with a background in the type deposit and applicable mining methods contemplated.

Sufficient information should be gathered to make a quick preliminary feasible and economic decision among alternative prospects. Table 1 indicates the type of information required and Tables 2 and 3 indicate a method of utilizing the information on a ranked basis to select the top prospects for further investigation. They can be roughly ranked on the mineral value and deposit location. Ranking is based on the more complex the ore and the more difficult the access to the deposit the more costly a project will be.

MINERALOGY		DEPOSIT
METALS	MINERALS	SURFACE
One metal	Native metal	Previously mined material
More than one	Straight oxides	In place ore
	Straight sulphides	Underground
	Oxides & sulphides	Horizontally accessible
		Vertically accessible

Preliminary Feasibility

Many small deposits mined at 100 to 200 tpd may only last for one to two years but if they are accessable, of sufficient grade and do not required much additional exploration or development, they can be seriously considered. Prior to undertaking exploration, a preliminary feasibility should be conducted to determine at what minimum grade, minable thickness and tonnage the project will at least break even. Subsequentially, exploration can be guided with these limitations in mind and a decision can be made in the early stage of exploration whether the deposit does or does not meet this minimum criteria.

Exploration and Economic Feasibility

Exploration typically involves geophysics, channel sampling, trenching, pits, drilling and mapping to achieve a level of confidence in a deposit. The average grade, average thickness, tonnage, depth, structure, mineralogy and ground conditions should be determined. Average grade and thickness are the average of those portions of the ore body which are above break even. A composite bulk sample, representative of the deposit, should be taken for beneficiation testing to determine the method and parameters of metal recovery. Once the desired level of confidence is reached further exploration should be discontinued and a detailed economic feasibility study should be prepared.

Economic feasibility involves formulating a detailed schedule of development and production expenditures, then a cash flow projection. From

TABLE 1. PRELIMINARY PROPERTY EXAMINATION

Prospect Name:
Town: Mine Dist:
Sec T R County
Long. ° " ' Latit. ° " '
State (Prov.) Country
No. Fee Claims Acreage
No. Located Claims Acreage
Map Name
Submit. By

Phone( )
Owner

Phone( )
Terms

Assessment or Tax Status

Claim Confl., Suits, Litig.

Sale
J.V.
Lease
Altitude:

Surface
Underground
Dump
Tailing
Ore Depth
Length
Width
Thickness

File No.:
Date:
Mountains
Plains
Placer
Lode
Vein
Dissem.
Strat fm.
Other

Mineral		Grade	Tonnage
	Mined		
	Reserve		
	Mined		
	Reserve		
	Mined		
	Reserve		

Current Lease(s) Royalties
Previous Expl.
Previous Prod. Years Ton Grade/$/Ton
Shutdowns & Reasons
Reports Available: Geologic, engineering, drillholes, assays, mine maps, reserve estimates, metallurgical, production records, geochem, geophy, photos, cost estimates, feasibility study (by:)
Other
Profit Financial Standing
No. Management Employees Quality
No. Hourly Employees Quality
Distance to Boat Rail Comm. Air Town/City
How to get to property:
Dist. to Comm. Power: Capacity Company
Climate: Winter(mild-moderate-severe)min °F/C Summer max °F/C Precip "/yr
Water Rights Acre/ft. Date of Priority
Prev. Develop. method, length/depth & cond.
Potential Develop. method, length/depth
Regional Geology

Prospect Geology

Mineralization-Alteration

Geochem/Geophy Results

Ore Reserves: Geologic Mineable
Property/District Potential
Type of Sampling Type of Assaying
Prev. Mine Method, Width/Height, Stope Cond. & TPD

Curr./Pot. Mine Method, Width/Height, TPD & Equip. Cond.

Prev. Mill Method, TPD & % Recov.

Curr./Pot. Mill Method, TPD & % Recov.

Tails Disp:Prev Curr/Pot.
Salable/Recoverable Co-Products or Real Estate

Smelter or Sales Outlet
Labor Market Local Attitudes
Ecology/Reclamation

Attachments/Outside Refer.

TABLE 2. SMALL DEPOSIT - PROSPECT RANKING CRITERIA

QTY	METALS	QTY	MINERALOGY MINERALS	QTY	GRADE ($/TON) Undgnd	GRADE ($/TON) Surface
10	Gold	10	Free metal	25	+$120	+$10
9	Silver	9	Straight oxides	20	90 - 120	7.50 - 10
8	Gold & Silver	8	Straight sulphides	15	60 - 90	5 - 7.50
7	Gold, Silver & Lead	7	Oxides & sulphides	10	30 - 60	2.50 - 5
6	Gold, Silver, Lead & Zinc			5	0 - 30	0 - 2.50

Metals					
$/Unit					

SURFACE DEPOSIT

QTY	RESERVE ACCESSABILITY MINED	UNMINED	QTY	TYPE RESERVE MINED	UNMINED	QTY	TYPE EXPOSURE/ SAMPLING MINED	UNMINED	QTY	OREBODY SIZE & CONFIGURATION MINED	UNMINED
10	Concentrates	--	9	High grade throughout	--	9	Trench &/ or drilled	--	10	+100,000T	+ 20' wide/thick
9	Tailings Placer	-- Placer	8	Medium grade	High grade throughout	8	Surface sampled	--	9	50to100,000	10-20' wide/thick
8	Dumps	--	7	Low grade	Medium grade	7	Unsampled	Surface sampled, trench/drill	8	25to50,000	5-10' wide/thick
7	--	Cut/Pit	6	--	Low grade dissem.	6	--	Trench/drill	7	10to25,000	< 5' wide/thick
6	--	Strip				5	--	Surface sampled	6	< 10,000T	--
						4	--	Unsampled			

UNDERGROUND DEPOSIT

QTY	OPEN	WATER FILLED	CAVED	QTY	MINED	UNMINED	QTY	MINED	UNMINED	QTY	MINED, FINES & FILL	UNMINED
7	Adit	--	--	8	Fines	--	8	Fines sampled	--	8	+100,000T	+ 20' wide/thick
6	Decline	--	--	7	Fill	Continuous lode/vein	7	Fill sampled	2 sides incl lowest side	7	50to100,000	10-20' wide/thick
5	Shaft	Decline	--	6	--	Contiguous small multi-veins	6	Unsampled	Across bottom	6	25to50,000	4-6' vert to 45° incl
4	--	Shaft	Adit	5	--	Discontinuous lode/vein	5	--	One side	5	10to25,000	<6' flat to 18% incl
3	--	--	Decline				4	--	Across top or entry to bottom	4	< 10,000T	--
2	--	--	Shaft				3	--	Entry to Top			
1	--	--	Winze				2	--	Drilled only			

TABLE 3. DEPOSIT - PROSPECT RANKINGS

STATE: COUNTY:

			* GRADE		SURFACE RESERVE ACCESS		TYPE RESERVE		TYPE EXPOSURE		OREBODY		UNDERGROUND RESERVE ACCESSABILITY			TYPE RESERVE		TYPE EXPOSURE		OREBODY	
NO RATING	METALS	MINER.	UND	SUR	MINED	UNMND	MINED	UNMND	MINED	UNMND	MINED	UNMND	OPEN	WATER FILLED	CAVED	MINED	UNMND	MINED	UNMND	MINED	UNMND

* A ( ) indicates the rank is based on old $/ton values and thus is one rank lower than if based on current value.

these projections the operator then has some confidence in knowing when positive cash flow will occur and what the project rate of return will be. Many small mine ventures fail because they do not take this one critical step. They do an injustice to themselves and their funding source. There is one other reason this step is so critical. Without a cash flow projection there is no guidance for developing and fine tuning mine planning or scheduling.

Mine Planning, Development and Operation

If the economic feasibility has had enough thought put into it, a detailed mine plan can be engineered around that study. Pre-production development and operation are but an implementation of that detailed plan. If additional information is found during development or initial production which alters information used in the feasibility study the economics should be immediately reassessed. To not do so is courting failure.

Management and Operating Philosophy Under the deposit selection section of this paper it was mentioned that most successful small operators are slightly conservative. The same also holds true through every other phase leading up to and including production. This attitude is best maintained by a well trained competent staff with key individuals who have a hands-on track record relating to the business at hand. Mining probably requires a wider variety of disciplines than any other natural resource industry. Since the small operation does not need many disciplines full time, like larger mines, it is beneficial to retain those individuals with a special talent only on an as needed basis. Those typically considered are geologists, surveyors, safety specialists, training specialists, mining engineers, metallurgists, mine contractors, mineral lawyers, mine accountants, hydrologists, electrical, mechanical and civil engineers. With recent tight economic conditions those still in business are generally competitive, motivated and well equipped to put the small operator on the right track. They provide the small company with the facilities, materials and technology that have been available to the major companies, (Bruce, 1972). A number of marginal small operators called the do-it-yourselfers have the attitude that this type of talent is too expensive and not necessary. Just the opposite is true. A well planned venture can earn back the cost of good advice many times over, but a misguided venture may never succeed at all. (Cope, 1981).

Small mining operations are well advised to adopt the operating practices used by mine contractors. They generally hire competent aggressive people, are quick to set up on a job, do completions in record time, and quick out with little wasted time. They employ very productive mobile equipment, easily erected and dismantled structures, and maintain close control on costs and schedules.

Optimum Mine-Mill Combinations In many older districts, such as Leadville and Cripple Creek, Colorado, the small mines operated while there was a central mill. Once it closed down, transportation to a more distant mill or smelter became to costly. Market conditions have made some of these districts attractive again, but without nearby milling facilities and with most of the larger smelters closed down the small mine operators have had to look at other alternatives. Generally a small mine can not justify its own mill but a combination of small mines in an area may justify a central mill.

MULTI SOURCE-CENTRAL PROCESS FACILITY: Several companies in the US have tried to establish central toll milling in older districts but have been unsuccessful primarily because small operations, on their own, can not be relied on to supply consistent feed. The only way most successful central facilities have assured consistent feed is to have control of the feed sources. Examples are the approach taken by uranium companies during the 1950's to 1970's on the Colorado Plateau. Companies such as Climax Uranium, Energy Fuels and Union Carbide acquired large claim blocks containing numerous small uranium deposits. They then subcontracted or sublet these small deposits to as many as 150 individual small operators. Some of these small operations produced as little as 50 ton. The ore was hauled on contract to central mills which were individually rated at from 350 to 1250 stpd. In some cases the mills were dependent on as much as 80% of the feed coming from small operators and the balance from company mines (Anon., 1966; White, 1978). Aside from controlling the feed sources the companies provided technical and some material-equipment assistance on a deductible basis. This setup assured there was consistent quality feed and allowed the small operators to profit from their own efforts. The multi small mine - central mill concept is presently more feasible in underdeveloped Asian and South American countries where the mining, socio and political climate is more attuned to the concept.

Another opportunity for the small miner is to seek out small high grade deposits within or near a currently active larger mining companies operations. The time and cost of discovery and exploration might be substantially reduced because current information may be readily available. The larger company may sublet or contract these deposits and mill the ore in their facility. Complex ores could also be considered because most large mills are not sensitive to small lots of complex ore mixed with their daily feed.

ON SITE SMALL PROCESS FACILITY: Locating a small concentrator near the deposit is only justified if the economics are better than transporting to and processing in another facility. Several factors make a small facility cost prohibitive in comparison to a larger one:

1. Per ton operating costs are higher.
2. Recovery is usually lower.

3. Milling complex ores is more costly.
4. Capital cost per ton milled is higher.
5. Require more permitting in comparison to size.
6. Requires substantial quantities of water and power.
7. Difficulties in finding a suitable tailings site near the mine and mill.

Before totally ruling out a mill on a small deposit there are other factors that a prudent small operator should consider. Sometimes a good balance can be achieved by transporting upgraded material to an established facility thus reducing transportation costs. Some ores can be upgraded by screening, ore sorting equipment, or rough gravity concentration. Other considerations are in-situ and heap leaching, column flotation, or mobile "caravan mills".

Smelting and Refining with fewer smelters in operation it has become more difficult for the small operation to sell their concentrates. Recent improvements in small scale precious metals leaching, smelting and refining have given the small operator the opportunity to produce a more saleable finished product. Also, some smaller precious metals refineries will take small lots of high grade concentrate and ore.

Marketing It is common irony that companies are often spending a great deal of time and money to save a few cents per ton on operating costs while at the same time overlooking equal or greater savings on marketing opportunities. It is quite prudent for the small operator to acquire as much market information and expertise as he can. One way to aid in doing this is to develop a formal or informal association with a group that can help him deal with some or all of the marketing problems a small operator does not have the time nor talent to address. Ironically, the merchant knows the market place and may often be able to get a better price from smelters or buyers than the small operator himself can get (Anon., 1970; Chender, 1983).

Equipment Leasing Capital equipment expenditures early in a project can be a heavy burden on financing sources. Equipment leasing is an attractive alternative to the small operator who needs flexibility and as little cash outlay as possible. Since new equipment sales are down many dealers are looking at other alternatives. Some are offering attractive lease deals. Equipment from different manufacturers can also be combined into one lease package through a leasing company. As long as the items considered are semi- permanent or mobile they will likely fit into a lease deal. Many advantages of leasing are:

1. Consistent use of newer equipment without tendency to keep equipment till it is unserviceable.
2. If mining or milling conditions change, equipment can be traded out for needed equipment rather than being an additional burden.
3. Problems of availability, maintenance and replacement are left to an agent or supplier.
4. Spare parts inventories are substantially reduced.
5. Package consolidations are more desirable than multi loan purchase agreements.
6. Non financable costs, such as freight and installation can be built into the lease (Anon., 1970).

Environmental and Regulatory One advantage a small operation has over a larger one is less visibility which does not cause as great an environmental concern. Determining permit and compliance requirements early on in a venture should still be top priority. Many small operations have been delayed months and some for years because they did not allow enough lead time for approval. Principal regulatory agencies to be concerned with in the US are: US Forest Service; Bureau of Land Management; Alcohol, Tobacco and Firearms (explosives); Nuclear Regulatory Commission (base metals radioactive emissions); Environmental Protection Agency; Mine Safety and Health Administration; Occupational Safety and Health Administration; Army Corp. of Engineers; and State Departments of Land Use, Fish, Game, Timber, and Water Resources. Local organizations to be concerned with are conservation groups, historical societies and county planning and zoning commissions.

Cooperatives and Associations The overhead cost and time to handle some administrative needs can be quite a burden to a small operator. If there are several small mining operations in a general region they can all benefit by forming a cooperative or association. Typically an organization can consolidate such administrative and operational functions as group insurance, workman's compensation, payroll, safety, bonding, financing, legal, accounting, transportation, milling, technical services and a multitude of other possible needs. One such association in the Eastern and Midwest US is the Small Coal Mine Operators Association, which is principally involved in group insurance.

Conclusions

Small mining ventures in precious metals take a different breed of mining firm than is seen in many larger mining companies. Those who run a successful venture have a "Think Smart" approach and use all the resources available to them. The little operator can not afford the time nor moneys on extensive exploration, development or unproductive programs but must get in, get the metal desired and get out as quickly and economically as possible. To accomplish this it sometimes becomes a business of "not what you know but who you know".

References

Anon., 1966, "Uranium Mining Starts a Comeback," Engineering and Mining Journal, Dec., pp. 79-88.

Anon., 1970, "Leasing v. Buying," Mining Magazine, Vol. 122, no.6, June, pp. 437.

Anon., 1970, "The Role of the Metal Merchant," Engineering and Mining Journal, Feb., pp. 90-92.

Bruce, C., 1972, "The Benefits of Using Outside Consultants," The Mines Magazine, Mar., pp. 22-25.

Chender, M., 1983, "Problems and Potentials of Small Mine Marketing." Second International Symposium on Small Mine Economics and Expansion, pp. 401-403.

Cope, L.W., 1981, "Pitfalls of Small Mine Development," 84th National Western Mining Conference, Feb., pp. 116-118.

Gardner, J.D., 1981, "Improving Small Mine Productivity," 188th Annual Northwest Mining Association Convention, Dec.

White, L., 1978, "Wide-Ranging Activity Marks Uranium Boom in Colorado and Utah," Engineering and Mining Journal, Vol. 179, no. 11, pp. 125-128.

Chapter 5

MARKETING FUNDAMENTALS THAT WORK FOR THE SMALL MINER

William Mayrsohn

Asamera Minerals (U.S.) Inc.

INTRODUCTION

The marketing aspects of a new precious metals property is often not considered until the mine actually begins production. At this point in time it is often too late or too expensive to change the mine's final product and not enough time to do an effective marketing program.

MINE PRODUCT

The rule for determining the mine's final product: The higher the gold and silver content, the greater return of metal at the lowest post-mining cost.

Dore Bullion: Most advantageous

Advantages:

- Highest return of metal.
- Lowest refining cost per ounce of gold and silver.
- Very compact and easy to ship in small quantities.
- Easy to assay.
- Assays between mine and refiner are comparable.
- Melt losses are very small at the refiner.
- Outturn of gold and silver 20-30 days for quick cash flow.
- Gold/silver refiners can return or purchase metal soon after arrival at a nominal financing cost.

GOLD DORE BULLION REFINING TERMS

Gold Content%	Treatment Charge/oz	Payable Gold %	Payable Silver
80-90+	$0.70-.90	99.9-99.8	96-98
70-80	.60-.70	99.8	96-98
30-70	.50-.60	99.7	96-98
10-30	.30-.50	99.4	96-98

Figure 1:

Above assumes balance of dore metal silver with minor impurities ie., copper under 5%. Minimum charges per lot: $175-275. Some have assay charges $50-60 per small lot. Metal advances vary with refiner.

SILVER DORE BULLION REFINING TERMS

000's Ounces/Lot	Silver Content%	Treatment Charge/oz	Payable Silver%	Payable Gold%
5	90-98	$0.22-.35	98	97-98
5 - 15	70-90	$0.20-.30	97.5	97-98
18+	50-70	$0.18	96.5	97-98

Figure 2:

Above assumes balance is gold, copper. Refiners have a low tolerance for Ni, Se above 5%, As, Bi, Al_2O_3, Sb, Cd.

Minimum charges same as Figure 1. Metal advances vary with refiner.

Figure 3 is a list of the more prominent refiners who tend to offer a wide variety of services. The large refiners can recommend a representative to witness the melting and weighing process if desired. The Royal Canadian Mint will not permit witnessing and rely exclusively on their own assays. All others will return a sample to the mine.

LIST OF PROMINENT REFINERS

USA	Location
Handy & Harman	N. Attleboro, MA
Englehard West	Anaheim, CA.
Johnson Matthey Ref.	Salt Lake City, UT
Pease & Curren	Providence, R.I.
Leach & Garner Ref.	N. Attleboro, MA
CANADA	
Royal Canadian Mint	Vanier, Ont.
CCR-Canadian Copper	Montreal, Quebec
De Gussa	Richmond, B.C.
Johnson Matthey Ltd.	Toronto, Ont.
Inco (Platinum Group)	Toronto, Ont.

Figure 3:

Above are the largest refining companies. Smaller refiners have been omitted. European refiners have been omitted as it is usually too costly as ship and refining costs are higher due to the weakness of the dollar.

SHIPPING DORE BULLION

U.S. Mail

Dore bullion can be sent by first-class mail in 50-75 lb parcels at the lowest cost. Many post offices have a maximum value per shipment of $100,000-125,000. They insure up to $25,000 but excess coverage is available from insurers.

Air-Freight-Armored Truck

Bullion shipped as valuable cargo will save one to two days over parcel post, but an armored truck is necessary to get it to the airport. Some refiners rely entirely on armored truck.

Placer Gold

This product can be shipped as is to refiners, but it is best to melt it to a dore bullion to upgrade the purity.

Loaded Carbon, Precipitates

Most refiners accept this material provided the precious metal content is high ie., 100-200 ounces per ton. However, the cost can be high, $5 to $10 or more per ounce because they have to do the melting and preliminary fire-refining Mercury in the material can be a serious problem.

It is best to strip or burn the carbon reducing it to metallic dore or melt precipitates on site.

Concentrate

This is the least desirable product. It is bulky, hard to ship, difficult to assay, and has the highest refining cost per ounce of gold ($30-60).

Smelters pay/return from 91 to 98% for gold and 94 - 95% for silver. The losses due to weight loss, assays, and assay methods can range from one half to three percent in addition to the returnable metal deduction.

GOLD CONCENTRATE TERMS

000's Tons Annual Output	Treatment Charge/Ton	Ref. Gold/Oz	Charges Silv/Oz
1-3	$150-250	$5.00	$0.35-50
3-5	125-150	5.00	0.35-50
5-10	90-125	5.00	0.35-50
10+	60-90	5.00	0.35-50

Figure 4:

Smelters charge for assaying and sampling on lots under 100 tons. These can vary from $75-150. There are also minimum lot sizes accepted. Penalties can be expected for As, Sb. High mercury is unacceptable.

Figure 5 is a list of smelters that routinely take gold-bearing concentrates. Terms can vary considerably alond with the transportation costs. The closest smelter may not be the best choice when the comparisons are made. Containerized shipping is expensive but it does eliminate in-transit losses. High-grade material is often shipped in bags or drums.

GOLD CONCENTRATE SMELTERS	Office Location
Asarco	Ore Depart., NY
Cominco	Trail, B.C.
Noranda	Noranda, Quebec
Nippon Mining	Tokyo, Japan
Dowa Mining	Tokyo, Japan
Mitsubishi	Tokyo, Japan
Metallurgie Hoboken Overpelt	Hoboken, Belgium
Norddeutsche Affinerie	Hamburg, W.Germany
Korea Mining & Smelting	Seoul, Korea

The major disadvantage of concentrates is the smelting time, which varies from 90 to 135 days after arrival. Bulk shipments have to be accumulated.

MARKETING THE MINE PRODUCT

The choices:

- Sell to a merchant who arranges everything, including sales.
- Have an agency agreement with a merchant who arranges everything on your behalf, for a fee.
- Sell direct to a refiner who prices automatically and deducts charges from the proceeds.
- Toll refine, and sell the metal separately to a gold broker/merchant of your choice, using a pricing method suited to your needs.

Before an alternative can be considered, market information has to be obtained by offering the material to refiners or smelters on a direct basis.

In making such an offer to smelters-refiners the following information is required:

- A sample of dore or other mine product.
- Complete assay/analysis for all elements totalling 100%.
- Amount per shipment and frequency of shipment.

Dealing Direct with a Refiner-Smelter

Refiners can offer the following services:

- Purchase the contained returnable metals.
- Toll the material into saleable bars, if desired.
- Arrange transportation of concentrates and dore bullion.
- Insure the material while in transit.
- Arrange for early outturn of gold/silver (at a nominal cost).
- Some sell metal forward.

Once the refiner has given you an indication of terms and you have the other related costs of getting it to their plant, you can compare one refiner against another. Then you can do a little negotiating to improve terms.

Offer your material to a number of refiners/smelters. The mine should try to do its own thing first smelt-refine and sell its own metal or receive early metal outturn with each segment of the business priced out separately. However, there is a point at which it is undesirable to do this, at about 20,000 ounces a year.

Comparing Smelter-Refiner Terms - NSR:

To decide which refiner is best, all costs related to smelting-refining, transportation, insurance, early gold return costs and the time-value of the payment date from the smelter have to be calculated to obtain the net smelter return (NSR).

Don't be overawed by very low treatment charges or high metal return. The NSR is the only figure that counts.

For the finer points of doing these financial comparisons, you will have to attend session five, "Financial Considerations for Negotiating the Gold Sales Contract."

Gold Leasing - Early Gold Return Option:

To overcome the time lag between shipment arrival and gold outturn for payment, gold can be leased about three to five days after arrival at the refinery as follows:

- From the refiner at a cost per ounce (15-30 cents per ounce).
- From a gold broker usually at a competitive cost 1.0 to 2.0% a year.

For mines producing under 20,000 ounces a year, it is probably less trouble to obtain early gold return direct from the refiner.

Advantages of Dealing Direct with Smelters/Refiners

- The mine maintains control over assaying and sampling by talking directly to the refiner about assay/sampling technique.
- Unusual losses can be dealt with on a one-on-one basis.
- Refiners can give advice as to the methods of refining dore at the mine for a cleaner product.
- A relationship can be built up between refiner and mine.

Dealing with Merchants/Gold Brokers and Consultants

As previously discussed, know your smelting and refining alternatives before you enter any discussions with a merchant.

If your output is small and the merchants offer is very competitive, insist on having a direct line of communication between the mine and smelter-refiner.

Don't over emphasize a merchant who also is an investment banker house. There is practically no connection between the metal traders and investment bankers.

Obtain financial data from the merchant on his company (annual reports, etc.). If the merchant is small, insist on dealing through irrevocable letter of credit backed by a bank line of credit.

Merchant-brokers can play a useful role for the small miner provided the miner knows how to utilize their services.

For a sizable private trader, insist upon a bank letter of support in lieu of financial data. In this respect, take nothing for granted.

You will be approached by brokers willing to pay cash for gold, at a 3% or so discount. Avoid these, as there is _never_ any need to sell gold at a discount.

Before you utilize a consultants expertise, find out how much he has sold and check references. Insist on dealing direct with smelters-refiners with the consultant's assistance.

Agency Agreements with Merchants

Merchants will approach the mining company to represent you for a fee for marketing mine production. You have to know the market before any contracts are signed with smelters or refiners.

There is a lot of concentrate swapping, where a merchant offers your material to a smelter at high treatment charges while asking for an offsetting discount on his purchased material from the same smelter. This tends to happen when large amounts of material is to be produced, ie., 10,000 tons of concentrates or more per year.

CONCLUSIONS

Marketing dore, concentrates, precipitates or placer gold, can be approached in a logical manner to obtain the lowest cost, highest net smelter return to the mine.

Talk to other miners about their marketing strategy, but in the end, you will have to solicit smelters and refiners.

Make contacts, obtain terms, do an accurate comparison with the assistance of your accountant and, by all means, listen to the advice of refiners and smelters. You can rely on their transportation people for the lowest cost shipping alternatives.

In the end, you will find that your own research will give you ideas to improve your marketing scheme as the refiners, gold brokers can give good advice as to what others are doing and how it can be applied to your situation.

Chapter 6

PC-POWER FOR SMALL MINES

How They Work in Exploration, Mine Planning, Process Design, Maintenance, and Cost Controls

by

Betty L. Gibbs
Gibbs Associates
Boulder, Colorado

ABSTRACT

Using a PC can help an operator analyze data more effectively and respond quickly to changing conditions. The cost of microcomputers is low, and fit into the budget of even small mines. Many software packages are also low cost and general software such as spreadsheet, word processing, and data base management programs can help the small miner organize and evaluate information very quickly.

A basic hardware configuration is proposed, including hardware upgrades and general and technical mining software packages. The hardware and software suggested can be used by very small as well as larger operations. PC's can be incorporated into all phases of a mining operation at a minimal cost and capabilities can be expanded as the operation grows.

INTRODUCTION

Microcomputers are no longer a new and unknown factor in the mining business. In fact, the range of options is bewildering, and it is difficult to sort through what is best for any particular operation. When you read about the latest gadgets or someone tells you what a great deal this new hardware widget is, it is difficult to determine whether your particular needs can be helped by a hardware or software addition.

Whether microcomputers are new in your company or have been in use for some time, any new acquisitions require evaluation of the company's needs and how particular software or hardware can fill the needs.

Guidelines are proposed in this paper for acquisition of a computer for fundamental applications as well as more detailed and technical mining calculations. The guidelines are general because each operation has different hardware and software needs. The list of hardware and software is a framework which can absorb details as needed to customize a system for any operation.

REASONS TO USE A PC

A mining operation is frequently faced with choices requiring quick decisions. For a manager to make the appropriate and most effective decisions, current information about the operation must be available as well as the calculation tools to correctly interpret the information.

In an open pit mine, for example, information about the grade of material in an area blasted must be readily available to guide the loader operator in sending a truck to the waste dump or to the crusher. One blast may have as many as a hundred holes. Plotting the holes and re-evaluating the ore/waste interface sometimes cannot be completed using manual methods until after the blast has been loaded into trucks. A well planned information gathering and analysis system centered around a microcomputer can ease the effort and time required to obtain answers to many of the problems facing small mine operators.

A microcomputer can be useful from the very beginning exploration stages through final close-down procedures. Operation of a microcomputer is relatively simple and can be self-taught if the desire to learn is strong. Many of the general software packages such as spreadsheet and data base management include tutorials which guide a user through the initial stages of learning how to use the program. Introductory computer courses are also offered in most areas and are usually recommended to help a new user past the first stages.

The following is a brief summary of the most typical applications of a microcomputer and associated software in any size mine.

Exploration and Feasibility Phase

The exploration and evaluation phase of the search for ore deposits offers many opportunities to take advantage of the capabilities of a microcomputer. The current hardware technology supports a very simple configuration for uncomplicated tasks as well as larger systems which can handle the most sophisticated evaluation methods.

Examples of how a microcomputer can be used in exploration include the following:

- **Organize data**. During exploration, geological data is collected which may consist of outcrop samples, drill hole core samples, geochemical analyses, and geological descriptions. A microcomputer can provide data base functions such as storage of information, selective retrieval, and simple calculations on the data.
- **Statistical analysis**. Standard statistics is used for analyzing geochemical results as well as examining distributions of mineralization data. Geostatistics calculations are not practical without a computer, and geostatistics often provides more insight into the true interrelationships of geological data.
- **Mapping**. The ability to quickly plot base maps, contour maps, and cross sections is valuable to the expedient evaluation of a potential ore body. As new information is collected, existing maps can be updated quickly with minimal effort required by the geologist. Rather than spending time in the laborious re-drawing process, the geologist can focus on interpretation of the available information.
- **Ore body modeling**. Using computer tools (programs), a numerical and graphical model of an ore body can be produced in a relatively short time as compared to using only hand calculation methods. In addition to speed, a computer allows more detailed examination of the modeled deposit and adjustments based on judgment and mining constraints.
- **Reserves and mine plan**. Once an ore body is modeled on a computer, the reserves can be calculated quickly. Calculation of total reserves is an integral part of a mine modeling package. The mine plan can be developed by hand from cross sections output by the computer, then mineable reserves can be calculated when the outlines of the ore are entered into the program by coordinates or with a digitizer. Some programs offer mine planning directly on the computer monitor screen.
- **Economic evaluation**. If you have ever spent hours calculating one cash flow analysis, you can appreciate the ease and quickness of using a simple spreadsheet program on a microcomputer. Once the cash flow is designed using the spreadsheet, many different economic scenarios can be quickly calculated and evaluated.

Mining Phase

When mining begins, a wide variety of applications are available which run on a microcomputer. Small mine operators must be able to handle all phases of their business from the clerical paper work through monitoring and production control. The work load can be eased significantly by using even the simplest and most general software.

- **Continuing exploration**. Continue to evaluate exploration drilling and more detailed delineation of the ore body.
- **Blast hole data**. Evaluate blast hole assays in a timely fashion and guide daily pit operations. The blast hole data can also be used to verify that model-building methods were correctly applied and to suggest adjustments in the methods.
- **Scheduling**. Daily and weekly scheduling is accomplished much faster with the aid of a microcomputer and appropriate software.
- **Grade control**. The goal of any operation is to produce more ore and less waste. A computer can expedite the task by providing timely information and analysis tools to guide the mining process. Re-evaluation of a model as new information is received can often make the difference between higher and lower mill or crusher feeds.
- **Refine ore body model**. As new information is received, the ore body model can be readily updated and the geology can be refined quickly.
- **Cost monitoring**. Cost of the operation can be tracked readily and provide up-to-date reports to management. A frequent summary of costs can pinpoint trouble areas before they become unwieldy.
- **Accounting and payroll**. The obvious advantages of using a computer for accounting and payroll functions are increased speed of processing information. Regular accounting provides closer control of the operation and everyone appreciates getting their paychecks on time and accurately.
- **Maintenance management**. Small operations, especially, benefit by attention to equipment maintenance. Small operations can afford few mistakes caused by ignored maintenance.

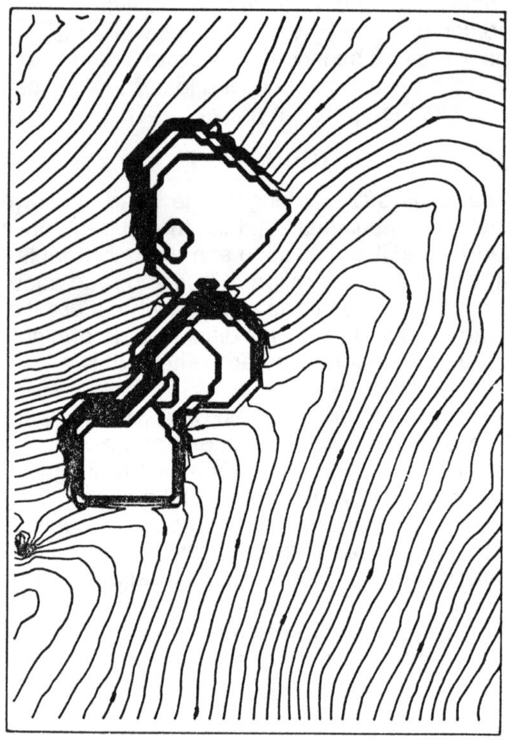

Figure 1. Example of a contour map with surface contours and a pit outline.

(Figure provided courtesy of MINEsoft.)

COMPUTER PROGRAMS

A wide selection of programs for microcomputers is available, and development of general software as well as mining programs is a continuing process. Programs for microcomputers are low cost compared to mainframe programs. Costs range from $50 to $1000 for general software and from $100 to $30,000 for geological and mining software.

Most software is designed to be easy to use for people who are not familiar with details of computer operations. The success of much of the software available is that the user does not have to be a programmer, but has access to powerful tools which are easily learned and have direct application to work they do on a regular basis.

Sources of Information

Information about word processing, spreadsheet and data base management programs can be found in computer magazines published bi-weekly or monthly. These magazines contain reviews of programs as well as articles of general interest related to microcomputers. The magazines are also a valuable resource for locating dealers in discount hardware and software.

Many of the mining magazines are incorporating periodic columns about computer applications specifically for mining. For example, the Mining Journal has started a quarterly column which discusses capabilities of selected computer programs.

Two other sources of information on geological and mining programs are the Computer Oriented Geological Society (COGS, P.O. Box 1317, Denver, CO 80201) and the Directory of Mining Programs (Gibbs Associates, P.O. Box 706, Boulder, CO 80306).

Universities, computer dealers, and individuals offer short courses in the basics of computers as well as advanced applications for particular industries. Some of the courses are available in the evenings, or may last for several days. These courses can be a good source of introductory information about microcomputers.

CONFIGURATIONS

A computer-based system is a combination of hardware and software. The goal of a new computer user is to build an effective system which makes work easier and faster. A good system contains a balance of computer equipment which will handle the volume of information and reporting required using software which is appropriate for the defined tasks. Often, this balance is easier to talk about than achieve. It is so easy to accept the popular beliefs about hardware or software packages and assume that what works for others will also work exactly the same in your operation.

To provide a starting point for small mine operators, hardware and software configurations are summarized below. The hardware is based on IBM PC(TM) and compatible computers, because of the variety and availability of software which can be used on these systems. All of the applications mentioned can be run with purchased software packages; no user programming is required.

A Minimum System (Level A)

Hardware: PC with monitor and two floppy disk drives. 512K RAM (working memory size). Cost: $800 and up. Printer, dot matrix (Cost: $250 up).

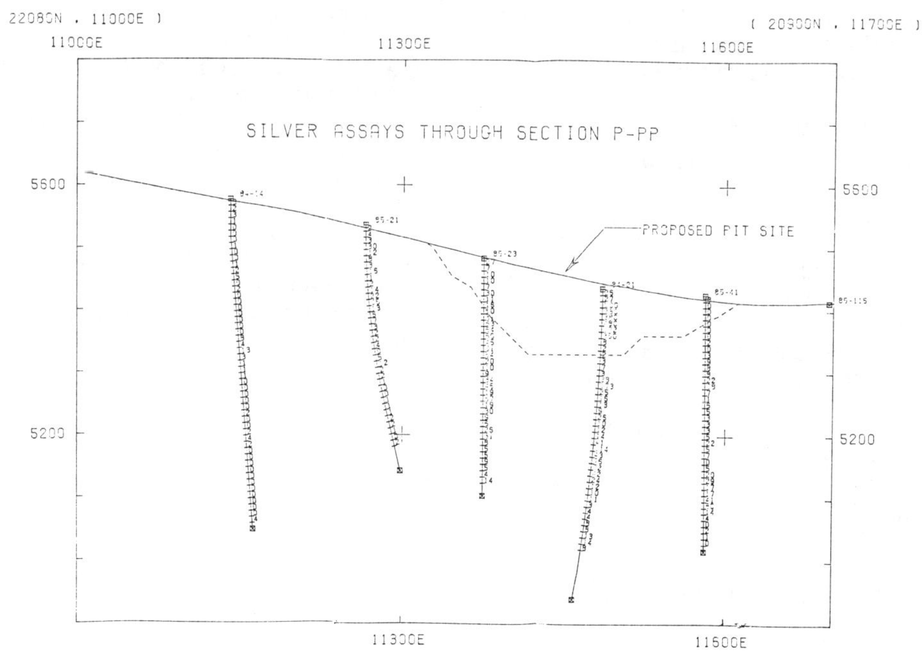

Figure 2. Cross section showing deviated drill holes, assays, and pit outline.

(Figure provided courtesy of MINEsoft.)

Software: DOS 3.2 operating system; word processing; spreadsheet program. Cost: $600.

What you can do with it:
- Write letters and reports
- Financial and economic evaluation
- Drill hole data calculations
- Scheduling
- Accounting and payroll
- Statistics
- Cost tracking

Spreadsheet programs can be used for a diversity of applications and are only limited by the user's imagination and need. For a small operation, spreadsheets can be used for a majority of the reporting and operations tracking needs. The list above is only a sampling of spreadsheet applications.

Add Data Base Management (Level B)

Hardware: No additional hardware is absolutely necessary at this point. However, if the data base you will be developing is very large, it is useful and inexpensive to add a hard disk at this point. A 30 Mb (megabyte) hard disk can be added for $450 and provide extra speed and sufficient storage space for most small mines.

Software: Database management software. Cost $250-$500.

What you can do with it:
- Organize and retrieve drill hole data, geological information, operations information.
- Maintenance tracking.
- Operations monitoring.

Data base management systems are flexible enough to handle almost any type of information. The best applications are for text and numerical data which require simple and minimal calculation. The data base allows quick access to information, and prints reports.

Add Graphics and Mapping (Level C)

Hardware: Add a graphics board (allows graphics to be output to the monitor screen). Cost: $100-500. If you have not added a hard disk yet, do so now. The calculations for mapping run significantly faster when a hard disk is available. A math co-processor added at this stage will also increase the run speed (Cost: $200). Output can be obtained on the dot matrix printer. Printer graphics are generally low quality but sufficient for some applications such as work maps. Simple graphs can be effectively produced on a dot matrix printer.

Software: Mapping program. Cost: $400-$5000.

What you can do with it:
- Base maps.
- Contour maps.
- Cross sections.
- Area calculations.
- Log plots (with some software packages).
- Blast hole maps.
- Mine plans.

Add a Plotter (Level D)

Hardware: Plotter. Many sizes are available from regular sheet size through plotters which can accommodate a 24"x36" sheet. Cost: $400-5000.

Software: No additional software is needed if the mapping program is selected carefully.

What you can do with it:
- Plot high quality maps.

Add Ore Reserves, Mine Planning (Level E)

Hardware: Additional hard disk capability can be added, but is related to the size of model you expect to create. Additional memory is also useful to speed up calculations if you have not already added it. One megabyte RAM or more can be added to a microcomputer. A math co-processor is a necessity at this level.

Software: Orebody modeling and mine planning packages. Cost: $2000-50,000.

What you can do with it:
- Complete ore body modeling.
- Reserve calculations.
- Mine planning and scheduling.

The above configurations are very general and do not include extras such as a higher quality monitor for better screen graphics or utility software which can increase the utility of the system. An option used frequently by owners of hard disks is a tape backup unit which will back up all the files on a hard disk onto a single tape. (A 30 Mb hard disk is equivalent to 84 floppy diskettes.) Prices range from $500 to $1500.

CONCLUSION

When an operation is very small, information can be handled effectively without a computer. However, as an operation grows, operators begin to feel a need for more efficient and speedy methods of monitoring the status of the operation. At this stage, introduction of a computer allows management to keep up with reporting functions and have sufficient time to keep the operation running smoothly.

The cost of a microcomputer system with software can be low or high depending on the components of hardware and software chosen. Even the lowest cost system with general software provides a useful tool to monitor costs, grades, and other operation parameters.

2

Understanding Geology and Ore Reserves

J. D. Lewis
Boise Cascade Corp.
Boise, ID

Chapter 7

LATERITIC GOLD: OPTION FOR SMALL MINE DEVELOPMENT

Prof. A.Bhaskara Rao & Prof. Maria S.Adusumilli

Dept. of Geosciences, University of Brasilia
70910 Brasilia-BRAZIL

ABSTRACT

The presence of gold in lateritic crusts over mafic and ultramafic volcanogenic complexes in / and greenstone belts, has called the attention in several parts of Brazil and also in Australia. Recent studies reveal the variety of structures and mineralogy of such crusts, and their importance in correlation with host(bed) rock types, as exploration guides, specially for greenstone belt metallogeny. The expected nature of growth of plates and/or nuggets in such crusts is under detailed study, and is considered as possibly due to dissolution and reprecipitation phenomenae of disseminated-fine-low grade ore in the host rock. In this investigation, laterite geochemistry, though incipiently discussed todate, is of fundamental importance and significance.

Lateritic gold can be classified texturally as: dusty and diffused, fine, thin platy,tabular, and nuggets. Such gold is pure, and is often restricted to nodular(pisoliths, oolitns, concretions) forms or the matrix, that constitute the crusts or lateritised gravel beds which have an essentially goethitic nature. Further this gold is recovered directly by mining the crusts or the adjoining placers with the metal derived from them.

The thickness and extension of laterite crust zones is extremely variable depending on the paleoclimatic conditions, and is conditioned to the favourable lithology and its structural expression. However, the "growth" of gold in these laterites will be conductive , without doubt, to several small scale mining activities, specially in the developing countries and regions limited to latitudes 30ºN and 30ºS.

INTRODUCTION

Extensive regions covering great part of the developing world with tropical, sub-tropical,rain forest and arid regions are often referred to as "lateritic". This term proposed by Sir Francis Buchanan in the early nineteenth century in India, has represented a great part of the bad land topography, and has been neglected for a long time. With the discovery of large quantities of bauxite deposits derived from bauxitisation process and/or lateritisation, and progressive addition of Ni,Mn, Fe, P and Nb to this list in recent years, laterites and lateritisation received more attention and International Seminars were held in Trivandrum in India(1979), São Paulo in Brazil(1982) and Tokyo in Japan(1985). Gold and platinum also found their deserved place in the laterite system.

LATERITE CRUSTS

Classification and Correlation.

Though laterite is a generic term used to charecterise ferriferous soil coverings and ferralites (55% of the Third World soil type coverage), yet the indurated and compact laterites are important to indicate original bed rock type. Thus classifications are proposed by Aleva(1979) and Schellman (1979) based on mineralogy and chemistry respectively.

Bhaskara Rao(1983, 1984) proposed a structural classification of the crusts based on the participation of fragments, nodules and bands; and correlated the crust types with possible bed-rock types of the greenstone belt units. The importance of this classification lies in its easy applicability in field and in exploratory campaigns.

It is necessary here to give the concept of a laterite crust(Bhaskara Rao, 1985 b): "Laterite crusts are indurated compact coverings on any bed rock type, with a deep brown colouration indicative of abundance of goethite, either dominating all over or as intergranular locking material between the rock fragments or bands or nodules constituted of varying mineralogy. Such crusts are usually autochthonous, eitherway indicative of the nature of bed rock type below or nearby". Studies usually are made on in situ crusts.

Common Types:

Among the crusts, the common types and their characterisation is given below, and illustrated in Plates I and II.

Nodular Crusts: These are constituted easily of about 50% of nodules or pisoliths or ooliths or concretionary forms. With the increase in the presence of oriented pisoliths and matrix inducing banding, varieties such as concretionary pisolithic (oolithic) laterite, and concretionary banded banded laterite could be detected. An increase in the fragments of rocks and/or minerals, would permit the denominator of breccioid nodular laterite or nodular laterite breccia.

The nodules are usually spherical,oval and rounded forms like pisoliths and ooliths. They are essentially constituted of goethite(about 80%) and some other components like gibbsite, clay,hematite, quartz and sulphides and occasional gold dust.

Frequently nodular crusts represent a bed rock type rich in iron, and geologically a mafic to ultramafic complex, or a volcanic suite.

Laterite breccia: Laterite breccia is the term applied to represent a crust with abundant(80%) rock and mineral fragments. They are normally quartzites, vein quartz, chert or sandstone,with normal inter-fragmental material made up of goethite and some gibbsite and clay. The constitution of the matrix is extremely variable. Hematite goethite and silica are abundant in case of banded matrix, denominating the laterite as breccioid banded laterite. In case of nodular laterite breccia, the matrix has nodular forms mainly of goethite, and intra-nodular goethite or gibbsite or clay of a later phase.

These breccia types often indicate original quartzites, banded ferruginous quartzites, BICFs banded iron cherty formations, heterogeneous metamorphic suites or sedimentary suites They frequently cover the nodular types, in case of greenstone belt suites, where the volcanic ultramafic and quartzites association is known and is good for strategy for gold.

Cases of "coarse"(over 10 cm diameter fragments of sandstones) laterite breccias , with successive lateritisation processes in sedimentary sequences covering volcanics, should be given adequate attention in precious metal exploration, specially for gold nuggets.

Banded laterite: Banded laterite is typical over BIFs banded iron formations, and is dominated by hematite-goethite association with some silica. Often this type is cavernous or "vescicular" due to heavy leaching of pre-existing silica. A variety which is laterite with banding often represents a parent rock as gneiss, gneiss with BIF, or a well banded sedimentary formation with intense faciological variation.

Lateritised gravel bed:(Bhaskara Rao, 1985 a): Extensive, covering an area of some km, lateritised gravel beds are recorded showing similarities to laterite breccia, but with characteristic rounded and oval pebbles of rock fragments that have suffered fluvial transport. Such gravel beds form paleo-fluvial terraces. The main aspect is the inter-pebble material which is essentially brown in colour and is goethitic in nature, resultant of intense lateritisation. The underlying laterites are nodular crust type, mottled and with occasional cavities. Such an intense lateritisation is attributed to the presence of underlying volcanogenic mafic to ultramafic sequences. Lateritised gravel bed contains gold which due to recurrence of denudation and erosion, contributed gold and pebbles to the fluvial placers in the area.

Lateritised colluvium:(Bhaskara Rao & Borges,1987) Rain forest regions where cycles of erosion have dominated the resultant topography, present frequent and abundant colluvium. From the Brazilian Amazon region in gold prospects, lateritised colluvium is identified and is considered as a very useful guide in gold prospection.

Colluvium is " a general term applied to any loose, heterogeneous and incoherent mass of soil material and/or rock fragments deposited by rainwash, sheet wash, or slow continuous down slope creep, usually collecting at the base of gentle slopes or hillsides"(Glossory of Geology, 1980. AGI).Such a desintegrated material constituted of angular to partly sub-angular or sub-rounded fragments with medium to fine grain fragments as intergranular material, and compacted by deep brown goethite is considered as lateritised colluvium.

Such a formation is usually underlain by nodular or concretionary levels, indicative of an intensive lateritisation of the bed rock. The nature of bed rock in known cases is usually a fine textured porphyry volcanic suite and/or cataclasite or sheared suite of rocks. The fragments are often resistant vein quartz or quartzites or hard rocks of varied types.

Heavy minerals, with abundant magnetite, include recoverable gold and permit small scale mining. It is also an exploration guide to gold metallogeny, in the parent rock.

LATERITIC GOLD

The metallogeny and importance of gold in recent years has attained special standards. Any attempt to organise "genetic models" or "deposit types" may run the risk of possible omissions due to greatly increasing scientific litterature.

Lateritic gold could be conceived as "native gold occuring in lateritic profiles constituted of indurated compact aggregates with incrustations of rock and mineral fragments or pebbles". Such gold is often very pure(wt %: Au 99,46; Ag 0,64; Cu + Fe trace. Gibb Maitland. Cit. Wilson,1983,p.230), and is texturally classified as:

a) dusty, diffused and invisible.
b) fine, visible,grains, and equigranular.
c) filmy and as coatings.
d) thin platy.
e) tabular.
f) botryoidal.
g) nuggets.

Derby(1884) recorded films of gold on botryoidal limonite, as peculiar deposition, in Brazil. Brazilian lateritic gold is already well known in Goias, Amazonas, Pará, Mato Grosso, Bahia and

PLATE I

A. A typical laterite profile on a gneissic suite showing(from top to bottom): soil covering, compact ferralitic crust,loose lateritic soil with low induration and rich in clyey-gibbsite-goethite assemblage; fluvial gravel bed with pebbles, clay and goethite no so indurated; compact indurated laterite crust rich in gibbsite;altered bed rock. Locality: Riacho da Pena, 26 km Goianesia-Rianopolis Road, Goias State.

B. Lateritised gravel bed, with pebbles of 10 cm diameter of quartz, with a matrix rich in goethite plus some gibbsite, indicating lateritisation. Locality:13 km Carmo do Rio Verde to Rubiataba, Goias State.

C. Laterite breccia, with angular rock fragments of varying sizes, showing original bedding along which lateritisation dominated. The fragments and blocks are quartz - ite and ferruginous quartzites. Locality: 53 km Goias Velho to Itapirapuã, Goias State.

D. Cavernous (concretionary) laterite crust constituted of goethite, and also gibbsite and clays. Locality: 23 km Goianesia to Entroncamento, Goias State. On the top of the hill.

PLATE II

A: Nodular laterite crust, compact and hard containing nodules of centimetric dimensions, and mineralogically constituted of goethite and hematite. Locality: 53 km Goias Velho to Tapira, Goias State.On tabluar morphologies indicating extensive coverings.

B: Pisolithic laterite crust, showing desintegration into small pisolihts indurated into crusts by fine mass. Both the matrix and the pisoliths are dominated by goethite. Locality: Montes Claros de Goias, Goias State.

C: Ferralite crust, essentially constituted of hematite mass, almost with a bedded aspect showing kilometric extension. Normally overlies nodular pisolitnic laterite crust zone.Thickness totally inconsistent. Locality: Montes Claros de Goias, Goias State.

Minas Gerais States. Some localities in Mato Grosso and Goias States have produced excellent nuggets of varying sizes and weights in the lateritic zone. Unfortunately the mining companies that extract such gold have not yet shown any interest to permit the documentation of their sizes, growth forms, coverings, and also their characteristic associations which will be useful parameters for further exploration work. Since lateritic gold has not attracted much attention till recently, though known since last century, of conventional exploration geologists, the mining men couldnot be blamed for ignoring these details of lateritic gold. It is worth mentioning that nuggets of 800 gms at Barro Alto, Goias State; 40 gms, several cases in Mato Grosso State; 15 gms in Fazenda Nova Goias State; several of 1-2 gms in many localities in Amazon and Parā States; 70-90 gms/ton in lateritised fault collapsed amphibolites in Minas Gerais State, are all known. Further mention may be made of the famous gold rush to Serra Pelada in Parā State that has produced gold aggregates of several kgs, which geologically is also interpreted to have had the concentration due to lateritisation. In Australia 164 gm nugget is known from laterites(Wilson, 1983, p.231).

The gold nugget of Fazenda Nova is shown in Fig.1(After Bhaskara Rao et al, 1983) and another but well grown and visible gold crystal aggregate (Courtesy Geol. Potira) in laterite matrix is given in Fig.2. Both of them are of interest and are described . In Fig.1. the pisolith nugget is surrounded by compact "limonite" bands deposited all around the gold fragment, and inclusions of quartz fragments are frequently noted. Gold as the nugget of 15 gms shows on the exposed nodule surface, step-like pattern gold, suggestive of a possible growth on a "relict" grain. The "limonitic" filling in the fracture(?) is rather strange and certainly a possible "channel dissolution fracture filling" not much related with the growth pattern of gold. In Fig.2. the laterite crust is indurated nodular to pisolithic with interstictial growth of hematite-goethite. Fragments of massive "vein quartz" with compacted goethite with vescicles is seen with fillings of highly hydrated iron hydroxides in the cavities. The main mass is brownish yellow and is goethitic in nature. Gold is visible with bright yellow metallic shine and has the following aspects:

1) Elongate tabular forms of 1.5 X 0.1 mm representing a growth of three small crystals in an elongate form. The aspect suggests a filling of a fracture, and recrystallised gold during the process of lateritisation.

2) Gold films on vein quartz, with beautiful yellow metallic shine, similar to iridiscence but representing the precipitation of dissolved and recrystallised fine dusty gold, coincident with lateritisation process.

Gold nuggets from India in laterites(Nair et al, 1987) record the SEM photographs and observe botryoidal and filamental to dendritic types also, and call the attention to the "painted type" of adherence of Au nuggets with Fe-oxides. Such adherence with Fe-hydroxides is frequently noted in nuggets in Brazil, which cause problems in the gold recovery in a washing plan operation.

PROFILE TYPES

Characteristics:

In Table I lateritic profiles are resumed. Gold is known in them and are thus useful for exploration works in any region.(see next page)

In Table 2 conceptual parameters of a lateritic profile are included.

Table 2:Conceptual Parameters of a Profile.

Topographic:	Small elevations and valleys,with tabular hill tops elongated and controlled by structural and/or lithological units of interest.
Preferred Geological Conditioning:	*Greenstone belt units. a) Mafic to ultramafic volcanics. b) Banded iron formations. c) Volcano-sedimentary sequences. d) Acidic manifestations of granite types. e) Quartz vein systems. *Intermediate acid volcanic suites. a) Porphyry systems. b) Epithermal systems. *Sedimentary sequences. a) Transected by dykes/sills of sub volcanic types. b) Intercalated by swarms of volcanic suites. *Sheared zones, cataclasites of the above systems.
Idealised Preferred Profile:	(From top to bottom). *Breccioid laterite Blocky laterite. * Lateritised colluvium Lateritised gravel bed. * Nodular, patchy cavernous laterite. * Concreationary laterite Crust Pisolithic laterite Oolithic laterite Nodular laterite * Altered bed rock Mottled zone Saprolite * Bed rock.

It should be remembered that such a profile is idealised and is conceptual, based on experience in several occurrences. Such parameters are, thus, very useful to limit the search for gold in laterite crusts, and are useful for exploration campaigns, formulation of strategies and also models .

EXPLORATION MODEL

Lateritisation is well known as a secondary process of chemical weathering dominant in tropical regions involving chemical decomposition with the alteration of parent minerals and liberation

Table 1: Profiles of Laterites with Gold Occurrences

A: GREENSTONE BELT TYPE		B: MAFIC TO ULTRAMAFIC TYPE (Volcanogenic)		C: ACID VOLCANOGENIC TYPE (Schematic profile)	
Breccioid Laterite	Laterite crusts with fragments of quartzites and vein quartz; in situ types located at peak points in the topography, and transported types on the slopes and in depressions, as huge blocks.	Lateritised Gravel Bed	Extensive fluvial gravel bed, with coarse pebbles, intensely lateritised with fine botryoidal goethite growths. Presence of cavities.	Lateritised Colluvium	Colluvium with angular to subangular and sub-rounded fragments of vein quartz or quartzitic and gneissic(?) rocks, indurated by banded goethite due to intense lateritisation. Interstitial material and matrix often heterogeneous with constituents of parent rock material. Located usually on slopes and depressions.
Concretionary Laterite	Compact concretionary laterite usually extensive as a bed, with or without banding, located at the top of the hillocks. Occasionally with fragments and also coarse cavernous types.	Nodular Laterite	Extensive nodular laterite crust zone, with an upper transition zone containing some coarse pebbles, and a lower transition zone rich in cavities.	Nodular Lateritic Colluvium	Idem, with nodules; upper transition zone with pebbles and heterogeneous matriz; lower transition zone with more nodules.
Nodular Laterite	Compact laterite with coarse nodular masses, forming almost an extensive unit.	Mottled Laterite	Medium nodular laterite crust zone with mottles of gibbsite and clays; nodules of goethite, as well as abundant matrix.	Nodular Laterite	Coarse nodular laterite as an extensive body with irregular thickness and abundant goethite and gibbsite mixture, and intercalations of clays as masses. Some grains and small fragments of quartz also present.
Pisolithic Laterite	Pisolithic and oolithic compact laterite, often with varying thickness. Normally extensive.	Pisolithic Laterite	Fine nodular to pisolithic laterite crust zone, very compact, mineralogically constituted of goethite.	Pisolitnic Laterite	Pisolitnic and oolithic laterite much compacted and representing a definite zone.
Transition Zone	Altered bed rock with frequent ferruginous infiltrations as reddish patches and bands.	Altered Bed Rock	Zone of intensely altered bed rock with relict structures, and iron hydroxide fillings.	Altered Bed Rock	Altered bed rock with clays, gibbsite and some goethite, in varying proportions inducing mottled appearance.
Bed Rock	Greenstone belt units, mafic to ultramafic sequences.	Bed Rock	Mafic to ultramafic, volcanogenic suites, and similar complexes rich in iron silicates.	Bed Rock	Acid volcanic suite, porphyritic suite; sheared and catalastic.
Climate	Tropical, Sub-tropical, Arid and Sub-arid .	Climate	Sub-tropical to Arid.	Climate	Tropical rain forest regions.
Example	Fazenda Nova, Goias Velho.	Example	Mara Rosa, Goias State.	Example	Localities in Brazilian Amazon.
Source	Bhaskara Rao et al(1983) revised version.	Source	Bhaskara Rao(1985 a) condensed and modified.	Source	Bhaskara Rao & Borges(1987) adopted.

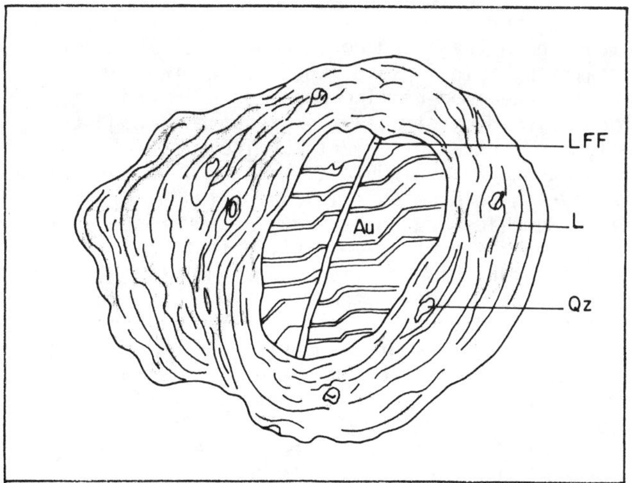

Fig.1: "Limonite" posolith with a small gold nugget, showing step-like over-growths on the surface of gold. LFF- Limonite filling the fracture. L- Finely banded "limonite". Qz- quartz fragments. Locality: Fazenda Nova, Goias State, Brazil. (Scale 5:1)

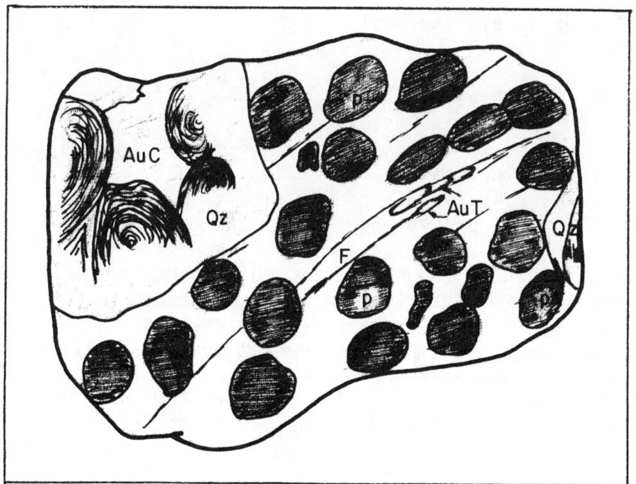

Fig.2: Sample of laterite crust with nodules(pisoliths-p) and matrix constituted of goethite. Quartz(Qz) fragments, with gold coating(AuC) and fine tabular gold crystal(AuT)aggregate along fractures(F)?.Locality: North of Goias State, Brazil. (Scale 5:1)

of chemical elements. The distribution of the dissolved mineral components is by vertical movement and transport, resulting in the deposition as enriched constituents. The metallic ions released as non-destroyable components are also enriched; further the metallic ions with lateral migration could be concentrated as metalliferous minerals or native metals in laterites.

Gold thus could form: 1) small nuggets either conserved from original rock, or chemically dissolved, mobilised and redeposited, or chemically grown due to accretion. 2) plates as fillings in a relict structure or fracture conserved from the parent rock. 3) fine and dusty or invisible type in the nodules, pisoliths and ooliths, due to dissolution, mobilisation and deposition with iron hydroxide gels and in colloidal form.

In view of this genetic model, an exploration model could thus be proposed taking into consideration the following typical laterite crusts as the best "hosts" of lateritic gold.

1) Nodular lateritic crust.
2) Pisolithic and oolithic lateritic crust.
3) Lateritised gravel bed.
4) Lateritised colluvium.

From table 2 it is observed that topographic features and geologic conditioning are important to seek the idealised laterite profile preferred to look for gold, and thus form part of an exploration model, as conceived in table 3.

SMALL SCALE MINING

The concept of small scale mining varies from one country to another, like from a developed USA or giant Brazil or populous India or small Sri Lanka. The parameters vary from place to place and from mineral to mineral.

Perhaps this is not so true in case of gold

Table 3: Conceptual Exploration Model for the Lateritic Gold

Bed rock		Types susceptible for chemical decomposition,with fine disseminated gold.
Climate		Rain forest,Tropical,Sub-tropical, Sub-arid, Arid regions.
Process	A	Chemical decomposition Dissolution Transport Oxidation
	B	Non dissolved minerals and metals left in situ(Gold)
Lateritisation	A	Mobilisation of components & element Fillings of fractures and open space (Gold) Precipitation Deposition
	B	Recycling of processes. Colloidal transport and deposition (Gold) Authigenic growth(Gold) Formation of nodular types
	C	Mobilisation and deposition of iron hydroxides(Gold) Filling of intra-fragmental spaces. Induration processes Dessication Formation of breccias, colluvium, and gravel bed- lateritised.
Erosion		Denudation and erosion Liberation of fragments and masses Contribution to placers(Gold) Formation of placers with coarse gravels.

or precious metals, Neverthless, for Brazil, a country of gigantic dimensions, all projects and plans are gigantic and "small is not so beautiful" or so "useful". This is neither a misconcept nor a correct concept. In one aspect it is not a correct concept because all small scale mining is a lead to regional scale exploration projects(Bhaskara Rao, 1982).

Small Mine Development for Gold

In every mining venture the most interesting aspect is the feasibility of the project, and this normally takes into account: 1) know-how, 2)reserves, 3) out-put, 4) operational costs, 5) investment, 6) returns, etc. In a small scale venture these could be reduced to a great extent.And in case of a surfacial deposit, these factors involve less time venture and reduced exploratory investments, and thus imply less operational costs. However, in case of precious metals such "reductions" need not be true and on the contrary could suffer an "increase". This is because, even the risk is increased in the determination of "tenors" and the parameters to "evaluate" the reserves.

From this analysis. the small mine development of lateritic gold is possible through:

1) Location of the presence of gold in fluvial placers, and checking of its possible lateritic derivation.

2) Location of goethite-rich nodular, pisolitnic, oolithic lateritic indurated units on a gold host rock type(see table 2).

3) Trenching and pitting in a systematic grid as part of an exploratory work in the laterite rich zone.

4) Total recovery of the lateritic material taken from the trenches and pits, for direct washing and recovery of gold, and/or assaying.

5) Preparation of the map of the mineralised area, and calculation of the potential gold content.

6) Determination of the size and distribution pattern of gold, in different zones and laterite horizons or episodes(see table 3).

7) Recovery of surface coverings with adequate machinery, and development of adequate method for washing and recovery of gold.

8) Operation and development of small mining for gold.

9) Identification and location of similar geological conditioning of laterite coverings, and its utilisation for:

a) regional exploration programmes in laterite horizons.

b) primary gold exploration programmes in type host rocks underlying mineralised laterite horizons, thus identified.

CONCLUSIONS.

Precious metals like gold and platinum are always sought and in varied geological conditions. The genetic variations that are being discovered permit new horizons like laterites where these metals could be concentrated.

Since laterites are surfacial expressions and are derived from other systems, and are indurated due to lateritisation, they are repositories of native metals and/or oxide minerals of economic significance.Thus the potential reserves are not usually high but could easily foster small scale mining.

However, as laterite is a system which is dominating in the Third World countries, the noble metal sought in it is of special interest and is of easy development. The experience in Brazilian localities demonstrates the presence of nuggets, which is a great stimulant for their search. Woodward, as early as 1906(cit. Wilson, 1983) described a rich gold deposit in "laterite" at Crusoe Hill, Menezies Region, Australia and wrote:" gold was found to be of an entirely different character and of a much higher value than that obtained from the oxidised zones in the mines, whilst subsequent trenching and cross-cutting beneath it failed to prove the presence of any auriferous body in the viscinity. The inference, therefore, is that this deposit results from the concentration of gold carried in small quantities in the country rock, during which process the silver and base metals were removed and the fine particles of gold united."

ACKNOWLEDGEMENTS

To National Research Council(CNPq)thanks are extended for the grant of Fellowships and research grants for laterite and gold studies.

REFERENCES

Aleva,G.J.J.,1979,"Laterite versus Saprolite".IGCP 129 Newsletter II, Hyderabad.June 1981,pp.30-34.

Bhaskara Rao, A.,1982,"Small Deposits as Guides to Regional Exploration Projects", 6th. IAGOD Symp. Tibilisi. Collected Abstr. pp.259-260.

Bhaskara Rao,A.,1983,Depositos Minerais: Ambientes de sua Formação e Modelos. Limited Circulation. University of Brasilia, Brasilia.202 pp.

Bhaskara Rao,A.,1984."Laterite Classification for Exploration." in: Applied Mineralogy. Metallurgical Soc. of AIME, Warrandale. pp.951-964.

Bhaskara Rao,A.,1985 a,"Lateritised Gravel Bed: A New Guide Horizon for Lateritic Gold ?"Internat. Sem. on Laterite, Tokyo,Proceedings.& Chem.Geol. 1987 . 60, pp.287-291.

Bhaskara Rao,A.,1985 b."Guide Horizons for Gold Mineralisation in Lateritic Crusts." Internat.Sem. on Laterite, Tokyo.Proceedings. & Chem.Geol.1987 60, pp.293-298.

Bhaskara Rao,A. & Barros, J.G.C.,1982,"Perspectives of Small Scale Mining in Developing Countries-Brazilian Example". AGID News ,Bangkok,30:p.19-25

Bhaskara Rao,A., Barros, J.G.C. & Adusumilli,M.S,, 1983,"Lateritic Gold Project" Proc. 2 nd. Int. Sem.Lat.Process, São Paulo.pp.159-176.

Bhaskara Rao,A., & Borges,M.R.,1987,"Lateritised Colluvium: Guide Horizon for Lateritic Gold in Tropical Rain Forest Regions". (in Press).

Nair,N.G.K., Santosh,M.,& Mahadevan,K.,1987."Lateritisation as a Possible Contributor to Gold Placers in Nilambur Valley,SW India".Chem.Geol.60: pp.309-315.

Wilson,A.F.,1983. "The Economic Significance of hydrothermal...Australia" Spl.Publ.Geol.Soc.South Africa, 7: pp. 229-234.

Chapter 8

PLACER SAMPLING, METHODS AND PITFALLS

Louis W. Cope

Placer Consultant

Abstract. The various types of placer deposits will be described as an introduction to the subject. The several methods and machines used for sample taking and processing will be discussed. The advantages and disadvantages of each will be covered. Included will be mistakes and pitfalls of placer sampling and sample reduction.

INTRODUCTION

Placer is a Spanish word with two meanings, "place" and "to please". What a delightful way to describe a mining operation--a pleasing place.

Because of the infinite diversity of placer conditions, sizes and locations, this paper is not intended to be a "how to ....." Instead, it will describe and compare concepts, and will refer only to gold. To knowledgeably work in placers requires a person with interdisciplinary experience of rudimentary geology, dirt-moving mining and gravity metallurgical knowledge. It is an art much more than a science. To read a book is not enough of a preparation; placer requires hands-on experience.

Gold placer deposits respect no international boundary nor climate. Placers are commonly located downstream from hard-rock mines. There are instances where deposits were formed from the erosion of an outcropping porphyry with a myriad of tiny spiderweb-like gold-bearing veinlets. And I have seen in several districts bedrock of metamorphosed slate and shale, classified as greenstones, with included auriferous quartz veins of up to 5 cm (2 in.) thickness, which obviously fed the placer.

PLACER DEPOSITS

To name the types of placer deposits is an imprecise exercise for three reasons. One, the names change in local useage; two, the deposits can merge one into another; and three, each author will define them differently.

In Place Placers

Eluvial. The first step in the formation of a placer is with the weathering and break-up of a gold-bearing vein outcrop, and the rubble is washed and creeps down the slope of a hillside. These deposits are sometimes called "colluvial". At times, a higher grade residual deposit is formed immediately below the outcrop.

River Formed Placers

River. As the name implies, a river placer is one which is in a river or creek, within the bounds of the flowing water. The rock and even the grains of sand are rounded by movement of the water flow.

Bar. There is a fine line between river and bar gravel. A bar is usually thought of as those gravel deposits above water level at times of low water. Bars can be lenticular-shaped islands, or attached to the bank on the inside bend of a curve. "Skim Bars" are a feature, particularly on rivers of high water velocity which rise and fall in relatively short periods of time. Fine gold is often carried and deposited on the cobbly surface of a bar. When the water drops, an area with easily recoverable gold remains on the bar.

Terrace. The low lying gravel on the valley floor is called a terrace. It can be flat and part of the flood plain, or gently sloping upward from the edge of the river. A terrace is often covered with soil and grass, or is farmed like bottom land.

Bench. These are placer deposits on hillsides well above the valley floor. They are remnants of terrace placers left after the river cut a deeper valley. In some districts, bench placers are prolific gold producers.

Wind Formed Placers

Aeolian. When wind removes the lighter and finer particles of a deposit, this allows the gold concentration to increase. These deposits occur in arid areas with a strong prevailing wind. They normally have only a superficial concentration.

Other Placers

Outwash. This type of deposit can be considered an incomplete placer. That is, the material is washing downhill as a piedmont or delta from a mountain range, but has not yet reached continuously running water. These placers seldom contain commercial quantities of gold in their entirely, but can have local concentrations.

Glacial. Glaciers can dig-up placers and gold-bearing rock and leave it deposited in moraines. Usually, this material is mixed with non-gold-bearing debris, which results in reduced grade. However, melting ice and precipitation can start to rework the moraines and reconcentrate the gold as in outwashes. There are instances of placers which have been covered by glacial debris, called "Paleo" placers.

Marine. The submerged, existing and uplifted beach placer deposits of Nome are well known. But this same feature occurs in other locations as well. The Spring and Fall equinoctial tides on the Pacific Coast of the Osa Peninsula, Costa Rica, cause a brief frenzy of work on the material pushed high onto the beach.

Tertiary. Placers which were buried by lava and ash in Tertiary geologic time carry this name. The deposits could be any of those previously described. Most are in California, but some occur in western Nevada, southern Oregon and other locations, New South Wales, Australia, for example. Modern California placers were formed by the west-flowing Feather, Yuba, American and other rivers incising the rich prehistoric river channels.

Dry. There are auriferous gravels in arid and semi-arid areas where there is little water. In treating these, air pulsing and vibrating machines to recover the gold require muck of two percent moisture, or less. For this reason, the scale of dry placer production is normally limited. Even in the desert, dirt a meter or more (several feet) below surface usually contains more than two percent moisture, particularly if clay is present.

SAMPLE METHODS

There may be signs that placer gold occurs in a locale, but the grade and reserves should be determined to be able to calculate economic feasibility of the property. Not to quantify the reserves will likely result in a plant being erected which is not profitable--another monument to someone's optimism. Proper sampling will also give information which is necessary for mine planning so that the correct type of equipment and mining sequence can be optimized for lowest cost operation.

Bulk Sampling

Large samples, such as those taken by the various bulk sampling methods have the two advantages that, one, a good inspection of the ore is made. That is, size distribution from the finest slimes to boulders is observed, as well as any stratification in the material. Secondly, enough sample is handled that it gives a better average gold grade for that specific area.

Inherent in the larger sample is one of the main disadvantages of bulk sampling. This is the larger equipment necessary to dig, haul and treat the sample. As handling takes more time, fewer samples can be taken and reduced for a given cost. With fewer samples, not as much is learned about the entire deposit.

As the bulk sample hole is dug, water running or seeping in effects sample integrety. It is not known if the run-in carries gold to raise the sample grade, or if only mud is carried in, lowering the grade. Careful hole measurements have to be taken so it will be known what volume is represented by the gold recovered in sample reduction. It is not advisable to use only part of the material of a pit because of the "nugget effect"--the considerable effect of one nugget falling into the sample or into the reject material.

Pitting. A pit can be dug by machine or by hand. If by hand, a 2.1 m (7 ft) depth is about the maximum for a worker to be able to throw the dirt out of the hole. Safety has to be taken into consideration as any pit deeper than one meter (3 ft) is a risk of caving and suffocating a worker. Some simple support can be used, but this gets in the way of digging.

Back-hoes are handy for digging sample pits and easily loading the gravel into a truck or processing machine. A back-hoe with a 0.4 cu m (0.5 cu yd) bucket usually has a 4.6 m (15 ft) boom with an effective digging depth of 4 m (13 ft), so is only useful in shallow placers.

Care has to be exercised when measuring a back-hoe-dug pit, because of the often curved front and back surfaces of such a hole. Any amount of sloughing of the top or sides of the pit will effect the final gold-per-cubic volume calculation.

Trenching. Pitting on a larger scale is to dig a trench, either with a back-hoe or a bulldozer. If the latter machine is used, deeper sample(s) can be taken. But again, the depth is limited by safety considerations for the machine and operator. Dozer-made trenches give an excellent view of the muck, but are expensive to dig. In trench sampling, the usual procedure is to take samples from the side of the trench.

A vertical cut is made in the wall of the trench of a predetermined size. The sample is caught in a bucket, drum or on a canvas, and taken to the process machine. Separate samples can and should be taken of the gravel when the character or color change, or divided merely by measurement. There are instances where all the material from a cut is taken for processing in a production-size process unit.

Limitations of trenching are the same as for pitting--large samples, few samples per dollar cost, a limit to the depth which can be dug, and water problems.

Operations. Production reports of previous operations on a property give a figure of gravel value. But only the gold recovered is reported, and this depends on appropriateness of the process and competence of the operator. An operation gives excellent data on characteristics of the deposit. However, it gives information on the area mined, which may or may not be the same in the unmined area.

Drill Sampling

By drilling, I am referring to cased holes in the 5 cm (2 in.) to 18 cm (7 in.) range, although there is equipment which will use a 76 cm (30 in.) or larger casing. By driving a pipe casing or using a drill stem the diameter of a drill hole preserves sample integrety from hole sloughing. But there can be run-in, particularly where a water-course under pressure is encountered.

Drilling should be of close-spaced holes in "fences" across the narrow dimension of a placer deposit. Several fences are made, with a relatively large distance between the fences. This is to be sure to intersect all higher-grade channels or paystreaks. These might be missed when drilling on a grid pattern.

As drilling produces a small sample, it is cheaper and quicker to process, resulting in a very large number of sample holes, compared to bulk sampling. This gives an array of samples over a much larger area, resulting in more ore reserve knowledge for a given budget. Data on the cross-section and bedrock profile obtained from drilling are invaluable for developing a mining plan. Large placer mining companies have historically depended on drilling for delineation of their gravel reserves.

At the same time, the small sample obtained in drilling is a disadvantage, because only a small amount of gold nugget effect will seriously effect the value of a sample. But the much greater number of samples will statistically give a better average for the entire property. Modern drill rigs are mounted on large trucks, which require good access roads and large flat drill-pads prepared in advance. Drills can be mounted on crawler rather than wheeled vehicles. As will be described, there are small portable drill units, but these are limited to drill-hole diameters at the small end of the range.

Drilling does not give a good picture of the type of material in the deposit. There is always the question whether gold particles dislodged by vibrations imparted to the soil will drop along the outside of a casing to "salt" the sample.

Churn Drilling. The method used for years in placer sampling was churn, sometimes called "cable tool", drilling. With this machinery, a pipe or casing is driven into the ground to rejection. The casing has a hardened steel ring of known diameter on the lower end called a drive shoe.

Drilling consists of dropping a chisel-point bit onto the core of material inside the casing. Enough water is added to maintain the ground muck in a slurry suspension. Figure 1 shows how the cable to the bit is threaded over a pulley on a derrick, then around a crank or eccentric mechanism to impart alternate pulling and relaxing of the cable. This results in a motion of lifting and dropping the bit.

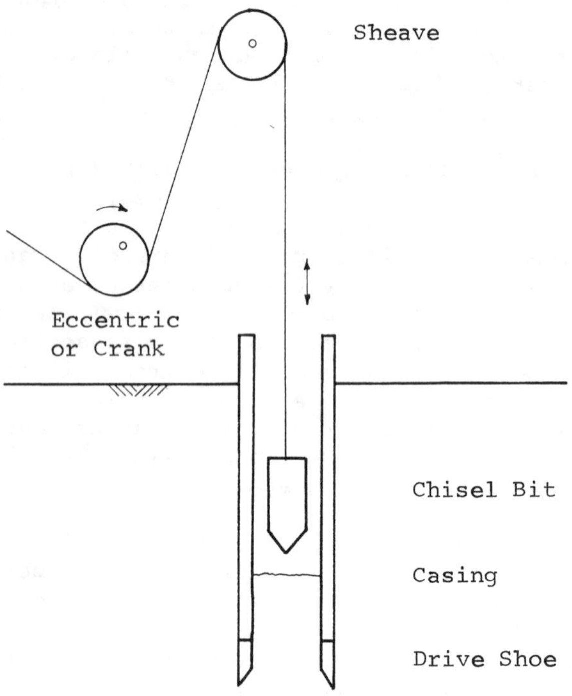

FIGURE 1. Churn Drill System

A frequency of 50 or 60 strokes per minute is necessary to keep the slurry in suspension. Care should be taken not to drill below the lower end of the casing. As the inside diameter of the casing is smaller than the crowd-in ring of the drive shoe, the core or slurry will rise in the casing. This should be measured and a volume factor applied. This is called a Radford Factor or Keystone Constant.

Removing the slurry, the sample, from the hole is called "pumping," but actually consists of "bailing". The bit is removed from the hole and a "sand pump" attached to a second cable called a sand line, is put into the casing. This is a circular container with a valve on the bottom, so that it will open on contact with the slurry and the bottom of the hole. Slurry fills the container and the valve closes when the sand line is actuated to quickly pull the container from the hole. Bailing is continued until all the sample is removed. The casing is driven deeper if necessary, and drilling continues.

A small, portable machine called an "airplane drill" is a light duty churn drill. It can easily be flown into a site or carried by pack animals.

Churn drilling has long been considered the industry standard exploration tool. However, it is slow, therefore considerably more costly than the next described reverse circulation method.

Reverse Circulation. Since about 1960, reverse circulation drilling has so replaced churn drilling for sampling that there are few placer-experienced cable-tool drillers available in the United States.

In reverse circulation drilling, a pipe within a pipe casing tipped by a drive shoe is driven into the ground by a pile driver-like mechanism. The pipe acts as the hole casing, carries the energy to the drive shoe and is the container through which the sample is recovered. Compressed air is delivered to the bottom of the casing through the annular space between the two pipes. This air picks up sand, gravel and rock chips which it returns to surface through the inner, smaller pipe. The air-gravel mixture goes to a cyclone where they are separated, and the sample drops into a container. (See Figure 2)

Drilling is more often than not done with water added along with the compressed air. The reason is that any moisture in the muck can cause some of the sample to hang-up in the return pipe, thereby

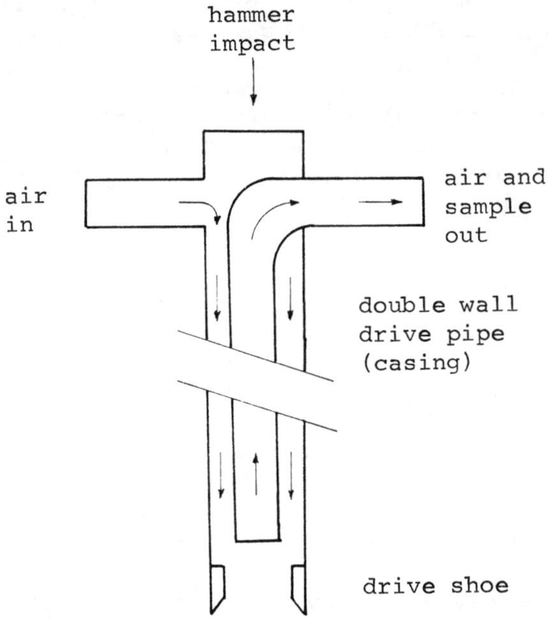

FIGURE 2. Reverse Circulation Drill System

affecting sample integrety. A wet sample keeps the system flushed out.

The main advantage of reverse circulation is the drill speed, which directly relates to lower cost. But at the same time, this speed can be a disadvantage because sample processing cannot keep up with the drilling. If results show a need to sample or resample a certain area, the drill is no longer at the site, and can only be returned at considerable expense. As with a churn drill, access roads and drill pads are required.

Vibratory Drill. A drill method which came available in the mid-1970's is vibratory drilling. In this method, a casing is pushed into the ground while having a high-frequency vibration imparted to it, and rotation if required. When the pipe has penetrated the deposit, it is pulled from the hole and the core which it has collected and brought to surface is taken from it. This is a true core of the gravel being sampled.

In favorable ground, the vibratory drill is even faster than reverse circulation, but information I have shows it to be more expensive per foot of hole. To be able to see a cross-section of the gravel in a core is very useful. However, a core of fines with little or no gravel will contract from the vibration, and boulders will not be broken as easily as with churn and reverse circulation drills.

Hand Drills. Empire and Banka are similar drills, and both depend on manpower. In each, the drill-stem is topped with a platform on which two to four men stand. They alternately hit a drive-head with a sledge hammer. Their weight plus the blows advance the drill. Meanwhile other men or animals are turning the pipe. The drive shoe has serrations to assist in breaking through boulders. Bailing is done as in churn drilling, but by manpower.

This system is little used nowadays, except in some remote foreign locations with very cheap labor costs. Despite this, hand drilling can end up to be high cost, due to the slowness of the method and the need to keep the crew in the field for a long period.

Other Exploration

Seismic. Determination of bedrock contours can often be made by high resolution seismic lines. This geophysical work gives no insight into the gravel grade.

Magnetic. I have heard of a few cases of magnetometer lines run across a placer deposit. This could be useful where there are significant quantities of magnetic minerals which occur with the alluvial gold, and where there is no interference from bedrock mineralization.

Comments

I consider the best sampling program to consist of a combination of drilling and pitting. Drilling gives a broad array of information for determining ore reserves, bedrock profile and water, all as described above. Then a few pits will give information on the material size, distribution, stratification, etc. Thus, combining the two methods utilizes the advantages of each.

SAMPLE PROCESSING

After samples are taken and the volume they represent is recorded, the contained gold has to be determined. It should be born in mind that an economically break-even placer will contain one part gold to about 5 million parts of waste! Sample processing has to be careful work, but results should be reproducible in a full scale plant, usually by the same or similar recovery method.

Placer concentrates should not be assayed, as this will report total gold, not just the portion which is recoverable to placer methods. Gold particles larger than ten mesh can easily be picked out of the black sand. Finer gold is commonly amalgamated with mercury to collect it from the heavy fraction. Mercury can be used with any of the sample reduction machines and methods described below. All of the following methods require screening to about minus 6.4 mm (1/4 in.), except hand panning and normal sluices.

Hand Reduction

Panning. Hand panning is the best known method for reducing a placer sample to its concentrate. Panning is used in reconnaissance exploration as well as the final step of most of the methods listed below. In remote, low wage areas, placer samples can be panned in their entirety.

Sluice Washers

Normal Sluice. In handling placer samples, a sluice box is the simplest and easiest method to use. However, sluices will commonly lose fine gold. A sluice 20 cm (8 in.) to 30 cm (12 in.) wide with sides 15 cm (6 in.) to 25 cm (10 in.) high and up to 3.6 m (12 ft) long are a common size for small samples. Larger samples will either have to be fed more slowly, or a wider sluice used.

Slope of a sluice is generally in the range of 1.3 cm (1/2 in.) to 3.8 cm (1-1/2 in.) per 30.5 cm (1 ft), depending on the size rock in the feed. To move the larger rocks down the sluice requires a steeper slope or more water, both of which cause greater gold losses.

Any number of types of collectors can be used in a sluice. Most common is the use of astroturf or indoor-outdoor carpeting under expanded metal. These are plastic products which are used in place of the traditional wool carpets, gunny sacks or cocoa matting. The plastic carpets are much easier to clean and if washed thoroughly, have no chance of carry-over from one sample to the next. These are lightweight and dry quickly.

Riffles can be of cross slats of wood or metal, either square wood or metal bars. An improvement on a common riffle is the Hungarian Riffle, which is one with an overhang, as shown in Figure 3. This over-

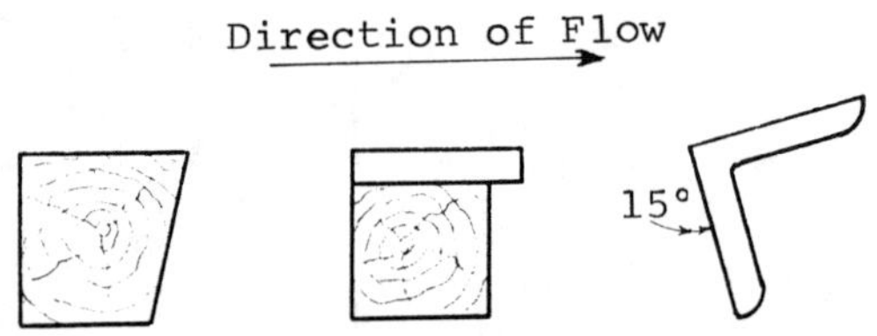

FIGURE 3. Examples of Hungarian Riffles

hang gives an eddy in the flow of pulp, thereby allowing the small gold particles a chance to settle. Other riffles can be of blocks of wood, round tree branches or rocks. I have seen the floor of a sluice with shallow, staggered 5 cm (2 in.) diameter holes and a little mercury in each. This method is particularly good for fine muck containing fine gold. In sum, types of riffles used are a matter of personal preference, or at times, limited by the materials at hand.

Long Tom. A Long Tom is actually a variety of a sluice. It differs in that it has a means of screening out gravel and cobble, thereby allowing less water and a flatter slope. Long Toms are of a wide, short configuration, and are normally a production machine fed by shovel or wheelbarrow. However, they can be any size, depending on the samples treated. (See Figure 4)

Motorized Sluices. There are any number of machines on the market with spray-bars over a vibrating screen. The screen, usually 6.4 mm (1/4 in.) or 9.5 mm (3/8 in.) mesh, feeds to a sluice, usually of carpeting and expanded metal. These machines reduce the work load and are faster than a common sluice, but still risk the loss of fine gold.

Rocker Washers

Cradle Type. This is a unit with a feed box, screen, canvas gold collecting cloth and a near-flat sluice made of wood. They

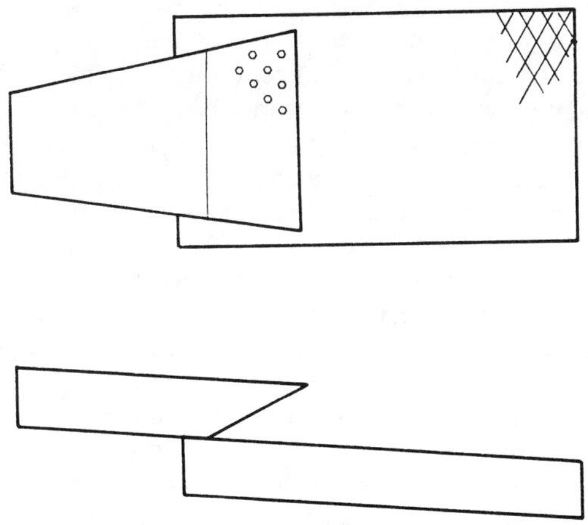

FIGURE 4. Long Tom

were rocked by hand, with gravel shoveled onto the feed box and water ladled on. Rockers such as this were common in the Gold Rush days, but are seldom used now.

Motorized Rockers. Just as with motorized sluices, rockers have been adapted to be motor driven. The most widely used machine has a revolving trommel with a spray bar. The material which goes through the trommel's punched plate screen drops onto a plastic molded riffle which reciprocates from side to side. This rocking motion causes the heavies to settle while the light fraction advances down the riffle to tailings.

Centrifugal Washers

Vertical Axis. Centrifugal clean-up and sample washing machines have been used for decades. The early ones, which are still a good machine, have a cast iron bowl with riffles cast onto the inside face. The bowl spins on a vertical axis while gravel is fed to the center. The mild centrifugal action forces the slurry to the sides, with the heavy fraction trapped in the riffles. The light material is washed upward and over the rim to tailings. An improvement is to cast a smooth bowl and insert replaceable rubber riffles.

Other manufacturers make similar machines, but with a different configuration. These are made of bent and welded plate. Another type has a water jacket with water ports into the riffles in order to maintain a live and self-cleaning concentrate.

Horizontal Axis. Another technology is a cylinder revolved on a horizontal axis. The interior of the drum has spiral riffles. A sample slurry is run into the drum. The heavier minerals settle into the riffles and are carried to one end while the lightweight waste minerals are washed out the opposite end.

Wheels. Still another type of machine to reduce a sample volume is a special kind of wheel, made by several manufacturers. They are made of a 41 cm (16 in.) to 51 cm (20 in.) diameter plate with a 10 cm (4 in.) high rim. The flat portion is faced with a spiral riffle which decreases in height from about 9.5 mm (3/8 in.) at the outside to about 1.6 mm (1/16 in.) in the center. The wheel is tipped to a convenient angle and slurry fed to the rim. The gold is carried by the riffles to a hole in the center of the wheel, while the waste overflows the rim as the wheel slowly turns.

Comments

By whatever means used to reduce samples, the supervisor should not allow loss of any black sand in the first step of processing. Gold would probably not be lost with the lightest portion of the heavies, but no one can be sure. In the subsequent stage, usually panning, the operator has better control due to working with a small volume, and can discard some of the heavy fraction under these controlled conditions. However, there is nothing wrong with amalgamating all the black sand produced.

DATA EVALUATION

After taking samples and processing them to determine the contained gold, calculations show the quantity of gold per unit volume. The numbers give the grade and quantity of gravel reserves.

Most gold balances use metric weights which is good. It is more precise to show, for example, 218 milligrams rather than 0.007 ounces of gold in a unit volume. Years ago, when the price of gold was fixed, results were usually given in cents of gold per cubic yard. In those days, however, some professionals reported gold content as pennyweight or grains of gold.

Sample results should be plotted on a map, both in plan and cross-section. Figure 5 is an example of presentation of drill or pit data. In this example, it can be seen that two fences of close-spaced holes are shown. Also sketched is a cross-section of one of the fences. Drill holes are indicated by dark lines from surface to bedrock. The little cross hatch marks are the limits of samples 2.4 m (8 ft) in length. The gold per cubic unit should be marked on the cross-section for each sample.

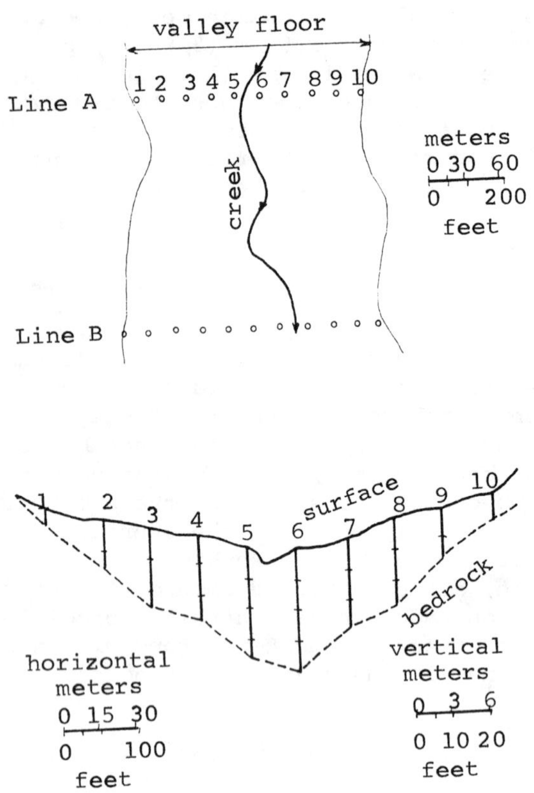

FIGURE 5. Presentation of Drill Data

PITFALLS

Sampling

No Sampling. Placer, just like all mining, is a high risk business, but a sample program is a way to reduce one of the major risks. People who buy life, home and car insurance will sometimes go into mining without spending the "insurance" money for sampling. Anyone who has been in placer consulting has heard statements like, "Gran'pappy panned gold all over the place," or "There's gold at surface and ten times as much at bedrock."

With luck, operations might start in good ground, but can run into barren ground so that the operator has to grope around for paying gravel. It is much cheaper to pay the cost of sampling than to have to treat 76 cu m (100 cu yds) or more per hour looking for the pay streak, or shut down while doing the sampling. No program will tell how much gold is in every cubic yard, but much more is known than if no sampling has been conducted. When sampling, always go to bedrock. To extrapolate values is very risky. If the sample taking equipment won't reach bedrock, go get some equipment that will.

Sample Integrety. The matter of sample integrety has been mentioned previously in this paper. After all, there is no use in taking samples if reasonable precautions are not taken to get a proper sample.

Sample Processing

Procedure. Logically, the processing machinery should be in good repair, and not be overloaded. A "white frock" mentality is not necessary in reducing the samples, nor are sloppy procedures condoned. There is a proper and practical mid-point.

CONCLUSIONS

An operator is asking for trouble if he does not do a sampling program to determine grade, quantity and location of reserves. Likewise, if the profile of bedrock is not known, operations are started at a handicap. Because someone is making money on a placer with a wheelbarrow and small sluice doesn't mean a large-scale operation would be profitable. But sampling will help answer the question.

There are enough problems in placering without injecting new ones by reinventing the wheel. That is, don't deviate from proven methods run by experienced people. I like the story that, when driving on ice, it isn't the tires on the car, but the "nut" behind the wheel that matters.

Chapter 9

THE ROLE OF GEOLOGY IN THE DESIGN OF DRILLING PROGRAMS

David R. Shaddrick

Rubicon Resources Company
Reno, Nevada

Abstract. Ore reserves are based on sampling developed by drilling programs. Correct choices concerning the type of drill to be used, depth of drilling, hole orientation and sample spacing are not only critical to the accuracy of the ore reserve but to the cost in time and money of the project. Appropriate interpretation and use of geologic information can significantly enhance the ability to make these decisions effectively. Geologic characteristics of mineralized rock that affect drilling and sample quality include: rock quality, openness, water content, mode of gold occurrence, internal gold distribution and grain of mineralization. Each of these characteristics and sources of data about them are revuewed, then related to the various drill program options. Finally, some procedures for checking the validity and appropriateness of the selected options are reviewed.

INTRODUCTION

Ore reserves in a mineral deposit are based on sampling. Fundamental concerns are the validity and accuracy of the samples, the adequacy of the sample distribution and the cost in both time and money of the sampling program. In virtually all cases today ore reserves and the samples on which they are based are developed through drilling. A number of drilling options are available and an understanding of the diverse factors affecting their selection is important to the success of a project. Program design options are: The types of drills to be used; the individual hole design including depth and orientation of drilling; the sample spacing which includes the down hole sample interval and length as well as the spacing between holes.

It is the intent of this paper to address five points which are believed to be critical in the early exploration and development drilling of gold deposits. First, the characteristics of mineralized rock that affect drilling decisions. Second, the impact of drill selection on sample validity and program cost. Third, the effect of drill hole design on the accuracy of tonnage estimates as well as sample validity. Fourth, the impact of the sample distribution on the accuracy of the tonnage estimates. Fifth, some procedures available to provide assurance that the program design decisions have been made properly.

Sample preparation and analysis procedures as well as geostatistics will not be addressed in this discussion. Although critical, these topics would be more properly treated in separate papers.

CRITICAL CHARACTERISTICS OF MINERALIZED ROCK

The many geologic properties of mineralized rock can be reduced to six critical characteristics which directly affect the drilling decisions outlined in the preceding section. These critical characteristics are: rock quality, openness, water content, mode of gold occurrence, internal gold distribution and grain of mineralization. Early identification and ongoing monitoring of these characteristics are important to the ultimate success of the drilling program. Under normal circumstances acquisition of this data should not require special studies or extra expense. The necessary observations, measurements and interpretations can generally be made in the normal course of the exploration and prefeasability work. The following paragraphs describe each of the six critical characteristics and present a brief discussion of how they may be documented or inferred.

Rock Quality

In this context the term rock quality refers to the ability of a drill to penetrate the rock cleanly and effectively and to return a good sample. This can be hampered by the presence of fracturing, shearing or alteration which can cause blocking of bits, binding of rods and hole caving or swelling. In addition to a significant reduction in drilling efficiency, poor rock quality can severly reduce recoveries and sample quality with some drilling methods.

The best early source of rock quality data is geologic mapping of rock exposures. Features to be aware of include: fracture density and orientation, shearing, faulting, geologic contacts and alteration zones. It is often possible to project or infer the presence of geologic features into a drill site from surrounding rock exposures, underground workings, air photos and geophysical surveys. Rock quality will commonly vary quite a bit across a deposit so adequate logging of cuttings or core as drilling progresses is important to early identification and correction of problems.

Openness

This characteristic reflects the extent to which fluid circulation can be maintained during drilling. Circulation can be lost to open fractures, faults, contacts, vugs, caves, underground workings.etc. Drilling fluids including mud, water and air are necessary for the effective operation of the drill and with some drilling methods as the means of sample recovery. The relative openness or tightness of the rock is therefore of real importance to the ultimate effectiveness and cost of the drilling as well as the validity of the sampling.

In addition to geologic mapping,as discussed in the preceding section, specific information on the potential for open ground in a project area can be derived from several sources. Published stratigraphic studies generally mention the occurrence of caves or vuggy units and some rock types, limestones or breccias for example, are more likely to have significant openings than others. Drilling in areas of past underground mining is always a cause for caution. Unfortunatly, very little can be done in the way of preventative measures. Old mine maps can be of some help and they are commonly available, either from government publications or local miners. However, these are usually not accurate enough to provide any assurance and constant preparedness remains the most prudent course of action.

Water Content

The presence or absence, as well as the quantity, of water in rocks to be drilled are of concern in the design of a drilling program. With some drilling methods water can directly affect the quality of the samples received by causing artificial concentration or dilution. When air is being used as the drilling fluid small amounts of water can make drilling very difficult and large amounts can shut down the drill. Prior knowledge of potential water problems will allow a more informed drill selection as well as contingency planning for alternate drills or drilling fluids.

Initially, the potential for significant water in the project area is inferred from geologic data developed during the early stages of exploration. Knowledge of local stratigraphy, structure and hydrology is developed from reconnaissance mapping and review of published reports and water well data. Projection of this information into the project area can provide a basis for early planning but the actual water content of the rocks on a hole by hole basis can only be determined as the hole is being drilled. Once drilling begins, careful monitoring of drill progress will identify water problems in time to take appropriate corrective measures.

Mode of Gold Occurrence

This refers to whether the gold is firmly bound to the rock and therefore easily retrieved with the sample, or is loose and subject to being separated. Gold occurring in situations such as, oxidized fractures, highly altered rocks, or open, vuggy veins is subject to being broken loose by the action of the drill or the drilling fluids. Due to its high density, gold has a strong tendency to differentiate from other rock materials. In tight ground with good circulation characteristics, this can produce an irregular series of artificially concentrated and diluted samples. In open ground the majority of the gold can be lost yielding erroneously low sample values. Early identification of a sampling problem and appropriate corrective measures can save substantial time and money and in extreme cases can mean the difference between having a mine or not.

The mode of gold occurrence can only be documented by careful sampling. In areas where sampling problems are suspected, comparison of selective versus bulk samples can provide early clues to the potential for loss or concentration of gold. The most effective method of identifying and correcting for sampling problems due to loose gold is careful monitoring of the drilling. Cuttings or core should be logged daily and special samples of drilling fluids taken on a regular basis. Finally, in cases where it is suspected that gold is being lost an alternate sampling method should be used to twin some of the holes, either another type of drill, a trench or an underground drift.

Internal Gold Distribution

A gold deposit can be represented graphically by an imaginary surface defined as the boundary between ore and waste under a specific set of conditions. The internal gold distribution then refers to the spatial distribution of gold within this imaginary surface. This distribution is never homogenous, although it can appear to be when viewed at certain scales, just as a wall can appear smooth when viewed from a distance and rough when viewed more closely. Most often, when viewed at the scale of the entire deposit, the internal gold distribution appears as a cluster of pods. If a single pod is viewed at an even smaller scale, it in turn, appears to consist of a cluster of smaller pods. At any scale these pods have the geometric variables of size, shape, orientation and inter-pod spacing, each of which varies widely and independently of the others. The degree to which the internal gold distribution varies in each of these geometric variables affects both hole design and sample spacing. If these variables are not properly matched, either the sampling will not be adequate and the resultant ore reserve invalid or the deposit will be over sampled with a significant waste of time and money.

The internal gold distribution is displayed by contouring gold values in both plan and section. Prior to drilling, careful sampling of ore exposures in outcrops, trenches or underground workings can provide data on the variability of the gold distribution. Study of similar deposits can help to develop an initial sampling plan. The early stage of drilling is the most critical time for the development of gold distribution data and for modifying preliminary sampling plans.

Grain of Mineralization

Quite often, the pods that make up the internal gold distribution are elongated or flattened and oriented subparallel to each other, thereby imparting a grain to the distribution. This grain can be seen at a large scale in the orientation of ore shoots in a vein system and at a small scale in the orientation of mineralization along foliation, cleavage or bedding planes in many deposits. A grain can be present at any or all scales and is not necessarily sympathetic with the overall orientation of the deposit. The presence of a grain imparts a directional dependence to the variability of the gold distribution. In other words, the gold values will vary more in one direction than in another. When this occurs the variability will be greatest across the grain and least along it. This in turn affects the validity of the samples and if not identified and accounted for, can result in an invalid reserve or a significant waste of time and money in resampling.

A grain of mineralization is almost always a result of the local ore controls. A knowledge of what has caused the gold concentration at the particular scale of interest is,therefore, the best source of information on the presence and geometry of a grain. Geologic mapping of exposed mineralization can provide early data on possible small scale ore controls such as specific sedimentary beds, shear or fault zones, cleavages, foliations, etc. Selective sampling across and along these features can help document directional sampling characteristics. At a larger,scale contouring of preliminary sampling in both plan and section, can help highlight a directional trend in larger pods or the deposit itself. Combining these contours with geologic mapping often helps to focus on deposit scale ore controls. As always, constant monitoring and testing in the early stages of the program are required for early identification of potential problems.

DRILL PROGRAM DESIGN

The end result of a drill program is a three dimensional array of samples. If done properly this array will accurately reflect the grade and tonnage of the deposit and will have been obtained at a reasonable cost in terms of both time and money. Significant errors occur when samples do not accurately reflect the volume of rock intended. These errors commonly arise from incorrect program design decisions such as: selecting an inappropriate drill for the job, taking non-representative samples or simply taking too few samples. The following paragraphs highlight some of the important program design options and some of the factors important in their selection.

Drilling Options

In most situations drill selection consists of a choice between core, reverse circulation rotary or standard rotary drilling. The fundamental question is: Which method will return the most appropriate sample with the best cost/time characteristics? It is seldom possible to accomplish all of the geologic, engineering and metallurgical goals with a single drill type. This will provide an opportunity to use another type of drill to test the primary drill selection.

Core Drilling. This is the most expensive and time consuming of the three drilling methods. It usually returns the best sample and has the added advantage of providing substantial pieces of solid rock for observation and testing. It is particularly appropriate for detailed studies of mineral distribution, geotechnical studies and metallurgical testing. In addition, it is the only really good method of drilling shallow angle, horizontal or up holes. Because of the ability to drill up holes core is the most commonly used underground exploration drilling method. Finally, core drilling is the only method not significantly hampered by excess water or loss of drill fluid circulation.

There are only a few, limited, situations where core drilling would not be appropriate from a sampling standpoint. These include: areas of broken ground such as open veins, shear zones and poorly cemented breccias, where the advantage of retrieving large pieces of rock is lost and good recoveries become difficult to achieve; and situations where the gold occurs in such a manner that the fluid flow necessary for core drilling will wash it away and invalidate the sample.

Because of the substantially higher cost per foot and longer drilling times,core should be used sparingly and only in those instances where its particular advantages are essential.

Reverse Circulation Rotary. This technique is significantly cheaper and faster than core drilling. The percentage of sample recovered, while generally not as high as core, is commonly quite good. The technique works well in broken ground and in situations where air or mud circulation is easily lost. Due to its good sample containment characteristics, reverse circulation rotary drilling is the preferred method in situations where gold values can be easily lost by separation from the surrounding rock.

Although reverse circulation drilling has, in recent years, become the method of choice in most cases, there are some specific situations where it would be inappropriate. First, and most important, in areas of tight ground with very good rock quality, it would be cheaper, faster and equally accurate to use standard or conventional rotary; second, in those instances where details of the gold distribution are critical to the accuracy of the ore estimate or for engineering purposes. In these cases core drilling would be the desired option. This is because the rotary drills tend to homogenize the sample over the sample length and small high grade zones or other distribution details can be obscured. Angle holes can be drilled with rotary drills and with care an adequate sample can be recovered but the quality decreases rapidly as the angle becomes shallower.

Standard Rotary. This method is the least expensive by a significant margin although the gap between standard and reverse circulation narrows each year. It is also normally faster than reverse circulation drilling. In areas of very good ground with few breaks or voids and good sampling characteristics this method is entirely satisfactory. Where angle drilling is required it is somewhat less desirable than reverse circulation drilling but with care the results from steeper angled holes can be quite good. As with reverse circulation drilling, standard rotary drills are not appropriate for shallow angle holes.

Drill Hole Design

Hole design consists of depth, angle and direction. The design question is: Will the hole cut the entire thickness of mineralization and sample a volume of rock that will be representative of the rock over the assigned area of influence?

In all cases hole depth should be sufficient to pass completely through the ore and into the surrounding waste. Because it is not uncommon for boundaries of gold deposits to be highly irregular it is also prudent to plan a number of deeper holes in selected areas across the deposit.

Selection of hole angle and direction is a matter of matching the hole orientation to the internal gold distribution. If no mineralization grain exists the variation in gold values will be the same in any direction and the hole angle and direction will make no difference to the validity of the sample recovered. If, on the other hand, a grain does exist at the scale of the drill hole the direction that the samples are taken becomes critical. In most situations, samples taken across the grain will be most representative and the hole orientation should be designed to approach that as closely as possible. Sampling along the grain is analogous to drilling down a vein and can produce completely biased results by sampling either all ore or all waste.

Sample Distribution

The sample distribution is a three dimensional array of sample points consisting of two elements: the spacing between holes and the down hole sample spacing. Although possible, a sample distribution with equal spacing for both of these elements is impractical. It is simply too expensive to make a drill pattern tight enough to match the down hole sample spacing or the down hole spacing broad enough to match the distance between holes. Additionally, tightening the drill hole pattern requires the addition of drill holes which is not only expensive but time consuming. Whereas, tightening the down hole sample density is much easier, faster and cheaper. Therefore, drill programs have a built-in directional bias with the down hole direction being significantly more appropriate for close spaced sampling.

In designing a drill program the question is: Will the sample distribution adequately reflect the variation of gold values so that the interpolation between sample points required for ore estimates is valid? A properly designed sampling program will include an adequate sample density in areas of high variability to insure the accuracy of tonnage and grade estimates but not waste time and money by unnecessarily sampling areas of low variability.

If no mineralized grain exists, the spacing of samples in all directions is determined by the scale of variations in the internal gold distribution. If gold values vary significantly over short distances a tight sample spacing will be required and vice versa. In most gold deposits a mineralized grain does exist. In these cases every attempt should be made to match as closely as possible the directional natures of both the gold distribution and the sample distribution. This is done by orienting the sample array such that the spacing between drill holes corresponds to the direction of least variability, and the down hole sample spacing corresponds to the direction of maximum variability, in the gold distribution. Even an imperfect match can save the cost of drilling a number of holes while retaining or even improving the accuracy of the sampling data.

ASSURANCE TECHNIQUES

Since the decisions and choices outlined above are commonly based on what is at best scant data it is essential that they be tested for validity early and monitored throughout the program. Also, since it is not uncommon for some of the physical characteristics to vary significantly over short distances, it is important to insure that the validation tests be done in several areas across a deposit. Common techniques for validating data and assumptions include: drilling twin holes, varying the hole design, trenching or drifting to expose the ore and changing sample spacing. The following paragraphs highlight some of the techniques that can be used to validate or confirm the data and assumptions discussed above.

Drill Selection

If the drill selection is not a clear cut case some of the required twin holes can be drilled using another type of drill. This should be done early in the program and the results critically compared. If rotary drilling is being used it is a good idea to drill some large diameter core for metallurgical testing and if planned properly this can serve the multiple purpose of checking the sample validity of the rotary drill and providing a wealth of data concerning the continuity, variability and physical characteristics of the rock. If core is being used without problems it might be helpful to try a few rotary holes to see if they could do the job as well.

Drill Hole Design

The validity of the hole angle and direction design can be checked by drilling check holes that cut the same rock from several directions and critically comparing the results. This should be done in several areas across the deposit. If at all possible a trench or drift should be made near some of the holes for detailed mapping and sampling. This information can confirm the basis of the original hole design or suggest changes. In addition information from trenching or drifting can greatly expand the data obtained from core drilling.

Sample Distribution

Continuity and variability data developed from core and trenching or drifting can be enhanced by drilling lines of closely spaced holes in selected areas. Also, the down hole sample interval should be reduced on a regular basis as an ongoing monitor of changes in the down hole variability. Sample maps and cross sections should be maintained and updated as assays are received in order to identify problems or opportunities as quickly as possible.

CONCLUSION

Knowing the geologic characteristics of a gold deposit and the use of this data in the design of a drilling program can significantly improve the validity of the resultant samples as well as the cost, in time and money, of the overall program. There are no pat answers or cookbook solutions to resolving such design questions as what type of drill to use or at what orientation and interval to take the samples. However, care and thought in the early design of a program, combined with ongoing review and modification, can help to avoid many potentially significant problems. The ability to recognize these problems early and correct them before they become critical can add an important incremnet of assurance to the final ore reserve.

Chapter 10

IMPROVING ORE RESERVE ESTIMATES FOR PRECIOUS METALS PROJECTS

J. Allen Hennessy

Bechtel, Inc. - San Francisco, California 94119

INTRODUCTION

The increasing price of gold and silver over the past 10 years has caused increasing activity in the exploration, evaluation and development of precious metals projects. Many of these deposits are relatively low grade and short lived which makes them economically sensitive to the major project variables such as metal price, capital cost, operating cost and of course, tonnage and grade. These deposits often have a high variability (a coefficient of variation -- σ/μ -- of 2 to 5) and an intermixing of ore and waste, making accurate ore reserve estimation difficult, especially when using traditional estimation techniques.

Often, the estimation errors are not the 10-20% that we see in base metal deposits, but the errors in grade and tonnage can be as high as 50%, which can obviously have a very substantial impact on project economics.

Recent developments in geostatistics when coupled with a sound geologic understanding of the deposit can significantly improve reserve estimates of these variable precious metals deposits.

Examples will be given of actual deposits where the traditional estimates were significantly different from the mining reality. An attempt will be made to explain the applicable geostatistical tools with a minimum of geostatistical jargon and without complicated formulas. The key to improving the reserve estimates is a clear understanding of the deposit geology and especially the structural controls, the use of local estimation techniques that recognize the intermixing of ore and waste and the effect of selective mining on recoverable reserves, and the use of a global reserve estimation technique to check the results of local estimation.

Geologic Controls

The vast majority of gold and silver deposits being mined or developed in the western United States would be classified as epithermal deposits. These are formed by hydrothermal fluids (hot water) flowing upward through the rocks and depositing the precious metals as the fluids cool or reach areas of lower pressure or favorable chemical sites. These environments are similar to present day hot springs-type environments with some deposits being clearly associated with surface hot springs such as the Homestake McLaughlin deposit in northern California.

These hydrothermal deposits can occur in sedimentary rocks resulting in deposits such as those of the Carlin district or the Jerritt Canyon deposit or the hydrothermal systems can occur in volcanic rocks forming deposits such as Round Mountain, Manhattan or Paradise Peak. The mechanisms of formation were similar in either case with the solution flow controlled by geologic structure and precious metals deposited when the conditions of temperataure, pressure and chemical environment were favorable.

The gold and silver bearing solutions travel through permeable faults and fracture systems depositing the metals along these channels. The grades are often highest in the central parts of these conduits (veins) with lower grades in the surrounding rocks. An understanding of the nature, distribution and orientation of these plumbing systems is important to a realistic reserve estimate. The point to be emphasized here is that, good, sound geology in the form of geologic mapping, logging of rotary chip samples and a few carefully logged core holes can significantly improve an ore reserve estimate.

Examples of Estimation Problems

Epithermal precious metals deposits with high grades concentrated in the central portions of permeable structures surrounded by lower grade ore and waste can lead to large estimation errors when techniques such as polygonal, inverse distance squared (IDS) or linear kriging are used. Examples of estimation problems are not usually publicized for obvious reasons. Some examples are presented here comparing estimates to actual production results, but the deposits names for the most part, must remain confidential.

With the availability of automatic computerized reserve software packages, it is easy to put the drill hole assay data in the computer and push the button, so to speak, and out comes a polygonal, IDS or kriging estimate. The problem with these types of conventional or linear estimates is that they often neglect the geological controls, must estimate blocks much larger than will be mined because of the exploration drill spacing, or can be unduly influenced by a few high grade assays. The frequent effect is that the polygonal estimate will overestimate the actual mined grade and underestimate the recovered tonnage while the IDS or kriging estimate will underestimate the grade and overestimate the tonnage.

A traditional geologically controlled cross section estimate often will be closer to the recovered grade, but is more difficult to computerize for use in time saving mine planning software programs. The cross sectional type estimate sometimes will overestimate grade and underestimate tonnage depending on how much mining dilution is included in the estimate.

Data has been accumulated to compare block kriging estimates with actual production for two structurally controlled gold deposits in volcanic rocks. The measured mined production was about 1 million and 3 million tons and in these two deposits the block kriging underestimated grade by 30% and overestimated tonnage by 25%. In a third case at a gold deposit in the western United States, the IDS method overestimated the tonnage in the first year's pit causing a significant modification of the mine plan in order to meet production requirements. In this instance, a few high grade samples influenced the grade estimate of a large area around them causing the problem in the estimated ore tonnage.

In the case of an epithermal gold deposit in volcanic and volcaniclastic rocks, the IDS estimate was 50% lower in grade and 30% higher in tonnage than an indicator kriging estimate which was checked by a global adjustment of the histogram. There is no production from this deposit as yet, but there is little doubt that either the IDS or linear kriging techniques will significantly underestimate the grade and overestimate the tonnage. This deposit has a coefficient of variation (s/m) of 3 to 5, and this extreme variability causes the severe problems experienced with the linear estimates (Journel and Arik, 1987).

An example of the overestimation of grade and underestimation of tonnage by the polygonal technique is found in the Pine Creek Gold Deposit of the Northern Territory, Australia (Schofield, 1987). This is a quartz vein saddle reef type deposit in Precambrian sediments. The deposit was estimated using the polygonal cross-sectional technique and the first years production indicates that the estimate was about 10% high in grade and 50% low in tonnage.

An interesting example of the application of nonlinear geostatistical techniques to the Jerritt Canyon gold deposit (Verly and Sullivan, 1985) indicates that these methods can estimate the grade and tonnage within about 10% for this deposit. Conventional estimation techniques were unable to accurately predict the grade or tonnage for a test area. The 2 million ton test block had a coefficient of variation of about 2.0 which is typical, but not particularly high, for these type deposits.

Geostatistical Techniques

Two geostatistical approaches can help in solving the estimation problem encountered in these precious metals deposits with a high variability of grade. The first would be the category of nonlinear local estimators such as conditional probability distributions, multigaussian kriging (MG), probability kriging (PK) and indicator kriging (IK). These methods have been discussed in previous papers by Parker and Journel and others (Parker, 1979; Journel, 1984). Only the simplest of these techniques, IK, will be described in this paper. The second useful technique is a global adjustment of the histogram of samples or composites to account for the difference in variability between the samples and the selective mining unit. The selective mining unit (SMU) is defined as the smallest block of material in the mine that can be classified as either ore or waste. For example, in an open pit gold mine with 20 ft benches and a

15 x 15 ft blasthole spacing, the selective mining unit would be on average 20 x 20 x 20 ft or possibly as small as 15 x 15 x 20 ft if the loading equipment were sufficiently small. If the same deposit was mined with larger equipment using a 40 ft bench with 20 x 20 ft blasthole spacing, the SMU would be about 30 x 30 x 40 ft and the recovered reserves would be lower in grade and higher in tonnage for the same cutoff.

The same volume variance principal applies when large blocks (say 100 x 100 x 20 ft) are estimated by IDS or linear kriging from exploration drill data spaced at 100 ft centers (Figure 1). The high grade and low grade assays are averaged together in these linear estimation techniques assigning an intermediate grade to the entire block. In actual mining practice, there would be about 40 blastholes to inform us about the grade of the large block and it would be mined in about 25 separate smaller blocks or SMUs, some being ore and some being waste. With exploration data spaced on 100 ft centers or even 50 ft centers, it is not possible to accurately estimate small blocks in this type variable deposit. The idea of indicator kriging and some of the other nonlinear estimators is to estimate the proportion of the larger block (100 x 100 x 20 ft) or the number of smaller blocks (20 x 20 x 20 ft) within a larger block that are ore and the grade of that ore.

Figure 1 depicts a case in which a large vein or a mineralized structural zone cuts across a 100 x 100 ft block. If the block is estimated by IDS or linear Kriging using the exploration drill holes spaced on 100 ft centers, the block would be 100% ore at a grade of 0.11 oz Au/t at a 0.05 oz Au/t cutoff. If the block is estimated by indicator kriging it would be 50% ore having a grade of 0.21 oz. Au/t and 50% waste with a grade of 0.02 oz Au/t. The estimate by indicator kriging will clearly be closer to actual mining production data for this case.

Another useful method in the assessment of global or overall reserves, either as a first look or as a check of a local reserve estimate, is the adjusted histogram of assays or samples. This estimate does not locate the individual blocks of ore but can give a realistic estimate of the total recoverable reserves.

A histogram of samples spaced evenly throughout the deposit is needed for this type of estimate. If the drill spacing is on a fairly regular grid then there is no problem, but if the drilling is more

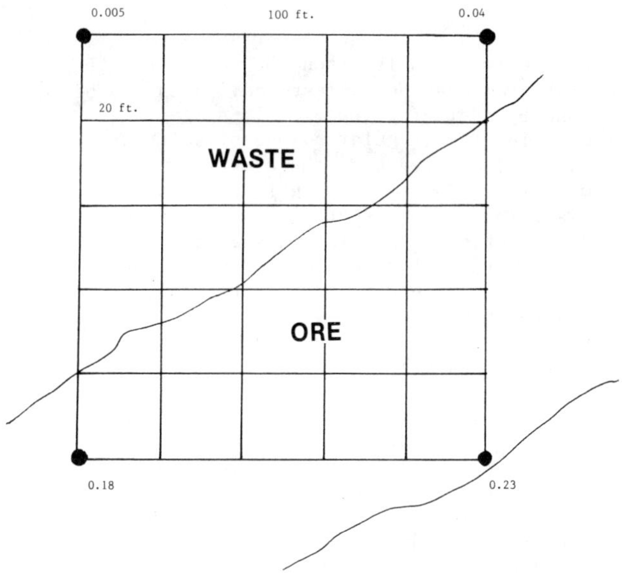

Linear Kriging or IDS	Indicator Kriging
Grade = 0.11 oz/t	Grade = 0.21 oz/t
100% Ore	50% Ore
	50% Waste

Figure 1
Indicator Kriging vs Linear Kriging

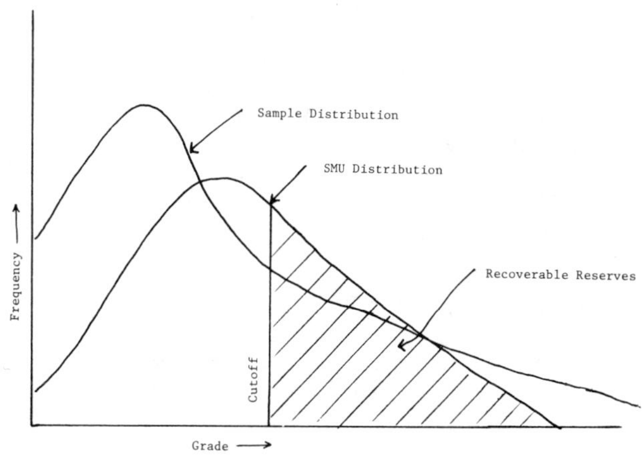

Figure 2
Histogram Adjustment

closely spaced in the higher grade part of the deposit, as is often the case, then the histogram must be declustered. This can be done by averaging the closely spaced drilling into regular polygons, so that each value represents the same volume of material. The assumption is that the histogram is representative of the true distribution of samples in the deposit. In other words, if a thousand or a million more holes were drilled the shape of the distribution would not change significantly, i.e., the mean and the variance would be about the same.

The next step is the adjustment of this histogram by a reduction in variance so that it will represent the histogram of SMUs that will actually be mined. This can be done using an affine correction which consists of adjusting each interval of the histogram without making an assumption regarding the type of distribution. The histogram also could be adjusted using a lognormal or a normal assumption, where the shape of the distribution is assumed. At any rate, the variance of the distribution is reduced by the amount of variance of a sample within the given SMU. The larger the SMU, the more the distribution's variance will be reduced. If the blocks are very large, as in the case of a block caving operation, the grades of the blocks will be almost the same and close to the average of the deposit but if they are very small, as in a small open pit gold mine, the difference (variance) between individual blocks can be very large. Figure 2 depicts an example of a histogram of samples and a histogram of SMUs with a reduced variance and the estimated recoverable reserves.

The variogram, a graph of variability vs separation distance is used to assess the amount of variance reduction appropriate for the particular deposit. The variance of samples in the entire deposit, a measure of the spread of the histogram, is reduced by the variance of samples within the SMU, referred to as the within block variance. The variogram is used to calculate this within block variance which is subtracted from the overall variance of samples in the deposit to yield the variance of SMUs in the deposit. This is Krige's relation (Journal and Huijbregts, 1978) written as:

$$D^2 (V/G) = D^2(v/G) - D^2 (v/V)$$

Where $D^2 (V/G)$ = dispersion variance of blocks (V) within the deposit (G)

Where $D^2 (v/G)$ = dispersion variance of samples (v) within the deposit (G)

Where $D^2 (v/V)$ = dispersion variance of samples (v) within a block (V)

The primary assumption in this method is regarding form of the distribution of SMUs. If the reduction in variance is less than about 30%, this method has been shown to give reliable global estimates.

With the high variability of some precious metals deposits, a few very high assays must be dealt with before the histogram adjustment can be applied. This can be done by treating the very high grade population separately. The high grade population can be assumed to represent a tonnage proportional to the proportion of high grade assays in the deposit. If 3 percent of the assays are above 0.3 oz/t, for example, it is assumed that 3 percent of the mined tonnage somewhere in the deposit will be above 0.3 oz/t with an average grade equal to the mean (or possibly the median to make it less sensitive to outliers) of the distributions above 0.3 oz/t.

The histogram adjustment can be applied to the remaining lower grade population and the high grade tonnage is added to the global estimate to calculate the final global reserve.

This method recognizes that the deposit will be mined as SMUs and not as large blocks nor as small drill core-sized samples. It has proven to be a very effective method for a first look at a grade-tonnage curve and as a check of local estimation methods.

Conclusion

The precious metals deposits being evaluated today can present very difficult ore reserve estimation problems because of their structural complexity and high variability. These problems can be at least partially overcome by a proper understanding of geology, by applying this geologic understanding in the ore reserve estimate, and by the use of nonlinear geostatistical techniques, such as indicator kriging, for local estimation and the histogram adjustment techniques for global estimations.

REFERENCES

Journel, A.G. and Huijbregts, Ch. J., 1978, "Mining Geostatistics," Academic Press, 600 p.

Journel, A.G., 1984, "Recoverable Reserves Estimation: The Geostatistical Approach", SME-AIME Annual Meeting, Los Angeles, CA, preprint 84-42.

Journel, A.G. and Arik, A., 1987, "Dealing with Outlier High Grade Data in Precious Metals Deposits", in press.

Parker, H.M., 1979, "The volume- variance relationship: a useful tool for mine planning," E&MJ, October, pp. 106-123.

Schofield, N., 1987, "Ore Reserve Estimation at the Enterprise Gold Mine, Pine Creek, Northern Territory, Australia; Part 2, The Multigaussian Kriging Model", in press.

Verly, G., and Sullivan, J., 1985, "Multigaussian and Probability Krigings - Application to the Jerritt Canyon Deposit," Mining Engineering, Vol. 37, No. 6, June, pp. 568-574.

Chapter 11

EVALUATION OF GOLD PROSPECTS: GEOLOGY INTERFACED WITH ORE RESERVE CALCULATION METHODS

Ron Jackson
District Geologist

Pathfinder Mines Corporation
San Francisco, California

Abstract. The report gives a general method for the geological evaluation of a gold deposit. Recognizable parameters associated with gold mineralization are used to develop a model of an orebody. The California Mother Lode is used as an example.

INTRODUCTION

Understanding the geologic controls of an ore deposit is important to accomplish at an early stage in the exploration program in order to know the tonnage-grade potential, and to properly design a drilling program.

Structure is often the principal control of ore bodies. Form a conceptual model of the area of interest to gain insight on the size and attitude of the deposit. Gold may be introduced up a fault and outward on favorable structural intersections. For example, a model may show mineralization moving up a strike slip fault and outward on a detachment bedding fault. Or, a model may show mineralization moving up a block fault and outward on a thrust fault. The model is refined for fault refraction in respect to variable rock competency and elaborated in respect to strain ellipsoid analysis. A detailed understanding of structural controls is often not gained until after mining has commenced. This is because surface exposures may not reveal all structures, and the expense does not allow for core drilling at close spacings.

Host rock lithology is described to determine its relationship to gold content. This is done by mapping surface geology, by a preliminary drilling program and by assaying for gold. Relevant literature research is done. It is important to separate facts from perceptions. Be willing to change your geologic model with the addition of new information.

An overview of the deposit can be obtained by looking for correlations between gold mineralization and host rock characteristics: alteration, protolith, fracture and veining characteristics, reactive chemistry, evidence of temperature/pressure conditions, depth of mineralization and paragenesis.

This further defines potential tonnage and grade of the deposit by knowing more about its geometry and if distinctive high grade zones exist. Assays may show that high grade gold associates with a more easily recognizable feature such as a particular structure (i.e., breccia zone or stockwork) or mineral (i.e., with sulfides adjacent quartz veins).

CALIFORNIA MOTHER LODE

The California Mother Lode is used as an example of the process of geologic modeling. For the literature review, I have selectively drawn from Knopf's excellent report on his observations and interpretations of gold mineralization in the Mother Lode. Some bias is used in this selection based on our own observations. Adolph Knopf, The Mother Lode System of California, USGS Prof. Paper 157, 1929:

Gold mineralization occurs at the end of the Nevadan orogeny after isoclinal folding and post Mariposa Formation, igneous intrusives which probably occurred by order of decreasing basicity: periodotites to hornblendite and gabbro to granodiorite to quartz diorite to albitite and albitite porphyry dikes. Serpentine was formed before gold depositing solutions. Carbonatization (ankerite) is strongest in serpentine and in augite rich greenstones.

Knopf recognizes that quartz veins can contain 1-2% sulfides which are pyrite, arsenopyrite, sphalerite, galena, chalcopyrite, tetrahedrite and rarely telluride petzite. Whereas, ankerite can contain 2 to 4% sulfides being restricted to pyrite and arsenopyrite. Galena and petzite are good indicators of high grade gold ore.

Alteration by decreasing abundance is given as ankerite, sericite, albite, quartz and pyrite. The low grade, ankerite 'gray' ore occurs most favorably by replacement of greenstone (especially augite rich), and black slate is unfavorable. Ankeritization has restricted permeation into slates outward from quartz veins.

High grade ore shoots contain a large portion of the gold previously mined from the Mother Lode. Ore shoots are persistent in dip direction and limited in strike length (i.e., ~5000 feet vertically and ~250 feet wide). Shoots are located at wedges between converging quartz veins, at the elbows in abrupt bends in quartz veins, at local expansions of quartz veins and at the transition zone between greenschist and slate.

Mother Lode quartz veins occupy fissures formed by reverse faulting. The veins are mesothermal with local hypothermal transition. The veins dip less steeply than the Mariposa Slate, but commonly will follow the lithologic contact. Quartz veins wholly enclosed in greenstone are likely to be low grade while veins in slate are likely to be high grade. Mother Lode quartz veins strike north to northwest and dip east. The veins commonly diverge from the northwest trending host rock by 15 to 20 degrees.

Knopf hypothesizes that gold is from a deep magma source after dacite had solidified; that gold is introduced during ankeritization; and that quartz vein origin is a result of the decomposition of silicates in the wall rocks by carbon dioxide.

PATHFINDER RESEARCH

Pathfinder Mines Corporation's business objective is the acquisition of proven gold reserves or advanced exploration projects. This often puts us in a position of having well established data bases to work with at the start of project evaluations.

An early exercise by Pathfinder in evaluating the Jamestown Mine in the Mother Lode was to reinterpret New Jersey Zinc's plan level geology maps to emphasize the distinct alteration phases and rock types. ABM Mining Group, Inc. (ABM)/Sonora Gold Corporation had composited gold assays and created a kriged, block model showing grade distribution. Normand Champigny of International Geosystems Corporation and Gordon Ford of ABM presented a paper titled "Geostatistics at the Mine Planning Stage: Harvard Deposit, Jamestown Mine, California" at the ICM Symposium in 1986. The paper describes their method of geological block model construction and geostatistical calculations.

We used ABM's data file to generate grade contour maps at the same scale as the geology maps to gain a general understanding of grade distribution with lithology/alteration. The geology map shows that serpentine and talc generally occur on the footwall side of the quartz vein. Mariposite closely halos the quartz veins. Ankerite preferentially occurs on the hanging well side over a wide zone adjacent to the Mother Lode veins. Slate/argillite form the outermost hanging wall. Additional dikes and tuff units are recognized. The grade distribution shows the highest grade zone follows the quartz-mariposite zone along strike and down dip. Ore does not penetrate the serpentine zone far and is not regulated by the ankerite zone. Cross structures cause local high grade trends.

The next stage in refinement of the model is petrographic analysis. Pathfinder's in-house petrographic studies are done by the mineralogy department of our parent company (COGEMA) in France. The major portion of this study is in progress. The studies are done by A. Pacquet and by J. Reyx. Their observations are summarized into the following:

PARAGENESIS

Metamorphism

- silicification
  - strongest in tuffs
- albitization
  - regionally intense to non-existent
- pyrite and chalcopyrite aligns with S_1 foliation;
- titano-magnetite alteration
- microfracturing
- vein fill: albite, apatite, quartz, pyrite - followed by chalcopyrite, pyrrhotite
- minor event: ankerite

Blastomylonitization

- fibrous quartz crystallizes in pressure shadows around pyrite, chalcopyrite and apatite crystals

- sulfides in pressure shadows and fracture fillings

Ferrostilpnomelane and talc

- microshears
- chalcopyrite can occur between stilphnomelane layers

Fe and Mg chlorite

- microshears

Mariposite

- locally white micas
- microfolding and microshears
- regional variation, event may be accompanied by sphalerite, chalcopyrite, arsenopyrite.
- mariposite can locally cement brecciated pyrite which often has gersdorffite (NiAsS) rims
- followed by minor fracturing event with quartz, vein fill.

Ankeritization

- a major, pervasive event
- strong corrosion of all previous minerals
- followed by minor magnesite

Their work is very detailed so this synopsis risks being an over simplification.

Gold can be found with all sulfide phases starting with the metamorphic phase as microinclusions in pyrite and chalcopyrite prior to microfracturing and pressure shadow deformations. Gold can occur with chalcopyrite cement of brecciated pyrite, with gersdorffite coating of pyrite, as free gold adjacent pyrite or quartz, in veinlets with sulfides, etc. Most gold occurs as microinclusions in sulfides. Chalcopyrite and other sulfides can occur as microinclusions in pyrite. Most important is recognizing the major phase of gold mineralization and recognizing what characterizes and spatially controls that phase.

A dominant alteration sequence is serpentine to talc to quartz, dolomite, magnesite. Volume change associated with alteration causes structural deformation which influences the circulation hydrothermal fluids which cause alteration. Deformation and alteration are interdependent.

After understanding the geological controls of gold mineralization, you must determine which controls can be identified by the exploration program and the significance of its impact on the orebody. As an example, one might suspect that there exists a ubiquitous, low grade dissemination of gold associated with sulfides identified in the metamorphic phase and that the geometry of this phase may be controlled by lithology. There might be overprints of gold mineralization related to later deformation and alteration phases which might be identified by rock type, mineral, or element. The important identifiers might be:

- existence of mariposite
- percent of silica
- percent of sulfides
- arsenic content
- Pb-Zn content

The drill sampling program makes certain to identify these factors as well as the lithology and alteration characteristics.

DRILLING PROGRAM

Optimum drill hole spacing might be determined by prior sampling, by assaying for gold, and by doing geostatistical analysis. Ideally, sampling is done of old underground workings. Samples should be taken in the strike direction, in the dip direction and perpendicular to the dip direction. Five foot channel samples should be taken over a couple hundred feet. The same assay techniques as planned for at the drilling stage should be used. Old assay records could be used, but with caution. Different sampling techniques, assay techniques, and cutoff levels may not allow for comparison of this data set and the later drill hole data set. Surface sampling can be useful, but sample only fresh rock. Also, a limited drilling program could help to define the geostatistical characteristics of the deposit. Drill holes would be oriented down dip, perpendicular to dip, and along strike direction. Samples collected along and immediately perpendicular to a high grade vein zone may not be representative of the lower grade, disseminated orebody. Separate variograms are produced for strike direction, dip direction, and perpendicular to dip direction.

Ideally, the drilling program would use a drill hole inclination perpendicular to dip. This direction will likely have the shortest range of continuity and the range interpreted from its semivariogram might represent the maximum sample size taken in the drill hole for assay. Drill hole fence lines should be oriented perpendicular to strike, and drill hole spacing along the fence line should allow the drill holes to intercept the vein in the down dip direction at a spacing equal to the range of the semivariogram derived in the dip direction. The spacing between drill hole fence lines should be equivalent to the range of the semivariogram derived in the strike direction.

Unfortunately for gold deposits, cost often prohibits drilling at the spacing specified by the semivariograms' ranges. You can start an initial program at some multiple of the range and then do infill drilling in the better areas.

A drilling program in the California Mother Lode used drill hole fence lines at right angles to the Mariposa slate contact and the Melones fault trend. Drill holes were spaced at 100 foot intervals, and fence lines were 100 feet apart. Samples of reverse circulation drill hole cuttings were taken at 5 foot intervals for assay and detailed lithologic description. All information was put into a computer data base and drill hole cross sectional data maps were produced showing assay results and lithology.

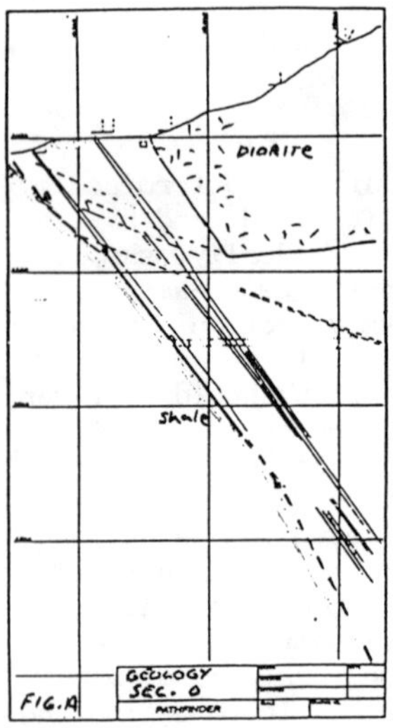

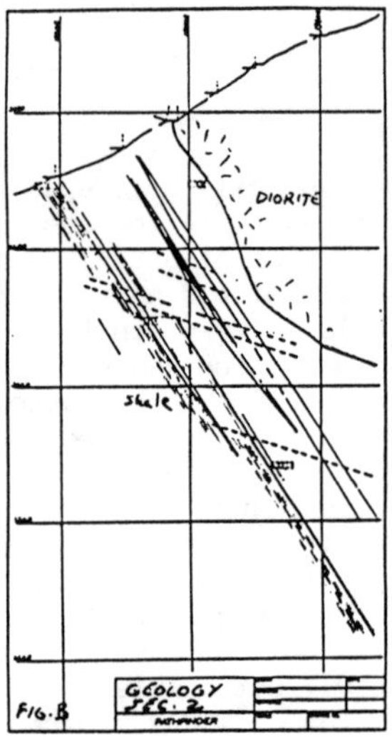

Geological cross sections were interpreted for each fence line (Fig. A and B). Interpretations used only distinctive rock types, which were black slate, quartz veins, diorite, and other. The other was a large category of various alteration types and a melange of protoliths usually not identifiable.

It was seen that the footwall Mariposa slate was surprisingly regular, and that Mother Lode quartz veins in the hanging wall, altered rock generally paralleled the slate contact. A quartz vein often occurred at the contact, and then one to three veins quartz might occur in the hanging wall up to 100 feet or more above the contact. Diorite often occurs in the outermost hanging wall.

It was observed that ore zones tended to parallel the Mother Lode quartz vein direction. This was true even for wide ore zones.

A separate set of cross sections was compiled to do cross sectional, polygonal ore reserve calculations. This is the best, manual method of ore reserve calculations. Minimum cutoff grade criteria was used. A computer data base was established in order to later vary thickness and grade cutoff parameters. Ore zones were always delineated by overlay reference to the geologic cross sections.

The diorite sill has been altered and is mineralized in places. There appears to be low angle faults which offset the slate contact, but not the quartz veins. Quartz veining tends to follow these low angle faults as does higher grade zones of gold mineralization. These faults may relate to the diorite sill emplacement and predate the Mother Lode quartz veins. A widen lens of quartz veining is often seen to occur at the intersection of the major Mother Lode veins and the low angle veins. When there are a multiple Mother Lode veins, an en echelon pattern of occurs.

High grade ore zones locally halo the quartz veins, and high grade zones locally straddle the Mariposa slate contact. The proximity of these two features is a recognizable control of higher grade zones. Longitudinally, the ore body is seen to trend approximately 20 to 25° north of the trend of the Mariposa slate contact.

A cumulative frequency curve of gold assay data was generated to determine the background level of gold content and to see if multiple populations of gold values exist.

A rock type model was digitized from the geologic cross sections. Separate classifications were made for (1) waste rock, (2) slate, (3) the slate/altered rock contact zone, (4) footwall alteration zone, (5) Mother Lode quartz vein zone, (6) hanging wall altered zone, and (7) diorite.

The model can be refined by interpreting trends between cross sections at different bench levels. Define barren interior waste rock, as well as exterior waste. As a final cross check, overlay your drill hole lithology and assay data on the model to verify that interior waste and rock types are correctly recognized.

There are several reasons to establish a rock classification inventory. Geostatistical tests might show that the zones have distinctive characteristics and should be treated separately. Rock density will differ for each classification, and adjustment must be made on tonnage figures. The metallurgical gold recovery from each zone may differ as well as the metallurgical process.

The average grade of each rock type did differ, the contact zone and the quartz vein zone being about equal, and the hanging wall and footwall alteration zones being about equal. But the semivariograms did not differ significantly.

A separate high grade ore model was developed (Fig. C). High grade zones which locally occur near to the slate contact, to quartz veins, and to secondary structures were segregated from the disseminated class of mineralization. The geologic cross sections were referred to in making these interpretations.

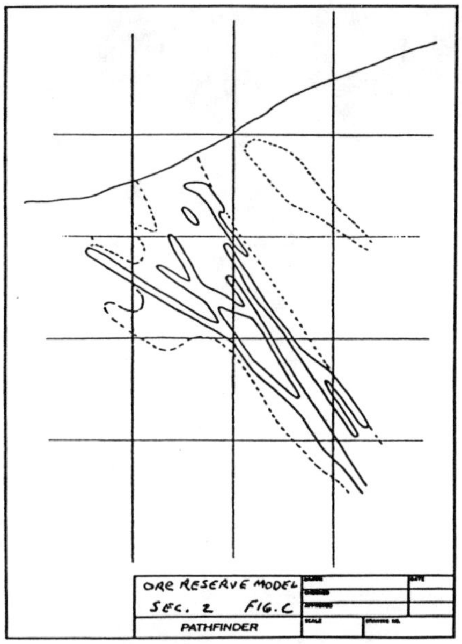

This method is flawed because of lack of detail. Minor structures were never identified by the geologist logging the drill hole cuttings. Significant quartz veins, as far as gold mineralization is concerned, that are less than 3 or 4 feet thick would not be identified on the lithology log. Gold is being used as the only identifier of gold mineralization, and gold has a high nugget characteristic, meaning that drilling might not intersect gold in what could be a high gold bearing zone. The most important flaw with this model is that the drill hole spacing was too wide to define intricacies of ore control. If we had taken more time, we could have tried to expand this model by using other mineral indicators of the high gold zones. We did use some different approaches to correct the model's shortcomings. The model did give us a better insight about ore distribution.

Chapter 12

GEOLOGICAL ASPECTS OF FEASIBILITY STUDIES

David S. Bolin
Principal Geologist

and

John W. Rozelle
Senior Geologist

1.Ø INTRODUCTION

During the normal course of a mining project, a feasibility study will be prepared that addresses all the aspects of the project including geology and mineral inventories, mining and geotechnical, metallurgical processing, infrastructure, environmental considerations and economic analyses. These studies are prepared for various reasons. Occasionally, a feasibility study will be prepared solely for use within a company as justification for development expenditures, however, these documents are frequently submitted outside the company either to raise funds from financial institutions or from public financing on one of the various securities exchanges.

Mineral industry consultants are often called upon to participate in the preparation of feasibility studies or to conduct an independent audit and review of an internally prepared study. As the recoverable mineral resources form the foundation upon which the balance of a feasibility study is based, including the economic analysis of the project, the resource aspect of the study is examined very carefully by financial and securities organizations. The balance of this discussion is directed toward the various geological and resource estimation aspects of feasibility studies that experience has shown to be of particular importance. Much of this discussion will be directed toward open pit gold operations, although the principles are applicable to any type of deposit.

2.Ø GEOLOGIC RESPONSIBILITIES IN FEASIBILITY STUDIES

The principal geological effort in the preparation of a feasibility study is generally dedicated to the following areas:

- Preparation of Geologic Reserve/Resource Estimates
- Condemnation of Sites for Facilities, Access Roads, Leach Pads, Waste Dumps and Tailings Disposal
- Collection of Geotechnical Data
- Provision of Mineralogical and Petrological Data Pertinent to Metallurgical Processing
- Selection of Appropriate Bulk Sampling Sites foR Collection of Metallurgical Samples

Each of these items will be addressed in detail in the following sections.

3.Ø RESERVE/RESOURCE ESTIMATION

An ore reserve estimate consists of the determination of a volume, or tonnage, and a grade for the deposit. There are numerous cases in the mining industry of projects that have experienced severe difficulties when operational ore grade, tonnage or metallurgical amenability have proven to be deficient compared to predicted estimates. An error in estimation of tonnage generally leads to more long term problems affecting mine life and production schedules. However, a significant error in the estimation of grade, particularly if mined grade is lower than anticipated, can have an immediate, and often disastrous, effect upon a project.

The geologic preparation leading to an ore reserve estimate generally follows a logical sequence familiar to economic geologists:

- Evaluation of the regional geology
- Definition of the geology of the deposit
- Specialized Studies on mineralogy, petrology, geochemistry and ore control, and other aspects of Mineralization
- Construction of a geologic model of the deposit

Depending on the commodity and type of deposit, reserve estimates can be prepared by a number of manual or computer assisted techniques. In recent years there has been an increased emphasis on computerized geostatistical techniques for a number of reasons:

- In many projects, the drill hole or other assay database is often kept on a computer system for ease in organization and manipulation of the data.
- A properly prepared and controlled geostatistical estimate can provide an optimum solution for the reserves.
- The block grade model containing the geologic reserves provides a good database for subsequent mine planning and estimation of minable reserves.

One aspect of a computerized reserve estimate that must be emphasized is that the law of "Garbage In-Garbage Out" applies every bit as much to reserve estimation as to other computer endeavors. The geologic rock model created for the control of the computer interpolations should be as accurate and detailed as possible.

3.1 Regional Geology

The regional geologic setting of the deposit should be examined in some detail. This information can provide substantial data with regard to district ore controls and predictability of mineralization, as well as, the potential for surrounding deposits. Often regional maps and geology can be obtained from state or federal surveys or graduate thesis, as well as reconnaissance work by project geologists.

3.2 Exploration/Evaluation Program

The subject of exploration program design is outside the scope of this document, however, a few points should be mentioned in that the careful design of the exploration/evaluation program can greatly facilitate the subsequent feasibility work.

Most deposits are evaluated by some combination of drilling; supported, in some cases, by varying amounts of underground exploration and sampling. The main purposes of this program are as follows:

- Determine the Limits of Mineralization.
- Provide Geological and Mineralogical Data.
- Determine the Distribution and Trends of Mineralization in the Deposit.
- Provide Samples for Assays and Analysis.

The spacing and orientation of drilling and sampling patterns should be determined by the geometry of the deposit. Usually, sampling on some regular grid pattern is preferable if conditions so permit. The general practice is to begin the program with a relatively wide spaced pattern to define the limits of the deposit and to continue infilling the grid until an acceptable confidence level with regard to geometry and grade of the deposit is obtained. If sufficient data is available such that a geostatistical variogram can be obtained, then an optimal sample spacing can be readily determined. The orientation of drilling patterns is a function of the known, or assumed geometry of the deposit. Whether the majority of the drilling is done with vertical or angle holes must also be determined based on local ore controls.

A final comment with respect to drilling programs concerns the advisability of core drilling. Current practice at many projects is to carry out an entire exploration/evaluation program using non-core methods, however, a strong case can usually be made for drilling a few carefully selected core holes for the following reasons:

- Many geological features such as structure and certain types of alteration are lost when observed in chip samples. In one case, a rock unit which has been logged as a quartzite for several drill campaigns proved to be a silicified siltstone when seen in core. Mineralization features such as veining and sulfide concentrations can also be better defined.

- Large diameter core (PQ-HQ) can provide samples for metallurgical testing at a relatively coarse particle size. Such samples are also available for bulk density tests.

- Features of geotechnical importance such as faults, fractures and bedding planes can be noted.

Finally, all drill hole collar locations should be surveyed. Downhole surveys for angle holes are highly desireable, particularly when vein or tabular deposits are being evaluated.

4.0 SAMPLING AND ASSAYING

The sample database forms the basis of all the subsequent reserve calculations, and it is imperative that this information be maintained in a logical and consistent manner. The process begins with collection of the sample. As the majority of deposits are evaluated by drilling, sample size for a given non-core hole diameter is a function of sample interval. In non-core drilling, accepted practice is to sample at intervals of five to 10 feet or occasionally shorter in some cases. Core drilling is sometimes sampled at a constant interval, or sometimes at irregular intervals based on geology. If drilling consists of more than one type or size, it is important that the samples be evaluated separately, at least in the initial phases of the work.

Sample recoveries can be estimated by length with some degree of precision with core, and to a somewhat lesser degree by weight with non-core drilling. This data should be obtained, presented, and discussed in the feasibility study.

In the urgency to proceed with an evaluation program there is a tendency to avoid detailed studies early in the program; this tendency should be avoided. Some detailed mineralogical and petrographic work early in the program can greatly facilitate the design of sampling and sample preparation programs as well as providing early information for the metallurgical planning. Often, this information is difficult to obtain from chip samples and rock samples of substantial size are required.

Sample preparation is often done by a commercial lab at the early stage of a project, however larger projects usually establish a sample preparation facility at the site. In either case, the preparation flowsheet should be reviewed carefully. Preparation flowsheets will usually be provided upon request by commercial labs, and custom procedures can normally be provided although additional cost may be involved. For any given ore type, an optimal procedure can usually be defined. Several references, such as the work by *Gy (1979) can be consulted for information. In any case, the sample preparation procedure, along with the history of any changes, should be incorporated in the feasibility study.

*Guy, P., 1979, Sampling of Particulate Materials: Elsevier Scientific Publishing Co., New York, Oxford, Amsterdam.

4.1 Assaying

Assaying for various commodities can be accomplished by a number of chemical, gravimetric or instrumental techniques. Most metals can be readily analyzed by atomic absorption techniques, and such methods are commonly accepted as standard engineering practice. Other methods such as fire assaying are commonly used for precious metals. Although there are arguments for both positions, most technical reviewers of feasibility studies like to see fire assay results used for reserve estimates for precious metals deposits rather than atomic absorption assays. If the ore is to be treated by cyanide extraction, some operators will also do cyanide extraction AA on all or a portion of the samples and use the ratio of cyanide extractable AA to fire assay as an indicator or metallurgical ore types.

In some deposits, there may exist a considerable bias between AA and fire assays, particularly at low grades, near the cutoff grade of the deposit. If this is the case, then it is important that the same method be used for reserve estimation as will be used for production ore control. For example, an overestimation of mineable tonnage above a cutoff grade can result if fire assays are used for reserve estimation and AA is used for ore control, if a bias of fire assay greater than AA assay exists near the cutoff grade.

Finally, it is important that a representative number of samples be checked by whole rock analyses and trace element analysis for any deleterious elements, for example mercury in some gold deposits. The presence of such elements may have an impact on metallurgical processing or waste disposal.

4.2 Check Programs

The sources of variability in samples are a function of the geologic variability of the ore, sampling errors, sample preparation and assaying. In order to maintain reasonable control on sampling precision and accuracy, a systematic check program should be established and maintained. Ideally, this check program should cover all stages of the sampling efforts. Checks can be established by taking multiple splits of the initial sample. Analytical precision can be established by multiple analysis of the same sample pulp while accuracy can confirmed by interlaboratory checks. Some samples are often checked by an alternative method such as neutron activation analysis. The cost of these analyses have dropped sufficiently in recent years to be available on a routine basis. The number of check samples/analyses run in a given program generally run from five to 20 percent.

The sample results can be analyzed in several ways such as a statistical "t" test on sample means or an "F" test of variance. Simple correlation diagrams provide a graphical display of the data (Figures 4.1 and 4.2).

4.3 Preparation of a Geologic Model

Whether the geologic reserve estimate is to be prepared manually or geostatistically, a geologic model should be constructed to be utilized in establishing shapes, trends and controls of the deposit. The model should be constructed initially in as much detail as the data permits. The specific items which can be modelled include, but are not limited to, the following:

- Alteration
- Structures (faults, veins, breccia bodies)
- Limits and boundaries of orebodies and ore types

When the model is completed, a statistical analysis of samples within various modelled rock units should be compiled, and it may be found that certain units are statistically indistinguishable, and the model can be simplified by combination of units.

Linear Correlation Of Check Assays

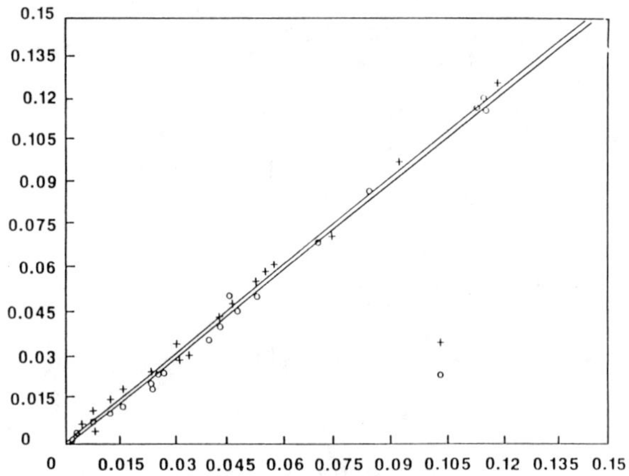

Figure 4.1

Log-normal Correlation Of Check Assays

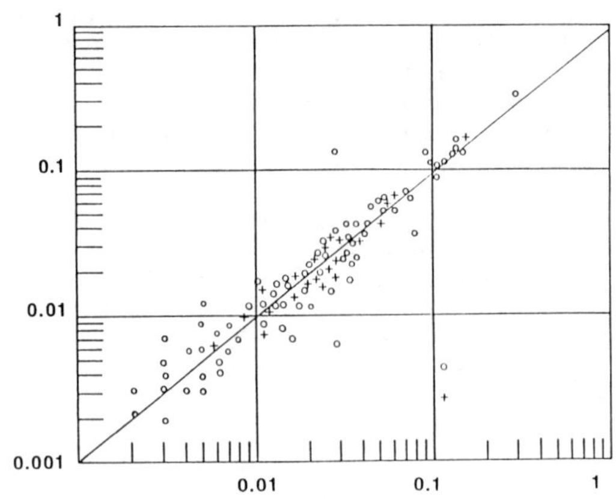

Figure 4.2

The shapes and boundaries of the mineralized units of the model should be defined with care, as these shapes will determine the volume, and therefore the tonnage, of the deposit. In some deposits, these boundaries will be sharp and defined by geologic contracts, in others, the limits must be defined by geologic interpretation.

Models can be constructed either in plan or in section, however, geologists have a tendency to work in sections. Examples of geologic rock model sections are shown in Figures 4.3 and 4.4.

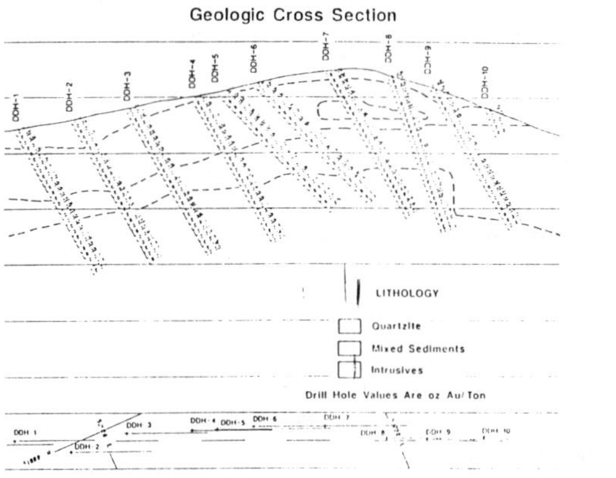

Figure 4.3

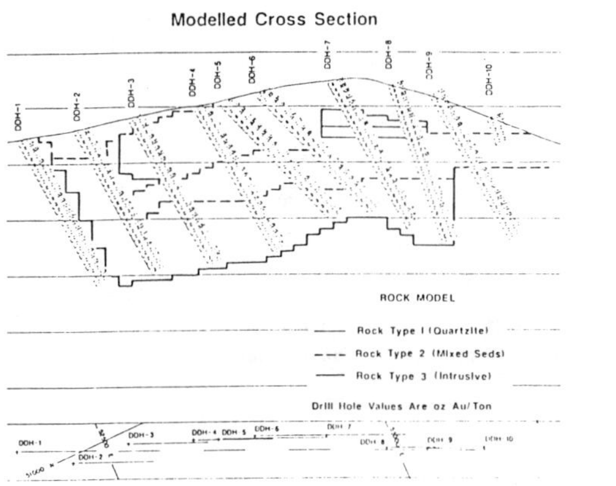

Figure 4.4

4.4 Resource Estimation

Properly prepared geostatistical resource estimates are generally accepted by most financial and securities institutions, and manually prepared estimates are also acceptable for many types of deposits. The method used is a function of cost and availability of facilities. In computerized systems, grade estimation can be done by relatively simple methods such as inverse distance to a power, or more sophisticated kriging techniques may be used. In some cases, specialized techniques such as log-normal kriging or indicator kriging may be used to provide a more refined estimate.

One aspect that is frequently overlooked or only given minimal attention is that of bulk density determinations. Relatively small differences in tonnage factors can have a large impact on reserve tonnage. Such factors should be determined for all material to be excavated, including waste.

When the calculations are completed, the resources are tabulated and categorized. At the stage of geologic resource estimation, cutoff grades selected are somewhat arbitrary, as no economic or engineering factors are incorporated. The range selected will, however, usually cover the range of expected economic cutoff grades.

Most financial organizations will require that resources be categorized by some measure of level of confidence. The USBM/USGS classification scheme is generally accepted by most financial institutions. For feasibility purposes, both measured (proven) and indicated (probable) resources can be included. These two classifications are often lumped as "demonstrated" resources. Inferred resources or geologic potential can be discussed, but quantitative estimates are not required. When economic parameters have been applied, and minability has been demonstrated, the resources become reserves. (Principles of the Mineral Resource Classification System of the USBM and the USGG; USGS Bulletin 1450A, 1976.)

5.0 GEOLOGY AND MINING

An effort should be made throughout the exploration/ evaluation program to collect data that are applicable to mining and geotechnical investigations. These data include, but are not limited to, the following features:

- Surface mapping of major faults, fractures and orientations of bedding and foliations.

- Noting of static water level in drill holes.

- Where permitting is not a problem leaving several holes cased and open for hydrodynamic investigations.

- Logging all core for geotechnical information such as shear zones and fracture density. This logging should be done before the core is split.

- All core should be photographed, in color, and in good light prior to being split.

This information should be filed in an organized archive until such time as it is required for detailed mine design.

6.0 METALLURGY

A rapport should be established at an early stage of the project between the geological and metallurgical personnel. Samples for metallurgical testing are required beginning early in the project. It is imperative that the samples are selected to accurately represent the types of ore expected to be processed. Critical features include major ore metals, degree of oxidation, associated gangue minerals and grade zones. The metallurgist is dependent on the geological staff for input in this regard.

Bulk samples will also be required for pilot scale testing as the program progresses. Obvious sources of such material include old mine dumps and existing underground workings. Although samples can usually be collected readily and cheaply from such locations, there are always questions with regard to representativeness of the sample.

Often, there is no ready access to major portions of a deposit, in which case if it is necessary to either drill large diameter core holes or to open new underground or surface workings for sampling. The metallurgical personnel can provide information as to quantities of material necessary for testing.

7.0 REPORT PREPARATION

All the information collected during the evaluation must be assembled into a concise, logical report for presentation. It is imperative that the report gives the impression that the project has the stated mineral resources available within acceptable limits of risk and that the economic projections of the project are realistic.

Shown in Figure 7.1 is a summary of the topics that should be addressed by the geological staff. Geological data is usually presented in expanded summary form. Great detail is not necessary. Sufficient data should be presented to convince the reader that a good understanding of geology and mineralization exists with respect to the resource base.

Finally, it is important to emphasize that all data be maintained in an organized, logical database. If a system is established early in the exploration phase and maintained throughout the program, the preparation of a professional feasibility study will be greatly facilitated.

Principal Geologic Participation

- Preparation Of Geologic Reserve/Resource Estimates
- Condemnation Drilling Of Sites For Facilities, Access Routes, Leach Pads, /Waste Dumps And Tailings Disposal Sites
- Providing Data For Geotechnical Studies
- Provision Of Petrological And Mineralogical Data
- Coordination With Metallurgical Staff In The Selection Of Bulk Sampling Sites
- Relevant To Metallurgy

Figure 7.1

Geologic Data Necessary For The Preparation Of A Feasibility Study

- Surface Geologic Map
- Drill Hole Location Map
- Location Of Underground Workings/Sampling
- Interpreted Geologic Cross Sections For The Deposit
- Summary Of Sample Collection And Preparation Procedures
- Summary Of Assaying And Check Assaying Procedures
- Assay Data

Figure 7.2

Chapter 13

SITE EVALUATION CRITERIA, EXPLORATION & DEVELOPMENT OF SMALL-SCALE PLACER GOLD MINES

by

Mr. Mortimer J. Richardson
Consolidated Placer Dredging, Inc.
President

Mr. Earl H. Malmstrom
Consolidated Placer Dredging, Inc.
Manager, Exploration

INTRODUCTION

The art of placer gold mine evaluation and mining has had an interesting evolution beginning in earliest times. DE RE METALLICA (c.1556), documents the use of rockers, circular sieves(jigs) and riffles; for classifying, concentrating and recovering placer gold, respectively. During the late 1800's in the heat of the international gold rush, a bucket ladder mining dredge containing its own processing plant was developed that enabled digging of placers to bedrock that were inaccessable to the hand miner. The power driven churn drill was developed at about the same time that facilitated more rapid and larger scale sampling of placers to bedrock. The mineral jig was applied in the early part of the 1900,s to begin the replacement of riffles on placer dredges. Mining dredges proliferated in California and New Zealand to mine the deeper placers and spread throughout the world. The companies that were leaders in dredge mining technology conducted experimental work that advanced the state of the art by processing of gravels with jigs and amalgam systems to recover gold on board dredges in a continuous process.

This paper will synthesize the technology of placer gold exploration, mine development and operation into guidelines for evaluation and development of small-scale prospects.

Prospect Reconnaissance

The location of a prospect usually results from some physical evidence of the presence of gold and/or past mining activity. Any earlier activity that may have produced gold from the prospect, should be investigated before making the trip. The initial reconnaissance should attempt to verify a near-surface presence and then make observations concerning the geomorphic characteristics that relate to placer gold accumulation. Estimates of the probable depth, breadth and length of mineable gravels can be made from the surface by the experienced placer mining engineer. With this information, he can speculate on whether there is a viable prospect to be pursued. The placer mining engineer should be experienced in panning samples found in shallow diggings or from existing pits. This will give him immediate indications of gold presence without fear of "salting" by an over-ambitious promoter of the project. The same capability of panning is important during the other sampling phases. A thorough knowledge of logging and calculating the results is essential to proper evaluation of the deposit's tenor.

Evaluation Methods[1]

Depending upon the character of the prospect such as river, benches, terraces, jungle, the type of equipment to conduct sampling of the deposit must be determined. In most cases, the drill has replaced bulk sampling. The churn drill has had the greatest amount of accumulated experience but is slow. The reverse circulation drill(RC), is limited by its lack of verification in subsequent mining in order to prove its reliability. However, in the hands of knowledgeable placer mining engineers experienced with its use and calibration, it may be used. The RC Hammer drill (such as the Becker), can be as much as ten times faster than the churn drill and compress the time necessary to prove a prospect, or, to disprove it. All costs in balance, it is often cost effective over the churn drill.

Drilling Equipment Since we consider that sampling of placer gold prospects can best be done with a drill, the following drills are described first.

HAND POWERED DRILLS: The earliest drill developed for placers is the Banka, developed in the 1800's in Indonesia for tin placers. Similar hand powered drills are the Empire and Ward. These are particularly useful for exploration in remote regions for shallow deposits and swampy conditions where low cost labor is available. The Banka drill can be purchased for about $10,000. plus the cost of pipe and extra tooling; complete for about $20,000. All can be transport- ed by canoe, beasts of burden or small aircraft. Many successful, placer dredging properties of considerable size were proven with this type of drill and hence can provide a high degree of reliability in the results when handled by experienced placer mining engineers.

CHURN DRILLS The best known of this type is the Keystone; developed and built in Pennsylvania in the mid 1800's. It first evolved as a water well drill and was then used in hard rock deposits in the mid west before being adapted to placer deposits in the west. This type of drill is built in various sizes from portable with 3-5 hp engines called an "Airplane" drill (because of its light weight for flying), up to 40 hp with correspondingly larger and heavier structures and drives. The casings used may vary from an I.D. of 3" to 8". The more commonly used size has been 5". They all use a simple eccentric drive that lifts and drops a weight that drives the casing. At periodic depths a sample is extracted from the pipe using a bailer.

Mounting this drill on tracks provides the greatest advantage for mobility in rough terrain. Where assessable by roads, a truck or skid mounted system is used.

REVERSE CIRCULATION DRILL(RC) The most common RC system that has been proven in placer mining, is the Becker Hammer RC drill. The casing used may vary from 3-1/8" to 4-1/8" I.D., thus producing a smaller sample than the churn drills. The Becker has been proven in its evaluation of offshore tin deposits in Thailand, where substantial drilling was followed by several years of dredge mining (Showing 90% R/E). In the case of gold we are aware of only one project, conducted (by our company) in New Zealand in 1981 drilling ahead of a large bucket ladder dredge then mining the same area, that has verified the results of the RC drill. Similar drills by other manufacturers are available. The advantage of the RC drill is its speed which is about ten times that of the churn drill. The RC's heavy weight prevents its use in some areas.

Bulk Sampling The earliest placer sampling was conducted by various methods of bulk excavation such as: Shafting, trenching, and pitting.

SHAFTING A shaft diameter of about 1.0 m is typical and may reach as deep as 30 m but usually 10 to 20 m. A wedge of 12 to 18" square is cut down one side of the shaft and sampled periodically. Seldom is the entire shaft sampled since the quantity of material to be handled is too large, unless a small trommel wash plant is used. Where low cost labor is available in remote regions, shafting is still used but is slow and not as efficient as the drill.

TRENCHING & PITTING Another form of bulk sampling is done by trenching or digging a large pit. As with shafting, a wedge is taken from a side of the pit and treated. Otherwise, by using a long riffle box or a trommel, the entire sample can be treated.

One problem with the above approach is that so much effort is expended on a few sites that there is a tendancy to limit the area covered and to extra- polate the results for the total area. This can produce distorted estimates of both the quantity of gold and the physical characteristics of the deposit since placers are typically erratic. This is where the drill is advantageous because of its mobility and less effort required in each location.

CAISSON SYSTEM The more advanced type of bulk sampling is with large diameter, caisson "drills" such as the German-made, Bade. This may have a casing with an I.D. of 1.0 m or less which is handled by a rolling crane. The surface interface equipment may have hydraulic rams that twist the casing while exerting a downward force to aid in the drive. A contoured clamshell just fitting the inside of the casing, is lowered periodically to remove the sample.

The caisson system requires a portable washing plant to process the large samples. Together with the rolling crane, the total cost of the system may exceed $1.3 million. For large and deep deposits such as tertiary gravels over 100 m in depth, this is sometimes the only effective way to conduct the sampling. With placer diamond deposits in Brazil and Africa, the caisson drill has proven to be essential. For small-scale deposits however the caisson drill is clearly too expensive.

Sampling Phases   Evaluation may be conducted by either drilling or bulk sampling, depending upon conditions. The following phases are described with drilling because of its dominant use but is applicable in principal to bulk methods as well.

SCOUT DRILLING: This initial sampling phase is intended to delimit the area and tenor level of gold in the deposit. While the Reconnaissance phase may have established the presence of gold in surface observations, it must then be determined if gold continues in commercial quantities down to bedrock. In most placers, the majority of gold is concentrated near or just above bed rock, and may be also locked in clay lenses above. If it is seen to continue and increase with depth, combined with some consistancy throughout the prospect, then some confidence can be developed in the potential for a mining project.

When the practical limits of the project have been established within the budget constraints of the investor, decisions can be made as to the area to be explored. The first holes are usually placed at the upper end across the river/stream or apparent flow of the deposit. If the results are low to nil this may terminate the evaluation. But if promising, random holes may be set at intervals down stream from one side to another to search for the limits of the deposit.

EXPLORATION DRILLING   Assuming positive results from the scout drilling, you proceed to set up a grid of holes on lines perpendicular to the flow starting at the up-stream point, spaced at fairly wide intervals from each side of the deposit to the other. Lines may be spaced 300 m or more apart depending upon the size of the prospect.Holes should be 50-100 m between centers.

The exploration phase is intended to either prove or disprove the concentration of gold in the deposit that will produce a reasonable level for economic exploitation. Once the drilling has been completed and the values achieve that level. the deposit can be classified as containing "Indicated Reserves".

DEVELOPMENT DRILLING: The spacing between holes and lines needs to be closed in, the ultimate objective being to have about one hole per 3 acres or less, depending upon the nature of the deposit. In deposits with more erratic dispersion or lower tenors of gold, a higher saturation may be necessary; such as one hole per acre. There are some deposits that have shown enough variation in values that a large bulk sample may be taken as a further verification of the gold quantity. When the sampling has been successfully completed, the deposit can be said to contain "Proven Reserves".

Feasibility Study After completion of the exploratory drilling phase, a preliminary feasibility study can be conducted. This will include a tentative design decision of the appropriate mining equipment that would be used, its estimated cost of operation and the amount of investment required along with the expected return. At that point, the decision can be made whether to continue with development drilling or to drop the project. Assuming a positive result from the study, the next step is to conduct the final drilling phase.

When the drilling phases have been completed and the tenor of the deposit established, final decisions are made concerning equipment design for the mining and processing system. From these decisions, the final feasibility study is conducted to determine the potential return and risks. Some of the factors to be considered are discussed in subsequent sections and a basis for parametric analysis presented.

Reserve Criteria

How does the investor or mining company evaluate whether a new prospect is a viable mining opportunity? In the majority of cases that we see, the promoter alleges that the tenor of the deposit is "high", using numbers that are usually associated with hard rock deposits. To the knowledgeable placer mining engineer, when figures like $10. to $50./m^3 are stated for a placer gold prospect, he knows that there is only a slim chance of it being valid.

Competent sampling of a prospect is essential to making investment decisions, if the investor decides to pursue it. When such a decision is made, how do you assess the prospect when tenor figures begin to emerge from a sampling program? To provide an answer to that question, we have developed a set of curves shown in Figure 1, "Reserve Criteria vs Tenor, Small Placer

Gold Mines", that are intended to provide a basis for decisions.

Criteria for Calculation of Reserves

In Table No. 1, we are assuming three levels of production that fall within the category of "small-scale placer gold mining". The corresponding capital investment for each size is also specified. It is assumed further that the mining will be accomplished using that equipment configuration but with differing sizes of components, specified in a subsequent paragraph.

Table No. 1

CAP M^3/HR	OPTG COST $/$M^3$	EQMT-COST	OTHER	TOTAL
		– $ MILLION –		
100	1.00	1.0	0.250	1.250
70	1.50	0.7	0.200	0.900
40	2.00	0.5	0.150	0.650

EQUATION FOR RESERVES (GOLD):

R = reserves – m^3
C = cost to produce – \$/m^3
T = tenor – \$/m^3 @ \$400./Troy oz.
I = capital investment – \$
P = profit objective = 4I

C X R = Total Cost
T X R = Gross Revenue

RESERVE CRITERIA vs. TENOR
SMALL PLACER GOLD MINES
@ $400/oz. TROY-Au.
FIGURE 1

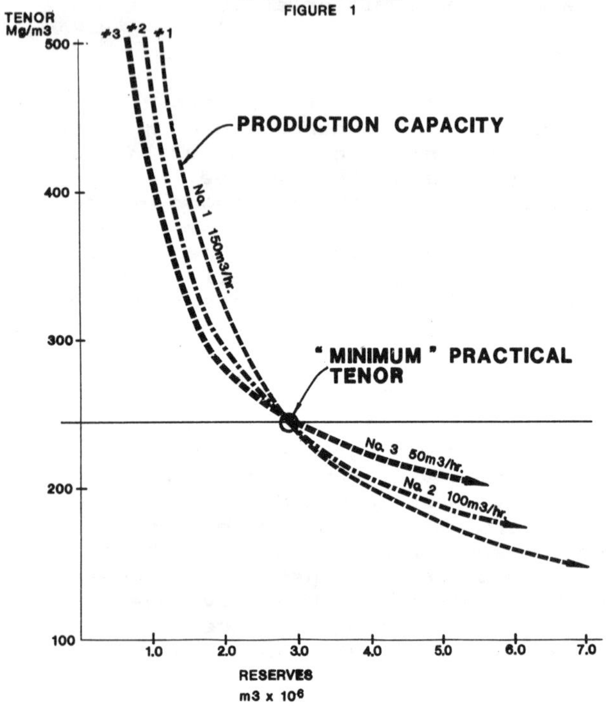

$$P = 400/31.1T \times R - C \times R - I = 4I$$
$$R(T-C) = 5I$$
$$R = 5\ I/12.86\ T-C$$

$$T = 5I - RC/R$$

$$R = 5I/(T-C)$$

$$I = R(T-C)/5$$

With the foregoing equation breakdown, the problem can be approached several ways. The curves resulting from this equation shown in Figure 1, are useful in establishing a rule-of-thumb for go-no-go decisions. The range of tenors shown up to 500 mg/m^3, are typical of placer deposits in volumes up to 7.0 million m^3. The three production ranges give logical dividing points for scaling the primary equipment sizes of trommel, jigs and backhoe; thus governing the amount of capital investment required. The "other" costs include exploration and development drilling; engineering design, feasibility and environmental investigations; obtaining permits and other associated costs.

The three production curves in Figure 1. coincide at a tenor of 245 mg/m^3 and reserve volume of 2.8 X 10^6 m^3. The curves are influenced by our arbitrarily selected profit goal for the project of 4 X Investment which we consider to be the minimum necessary to absorb risks and provide a reasonable return. This can be set higher depending upon the tenor and price of gold; not unusual to have 10 I or higher. It can be seen that as the total volume of the reserves exceeds about 3.0 m^3 X 10^6, the efficiency of return diminishes rapidly. This suggests that consideration should be given to installing higher production equipment such as a bucket ladder dredge.

MINING PLAN

When approaching the problem of mining a property while in the evaluation stage and confining the activity to a relatively small deposit, the difficulties become apparent. This is mainly due to the limitation of reserves and tenor, the consequently low potential revenue which in turn reduces the investment capital that is justified for allocation to the project. The foregoing example of placer mining equipment, is the most common system used for small scale placer gold mining.

The following discussions illustrate the basis for decisions that must be made to define the Mining Plan.

Equipment Selection[2]

In a small-scale placer gold property in the range of reserves illustrated above, it appears that a maximum for equipment costs should $1.5 million. The example shows total investments ranging from $650,000. to $1,250,000. including evaluation and permit acquisition. One of the pitfalls to be avoided is devising a method of mining that involves double or triple handling of material. This includes using front end loaders, skip loaders, trucks and multiple stacker-conveyors. How then to move material in the excavation and processing stages without engulfing the operation in its own tailings and staying within tolerable costs of operation? The following is an equipment configuration that we selected for a prospect whose general characteristics are in the small placer category:

SIZE OF DEPOSIT:

Volume:	4.0 million m^3
Tenor:	250 mg/m^3
Depth:	15-25 m

EQUIPMENT DESCRIPTIONS:

Excavator: Backhoe, Cat 225
Dozer D 7
(working from shore)

Floating Washing Plant

Revolving Trommel: 28'X5'
Screen holes: -1/2"
Stacker for oversize: 80'
Primary Jig, Circular: 9'
Secondary Jig, Circ: 3'
Amalgam system, 3 stage

Production: 150 m^3/hr, 70,000 m^3/mo

Excavation and Tailings Disposal

Given the foregoing configuration and other assumptions, we have developed a graph that demonstrates the consequences of a drop in production level that will degrade the cost of operation. This is shown in Figure 2, "Production vs Costs, Effect on Small-Scale Placer Gold Mines". In Figure 1 we showed the effects of different equipment sizes on production levels. In Figure 2, with the choice made in size of equipment, it can be seen that there is a range of tolerance of production loss before the cost of operation renders the project uneconomic. The margin of tolerance will vary, depending upon the price of gold and tenor of the deposit. But to achieve the higher production level the excavation and disposal methods must be worked out on the most efficient basis.

The cost of operation is shown to vary between $1.30/$m^3$ and $2.00/$m^3$ when production decreases. This will extend the length of time to deplete the reserves and have other financial consequences. Figure 2 also dramatizes the effect of a small operation in terms of fixed vs variable costs that interplay with production levels.

Excavator During the 1930's, California experienced a proliferation of what were then called "Doodle Bugs". These were placer gold mining plants using a dragline for excavation and feed to the plant which was on a barge floating in a pond. They usually used revolving trommels and tables or riffles for both concentration and recovery. In some published reports it has been said that there were over 300 "Doodle Bugs" operating in California during the 1930's.

The dragline was not an efficient tool for placer mining since it will not clean bedrock effectively. The backhoe had not yet been perfected but since WWII it has become the ideal tool for placer gold and tin mining on a small scale. In the latter case, a company in Brazil is reported to be operating up to 45 plants using in most cases a

PRODUCTION vs. COSTS
EFFECT ON SMALL SCALE PLACER GOLD MINES
FIGURE 2

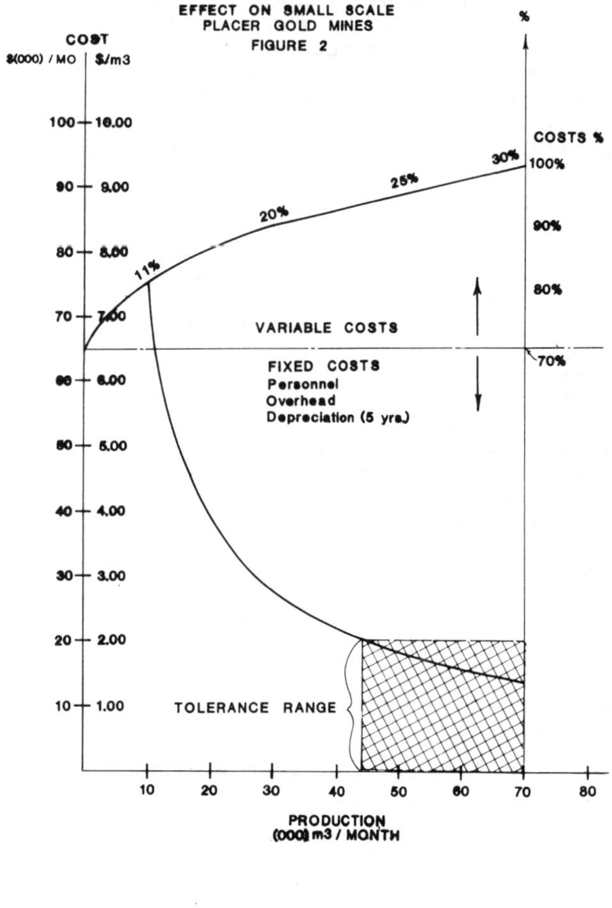

backhoe. In two plants they feed with a bucket wheel suction dredge but in those cases, the deposits are not considered "small-scale". With jungle operations and shallow ground, the backhoe can work effectively with little concern for tailings disposal. In the typical North American placer however you must deal with greater depths and earth moving becomes more of a problem. Disposal also involves environmental issues. We therefore see that the backhoe has become the preferable small-scale, placer gold mining tool. Its greatest limitation however is its maintenance and repair cycle that makes it difficult to keep operating around-the-clock.

Tailings Disposal How this problem is solved often spells the success or failure of a small project. The environmental requirements usually specify that the land contour must be restored and grasses and plants replaced. The operator should insist on the requirements not causing him to re-create more than the land had on it before mining takes place; and to only produce a contour that simulates what had previously existed and no more. In most cases that we have seen, a properly designed operation will in fact enhance the area.

Processing Plant

There has been a tendancy to use lower cost equipment in small scale placer gold washing/processing plants with little regard for its ability to recover the majority of gold. This includes using flat shaking screens for classificaton and riffles for concentration. Few have knowledge of amalgamation and thus will not have a continuous system for recovery.

The history of placer gold recovery systems shows a slow development of technology until the bucket ladder mining dredge was combined with a processing plant as a single system, in the 1880's in New Zealand. While mineral jigs were being actively developed for other minerals in the 1880's, no application was made for placer gold on a dredge until about 1914, in California. Until that time and for some years after, only riffles were used and records have shown that as much as 40% of the gold was lost to tailings. Adding jigs to the circuit recovered more and by the mid 1930's our company and a small dredge operator in Idaho, took the lead and replaced riffles with jigs on our dredges.

The solution to environmental problems with small scale placer gold mining, is to adopt more modern methods using jigs along with recirculation of processing water. The equipment described in this paper can offer a step forward in meeting those goals and to improved disposal reclamation methods.

CONCLUSION

Deciding to invest in and mobilize a small-scale placer gold mine requires professional, experienced evaluation. The essential criteria should be examined at the outset to see if the given parameters of size and tenor fit. Second, no final decision to mobilize the mining equipment should be made until the property has been competently evaluated. The selection of proper equipment and seeking professional help in running the operation, must be the next priorities if the investor/mining company is to make a profit in this scale of mining business.

REFERENCES

Richardson, M.J., 1986, "Evaluation/Decision Process for Small-Scale Placer Gold Mining," Mining, April.

Richardson, M.J., 1984, "Handbook of Alluvial Gold Evaluation Methods," Consolidated Placer Dredging, Inc., Irvine, California.

3

Mining/Production—Options to Maintain Competitiveness

W. J. Schlitt
Brown & Root USA, Inc.
Houston, TX

Chapter 14

CONTRACT MINING AND ITS ROLE IN GOLD MINE DEVELOPMENT

W. Joseph Schlitt, P.E.
Manager of Technology
Mineral & Metal Industries

Donald B. Swanson
Business Development Manager
Contract Mining/Heavy Civil Construction

Brown & Root U.S.A., Inc.
P.O. Box Three
Houston, Texas 77001

ABSTRACT

The paper provides an overview of the role that contract mining plays in the minerals industry. Particular emphasis is on contract mining and its place in the development and operation of small to medium sized precious metal mines in the western United States. The paper begins with a discussion of the mining options available to owners. This is followed by a description of contract mining organizations and the services they provide. In addition to various mining activities per se, these can include other processing functions such as crushing and agglomeration, various types of heavy civil construction, other types of technical and administrative support, and even assistance with financing. Typical equipment spreads and unit costs are presented. Finally, the advantages and disadvantages of contract mining are discussed. Liberal use is made of hypothetical examples and case histories.

INTRODUCTION

The contract miner has long played an important role in the minerals industry. Historically, he was an individual who offered his services as an underground miner. Payment was received either on a day rate basis or more often, on a per ton rate for rock broken and hauled. Today, this same principal remains in effect. In fact, individual contract miners still ply their trade in the industry. However, now the term contract miner usually refers to an independent contractor that supplies the people and equipment needed to handle open pit mining activities for the mine owner or lease holder.

Of course, use of a contractor is only one of the mining options available to the owner. Another is for the owner to purchase all necessary mining equipment, either in new or used condition, and then to operate it with his own people. Alternatively, the owner can lease equipment but still operate it using people on his own payroll. Finally, the owner may use a mix of his own people and equipment, plus those from a contractor. Under the latter scenario, the owner would keep his own equipment fleet fully utilized while using the contractor to handle overload situations or provide specialized services. These might include such activities as removal and stockpiling of topsoil, prestripping the ore body, or building access and haul roads.

The role of the contract miner and his relationship with the mine owner are discussed in more detail in the following sections. These cover contract mining organizations, scope of services, equipment spreads, costs, and the advantages and disadvantages of contract mining.

CONTRACT MINING ORGANIZATIONS, EQUIPMENT AND SERVICES

General Services

A contract miner functions as an independent contractor that offers to provide a variety of services at or around a mining site. These services may be limited primarily to mining activities such as the following:

- Constructing access and haul roads
- Stripping and stockpiling topsoil
- Prestripping barren overburden
- Drilling and blasting ore and waste
- Mining and hauling waste and ore

However, there is a trend for the larger contract mining organizations to provide additional heavy civil construction services to the mine owner. These may include:

- Exploration roads and drill pads
- Site preparation for surface facilities
- Earthen dams and diversion channels or pipelines for control of surface and subsurface water flows

- All types of back filling, recontouring and reclamation
- Other environmentally related activities
- Gold ore processing for heap leaching

The latter service could include leach pad construction, ore crushing, fines agglomeration (if necessary), ore stacking, and removal of leached material to a final tailings site. This site itself may be prepared and maintained by the contractor.

Contract Mining Organizations

The organizations that provide contract mining services range in size from large international engineering and construction firms down to small local contractors. The latter generally have small equipment spreads and only serve their own immediate areas. The intermediate to large contractors are active on a regional basis or even throughout the western United States. Some, but surely not all of these major operators are listed below:

- The Argee Corporation (A Groves company)
- Brown & Root U.S.A., Inc.
- N. A. Degerstrom, Inc.
- Gilbert Western Corporation (A Kiewit company)
- Granite Construction Company
- Lost Dutchman Contractors
- Morrison-Knudson Company, Inc.
- R. E. Vining Company
- Washington Corporation

Equipment Spreads

As mentioned above, all contract miners, large or small, function as independent contractors. They provide their own equipment spreads. The exact spread will be matched to the job, with equipment drawn from strategically located equipment yards. For a typical mining project, major equipment may include all or most of the following:

- Drills and Compressors
- Loaders and/or Shovels
- End Dump Haul Trucks
- Bottom Dump Scrapers
- Tractors with Dozers and Rippers

The loader buckets or shovel dippers will be sized to be compatible with the haulage truck capacity. The loading and hauling equipment will also be matched to the mine tonnage, haulage distance and number of ore and waste faces to be worked.

In addition, the contract miner will provide an array of ancillary equipment to support the mining operation. This may include such things a powder truck(s), mechanic's and lubrication trucks, pickups and a water wagon. Other ancillary equipment may include motor graders, backhoes, and pumps. The contract miner also normally provides trailers that serve as administrative and safety offices, as a parts warehouse and the like.

In cases where the contractor also handles the heap leaching, several additional types of portable or semiportable equipment will be provided. Included will be crushers, screens, conveyors and a loader for heap construction. Typically a jaw crusher will be used as the primary unit, while standard and short head cone crushers will be used as the secondary and tertiary units, respectively. The minimum particle size for the crusher circuit discharge will seldom be less than 3/8 to 1/2 in. If such a fine particle size is not required, the tertiary or even the secondary crushers may be eliminated. A mobile crane will be needed for crusher repairs. A spray system for dust supression will also be needed.

Where crusher fines or clayey ores must be accommodated, the contract miner may be asked to provide the agglomeration system. This will produce a feed that can be heap leached without solution percolation problems. As described in Chamberlin's recent article, this type of operation requires specialized equipment such as drum or disk agglomerators or a conveyor system set up for belt agglomeration. Provision must also be made to add moisture (usually as cyanide solution) and binder (lime and/or cement) during the agglomeration step.

Some of the larger contractors will also be able to provide other specialized equipment. This could include drag lines, gyratory crushers, diesel generators and the like. Under exceptional circumstances, even a contract miner may have to lease an unusual piece of equipment.

Table 1 provides a detailed listing of the equipment spread needed for a typical gold heap leaching plant that handles 2000 tpd of ore at an ore to waste strip ratio of 1 to 2.5. In addition to the mining, the example assumes that the ore will be crushed to minus 0.75 in. then stacked in 20 ft. high heaps for leaching. No agglomeration is included. The information in Table 1 includes mine and crusher parameters, principal mine equipment, required spares, support equipment and major ore processing equipment. Hypothetical layouts for the overall operation and the crusher circuit are shown in Figures 1 and 2, respectively.

Contract Mining Services

Obviously the principal services provided by the contract miner are to mine and haul ore

Table 1

EQUIPMENT SPREAD FOR TYPICAL SURFACE GOLD MINE IN THE WESTERN UNITED STATES

Mine Parameters

- 2000 tpd ore, with ore to waste strip ratio of 1 to 2.5.
- 5 mining shifts/week at 10 hrs/shift
- In-place rock bulk density, 4000 lb/cu yd (dry basis)
- Average round trip haul distance over life of mine, 5000 ft.
- Average haul grade under full load, +6%

Crusher Parameters

- Run-of-mine ore top size, 24 in.
- Crushed ore size, 100% minus 0.75 in.
- 10 crusher shifts/week at 8 hrs/shift
- Crusher plant availability, 80%

Major Mine Equipment

- Rotary drill, Drill Tech 25 or equivalent (one)
- Loader, 8 cu yd (one)
- End dump haul trucks, 50 ton (four)
- Tractor with dozer and ripper, D-8 or equivalent (one)
- Motor grader, 14 G or equivalent (one)

Spare Equipment for Project Continuity

- Loader (one)
- Haul truck (one)

Major Crushing Equipment

- Fixed grizzly, 16 in. slots (one)
- Rock breaker, pneumatic, 24 ft. reach (one)
- Apron feeder, 42 in. wide x 20 ft. long (one)
- Jaw crusher, 36 in. x 42 in. (one)
- Standard cone crusher, 5.5 ft. dia. (one)
- Vibrating double deck screen, 6 ft. x 20 ft., (one) with 2.5 in. and 0.75 in. openings
- Extendable conveyor system to heap leach site, 30 in. wide (one)
- Loader (at crusher plant), 8 cu yd (one)
- Loader (at heaps), 10 cu yd (one)

Support Equipment

- Water wagon, 10,000 gal (one)
- Flat bed truck for powder (one)
- Service truck with diesel and gas compartments, plus oil, grease and air (one)
- Mechanic's truck with welder (two)
- Tire truck (one)
- Parts trailer (one)
- Office trailer (one)
- Pickup trucks (three)
- Mobile crane, 25 ton (one)
- Portable flood light stands with gasoline generators (six)
- Buildings/trailers for offices, shops, etc.

Figure 1

LAYOUT OF OPEN PIT GOLD MINE WITH ORE CRUSHING AND HEAP LEACHING

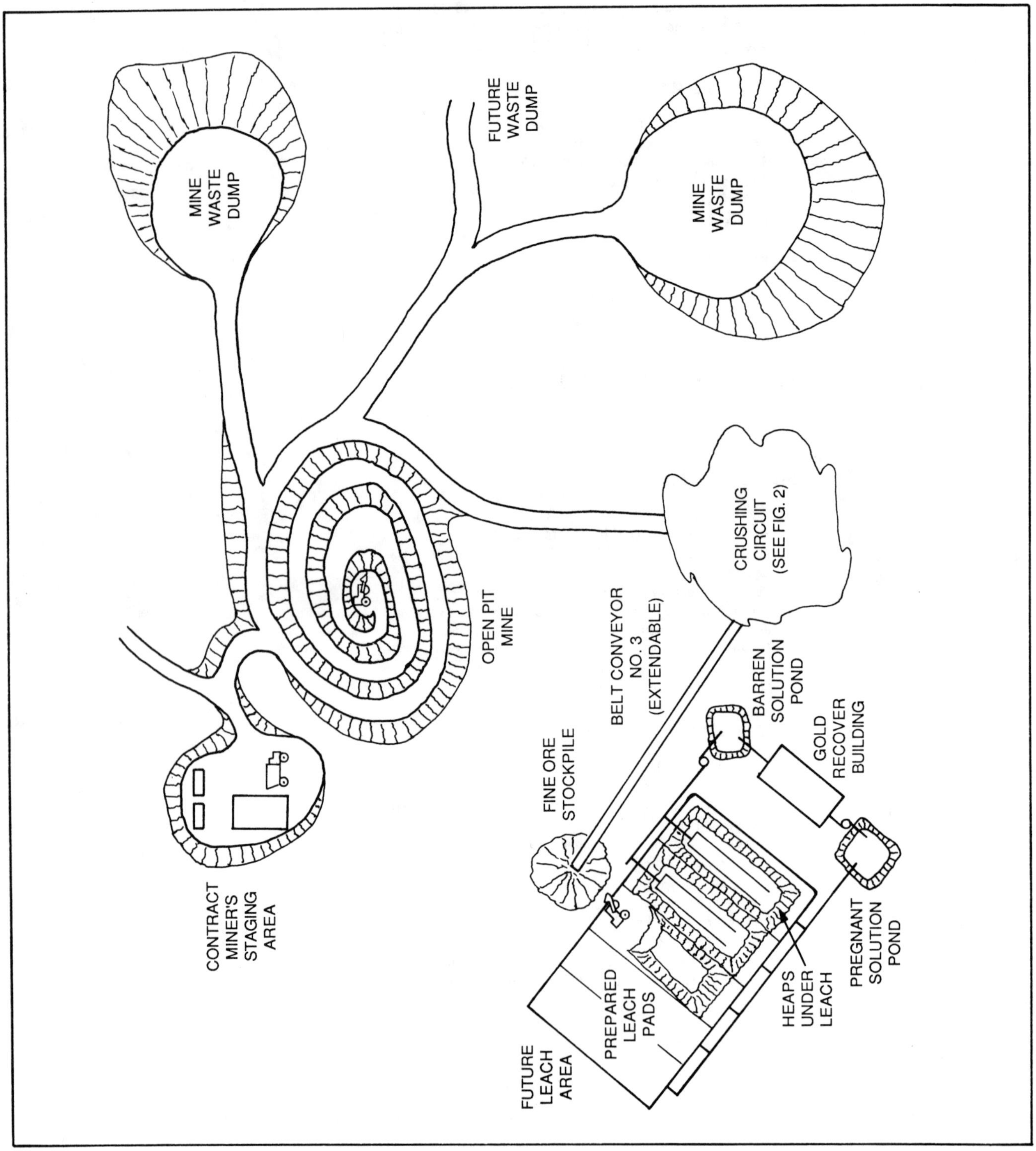

Figure 2

LAYOUT FOR CRUSHER CIRCUIT

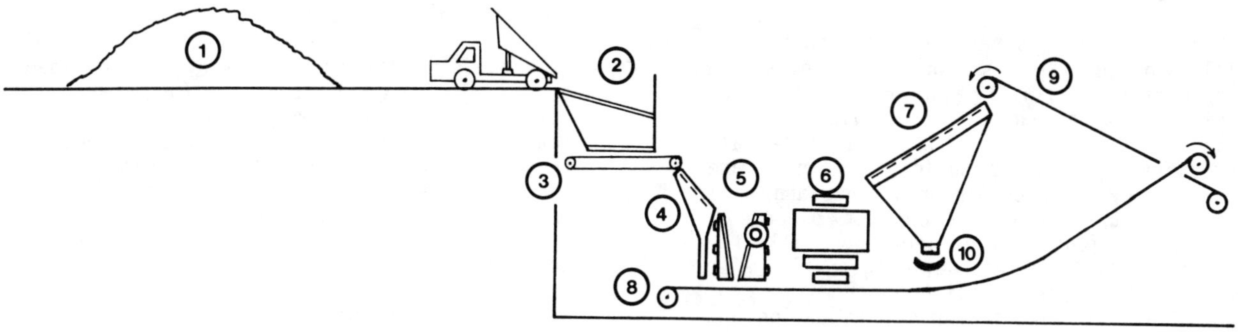

1	10,000 TON ORE STOCKPILE
2	100 TON HOPPER - 16" GRIZZLY
3	42" X 20' APRON FEEDER
4	4" FIXED GRIZZLY
5	36" X 42" JAW CRUSHER
6	5.5 'Ø SHORT HEAD CONE CRUSHER
7	6' X 20' VIBRATING SCREEN
8	BELT CONVEYOR NO 1
9	BELT CONVEYOR NO. 2
10	BELT CONVEYOR NO 3

and waste, and perhaps to process the ore for heap leaching. However, the contractor also provides many additional services. These free the owner to focus his attention on other matters and allow him to keep his own staffing to a minimum. Thus, the contractor provides a project or general manager; other operating, maintenance and administrative supervisors; emergency medical technicians; any field engineers; and his own operators and craft (maintenance) personnel. All these people will be on the contractor's payroll and will be his responsibility, not the mine owners. The staffing level will vary in proportion to the size, duration and complexity of the mining operation. However, over-staffing is seldom a problem due to the competitive bidding originally done to obtain the mining contract. The larger contractors can also provide additional home office support on an as-needed basis.

Along with the operations, the contractor will provide all his own fuels, oils and lubricants. He will also handle his own equipment maintenance and provide a full repair program. This will include all necessary maintenance buildings. These are usually of the preengineered type and will be sized to handle the largest pieces of mobile equipment in the fleet--generally the haulage trucks. The buildings will also reflect the expected climatic conditions at the site. For example, good insulation and heating will be used in cold weather locations. Necessary spare parts and supplies will be kept in the building, in separate trailers, or for items like diesel fuel and tires, in fenced yards.

Other services provided to the owner may include field engineering, permitting, and various administrative functions. All these will relieve the owner of various responsibilities and liabilities. The contractor will administer such areas as payroll, personnel and production records, insurance, and purchasing. Engineering can include day-to-day mine planning, plus any engineering needed to layout, install and operate ore processing equipment when this is the contractor's responsbility.

Permit engineering will also be required if the contractor is responsible for getting the necessary approval for his own activities. In this regard the contractor and owner are likely to coordinate their activities closely in order to present a united front to the appropriate federal, state and/or local officials and agencies. Where the contractor is responsible for his own permitting, he will also be directly responsible for inspections by MSHA personnel and those from other regulatory agencies.

A final administrative area covered by the contractor will be safety and training. He will be required to conduct a safety program that meets or exceeds applicable MSHA, OSHA and/or state and local requirements. Most of the larger contractors such as Brown & Root do have their own MSHA-certified instructors. These people provide on-site safety certification training and also conduct refresher training. The contractor will then maintain the necessary safety program and records.

On most projects only skilled operating and craft personnel will be hired. However, on large projects there may be opportunities to upgrade skill levels through training programs. Where these are warranted, the contractor will administer the training necessary. For most contractors, safety and training are areas of importance and they do receive considerable attention from the project management team.

Regardless of the scope of services provided there will always be one or more interfaces between the contract miner and the owner or leaseholder. These are important because they represent potential points of conflict. As noted above, permitting and various types of regulatory compliance are two areas where close coordination between both parties is required.

Two other areas where interfaces occur are in grade control and production accounting. Grade control is so important that the owner generally retains responsibility for this area. This includes assaying of exploration drill core and production (blasting) drill cuttings. Based on these results the owner then flags the bank run material as ore or waste. Finally, the owner should check periodically to insure that the contractor is properly observing the flagging. Thus, he verifies that ore is going to the mill and/or leaching heaps, or to a stockpile, while waste is going to the mine dumps.

Production accounting is vitally important to both the contract miner and the owner. On the one hand the quantity mined and processed represents the basis of payment to the contractor. For mining this is generally determined by yards (or occasionally tons) of bank run material moved. This is based on surveyed cross sections which are used to corroborate the load count. Surveying is usually done monthly and is often verified by aerial surveys taken at the beginning of the work and then periodically during the job. For crushing, agglomerating and stacking, payment is commonly based on tons or loose cubic yards handled. This is tied to the ore mining rate as adjusted for surveyed quantities in any ore stockpiles.

Obviously, these same quantities are also important to the owner. On the one hand they represent a major expense, but they also become the basis for the gold and silver production that generates the cash flow to the operation. In summary, both grade control and production accounting are so critical to the owner and contractor that the terms and conditions covering there areas are carefully set forth in the contract.

Quite often the contract miner will have more personnel on site than the owner. This is because mining and maintenance are more labor intensive then the leaching and metal production operations normally run by the owner. Thus, once the owner and contract miner have developed a good working relationship, the owner may have the contractor provide additional services on an as-needed basis. These will often be paid as extras on an hourly or day rate basis. Typical services handled in this way might include snow removal, site access road maintenance, dust control on access and haul roads, and maintenance on any of the owner's mobile equipment. The contractor may also handle small capital projects and periodic heavy civil construction. Examples could include construction of additional heap leach pads, expansion of a tailings area, or drainage control. Full service contractors may also supplement their normal complement of personnel in order to handle other types of small projects. These would generally cover limited work for the owner that does not justify a separate competitive bidding approach. Examples might include installation of a new piece of equipment or modification to an area in the process plant. If the contractor has a strong minerals engineering group, he may even provide occasional engineering or metallurgical support in addition to construction oriented services.

The ultimate in contractor service would be to run the entire operation from mining through gold/silver production. While not unheard of, this type of agreement seems to be more common for cabonaceous fuels such as coal and lignite. However, where the owners or leaseholders are entrepreneurs or venture capitalists with little or no mine operating experience, a contractor may agree to run the entire precious metal operation for them.

COMMERCIAL CONSIDERATIONS

Contractual Matters

Most mining contracts are awarded on the basis of competitive bidding. The bids are normally prepared in response to a specific scope of work issued by the mine owner or lease holder. The bid document defines such things as the schedule and tonnage requirements. The prospective contractors then supplement the information in the scope documents by visiting the mining site. This allows firsthand assessment of the site-specific conditions that could impact either positively or negatively on the cost of a mining operation. Then a bid is prepared. A partial list of the many items to be considered in doing this is as follows:

- Ore and waste tonnages and the schedule for each
- Haul distances for ore and waste
- The nature of the material to be mined, e.g. hard and abrasive, or soft and friable
- The area topography, e.g. flat, rolling hills, or steep canyons with little space for waste dumps
- The climatic conditions, e.g. extended cold snaps, heavy snow or rain, high winds or extreme heat
- Utilities, specifically power and water
- Site accessibility, e.g. remote or readily accessible to populated areas that can supply people and services
- Environmental and regulatory restrictions, e.g. limitations on dust or noise, or the need to protect surface or ground water sources
- Mobilization and demobilization requirements
- Special hiring restrictions, e.g. the need to use Indians for work to be done on a reservation
- The required size distribution for ore, if crushing is required

Once the contract miner is successful in being awarded a project, he will meet with the owner and finalize the contract. This is not always easy as there are almost as many contract variations as there are jobs. For example, the contractual time frame may cover the available mining season, a fixed period such as a year, or the life of the operation. The contract may also cover a specific tonnage or may be an evergreen type that can be extended under a set of mutually agreed-to conditions. Particularly for long term contracts, there should be reopening clauses to protect both parties. These would allow for adjustments in such things as fuel prices, insurance requirements or mandatory regulatory changes.

Depending on the type of contract used, pricing will usually be specified as either a firm fixed price for the job or as unit rates per yard or ton for different activities. Other types of contracts such as cost plus a fixed fee are not common. In any case, the prices will be all inclusive and will cover the contractor's mobilization, maintenance, overhead, demobilization and profit, in addition to his direct operating costs. A single weighted figure for ore and waste mining and hauling is commonly used to avoid accounting problems. Finally, any bonus/penalty arrangements will be clearly specified.

The price for the services identified in Table 1 will vary with such factors as the haul distance, the nature of the material, and the mining method. However, the price to drill, blast, load and haul will be in the range of $2.00 to $3.00 per bank run cubic yard. The price for crushing and stacking the ore wll also depend on many variables but generally the price will be between $1.50 and $3.00 per ton. These costs assume no agglomeration of the ore and no unusual site-specific conditions at a location in the western United States.

As with any contractual agreement, the mining contract must clearly define the responsibilities of both the contractor and the owner. Some of the key contractual points have already been mentioned above, i.e. the basis of pricing (compensation), billing and payment, method(s) of production accounting, and definition of such important areas as grade control where there is an owner-contractor interface. Other important contractual areas will include the following:

° Definition and pricing of extra services
° Safeguarding of owner's proprietary information
° Standards of performance and warranty
° Insurance and allocation of risks
° Remedies
° Termination of services

Project Financing

Quite aside from the normal contractual considerations, some contract mining firms, particularly the larger ones, offer one or more types of project financing. Such a service would be particularly valuable to small or new mine operations. Following their exploration and development programs these companies are often short of cash needed to cover fixed and working capital requirements until a positive cash flow is established from the operations. Thus, in extending some type of financing, the contractor can help the mine owner by allowing the operation to commence without the need to secure additional loans or to go into the equity markets. Of course, financing may also assure the contractor of the job. However, the contractor will certainly receive some compensation for assuming the additional risks associated with the financing package being provided.

The contract miner may participate in project financing in several ways. The simplest probably involves deferred payments. Under this approach the contractor begins prestripping the deposit and then begins mining the ore body. Each month the owner will be billed for the current month's work, plus an agreed-to fee or interest charge for overdue invoices. Once refinery receipts start coming in and a positive cash flow is established, the owner begins meeting current invoices and paying off the overdue ones. One variation on this theme is for the contractor to receive all or a large portion of the net profit until the owner is current on his payments.

Another approach is for the contractor to take an equity position in the mine or mining company. Under such an arrangement the contractor may receive stock, a part ownership in the claims, receipts for forward gold sales, or a share of the net profit or gross revenue from the mine. Such options can permit some rather creative financing arrangements.

A third option is for the contractor to simply co-sign or guarantee loans from a bank or other investment group. The mine owner can then keep his payments to the contractor current. However, to protect his interests the contractor may require some collateral, e.g. a lien against the claims or some other asset of the mine or company.

Any time financing is involved, the contract miner is entitled to considerably more information than he would otherwise need. This will allow him to properly assess the projected mine cash flow and the added risk he takes when he provides financing. Examples of such additional information include the following:

° Tonnage of the deposit
° Average grade of the deposit
° Proposed mine plan and initial ore grade by week or month
° Metallurgical results from laboratory and/or pilot tests
° Design basis for the plant
° Financial condition of the mine owner or lease holder
° Any restrictions or other liens against the mine claims

Once all this information is in hand, the contract miner will be able to judge the mine's profitability (or lack thereof). He can then decide how much risk to take and at what price when offering project financing.

ADVANTAGES AND DISADVANTAGES OF CONTRACT MINING

The previous sections have defined contract miners, the equipment and services they provide, and the commercial conditions underwhich they do their work. Some of the advantages and disadvantages of contract mining have also been mentioned in passing. In general, the advantages of contract mining, particularly the economic ones, decrease as the size of an operation and the expected mine life increase. However, the new tax laws favor the contract miner over an owner operation. Factors include elimination of the 10% investment tax credit and a decrease in the rate at which equipment can be depreciated. Both these reduce the incentive for the owner to purchase his own equipment.

Reserves are a key factor in determining whether an owner should do his own mining or contract this out. For the case described in Table 1, one with an ore mining rate 2000 tpd, minable reserves typically run about 7 million tons. This gives a 10-year mine life so that equipment can be written off over a reasonable period of time. If reserves are substantially less, then contract mining will become more attractive as there will be insufficient production to pay off all capital costs, meet operating expenses and still leave enough cash flow to adequately compensate the owner for taking the risk of opening a new mine.

To be more specific, the equipment spread shown in Table 1 is estimated to have a capital cost of $6.5 million (new equipment--1986 4th quarter basis). At a gold price of $400/oz and cash costs of $200/oz the profit from some 32,500 ounces of production would be needed just to cover the costs of the mining and crushing equipment. Assuming 0.03 oz of recoverable gold per ton (0.05 oz/ton at 60% recovery), this production level requires more than one million tons of ore or one and a half years of operation. Total capital requirements for the project, including such items as land acquisition, exploration, development, mining and processing equipment, ancillary facilities, and working capital, may easily be two to three times the $6.5 million figure, particularly if the time value of money is taken into account. This also doubles or triples the reserves that must be mined just to cover the capital costs for the project. Finally, if total operating costs, including the noncash items, are also to be covered then more than half the reserves would have to be mined in order to cover all costs. Clearly then, if reserves are much less than the 7 million ton figure mentioned earlier, contract mining would be a more economic choice for the owner than buying the equipment and doing the mining himself.

Overall, the advantages and disadvantages of contract mining can be categorized into two classes - those that are primarily economic and those that relate to operating flexibility and control. In each case, the owner or leaseholder must carefully evaluate his own situation to determine whether or not he should contract for mining services or purchase the equipment and do his own work. If the situation is not easily assessed, the owner can estimate the cost of doing his own work, then solicit a bid from a contract miner for comparison.

Some of the principal advantages of using a contract miner are as follows:

- A reduction in capital requirements for mining and also ore beneficiation equipment
- A reduction in the period required to pay back the capital investment
- A reduction in direct payroll and administrative expenses
- A reduction in costs for safety and training programs
- A reduction in some engineering costs, e.g. day-to-day mine planning or engineering on a contractor-operated crusher circuit
- A clearly defined, fixed cost for mining and any other contracted services
- Relief from certain permitting requirements and regulatory inspections
- Increased operating flexibility, particularly with regard to changes in production schedules
- Specialized services and equipment available and easily accessible, e.g. use of bottom dump scrapers to strip and stockpile topsoil at the start of operations

In addition to the advantages, there will also be some disadvantages to using a contract miner. The principal one is that the unit operating costs will be higher. The owner may also have to pay a termination fee or a special demobilization charge if he shuts down for some reason outside the contractor's control. Examples might include falling metal prices, poor metallurgical recovery, or lower than expected ore grades. The owner also faces the risk that the low bidder may be financially weak or may not actually have the equipment needed to maintain the required schedule. Thus, the bidders' qualifications must be carefully considered. Finally, the owner will have less direct control over a contractor than he would over his own personnel. Such a factor is particularly important when it comes to maintaining close grade control. This is because a contractor will have less incentive than the owner to see that ore and waste are sent to their appropriate destinations or that all ore is crushed, screened and stacked properly for leaching.

REFERENCE

Chamberlin, P.D., 1986, "Agglomeration: Cheap Insurance for Good Recovery When Heap Leaching Gold and Silver Ores", Mining Engineering, December, pp. 1105-1109.

Chapter 15

THE KINGSTON PROJECT
NEVADA GOLDFIELDS' INITIAL MINING OPERATION

by

Mr. Brian P. Micke
Operations Manager
Nevada Goldfields Inc.
Reno, Nevada

Mr. A. E. Jack Swanson
Project Manager
Nevada Goldfields Inc.
Kingston, Nevada

INTRODUCTION

Nevada Goldfields began underground mining development at the Kingston Project in September, 1986, followed by mill start-up in March, 1987. The first gold pour in April followed three years of preparatory work in establishing a North American operational base for the production of precious metals by Golconda Minerals N.L., an Australian mining firm.

GENERAL CORPORATE BACKGROUND

The Golconda Group of companies was established in early 1983 in Australia. It began operations by successfully retreating tailings at several old gold mining sites throughout Australia. With the limited reserves available for tailings retreatment, Golconda began to look toward the future by conducting a gold exploration program in Australia. However, the continued high price of gold in Australia had limited the possibilities of acquiring good properties at a reasonable cost and consideration was given to looking outside Australia in addition to Golconda's domestic exploration activities. North America was favored because of the proven potential for substantial gold deposits, the established infrastructure which would allow for rapid and cost-effective development of new projects and the advantages of similar political and economic climates.

In mid-1984, Golconda established an office in California and began an extensive study of over 100 prospects throughout the United States. Two prospects were identified as a result of this study whch showed merit; one was located at Empire, Colorado and the other at Aurora, Nevada. Negotiations were then successfully undertaken to acquire these two properties. Golconda's U.S. offices were moved to Reno, Nevada in early 1985 to facilitate management of its projects.

To operate and manage these projects, a small Canadian company listed on the Vancouver Stock Exchange, Cannon Minerals, was acquired by Golconda and renamed Nevada Goldfields Corporation. The current structure of the company is shown below in Figure 1.

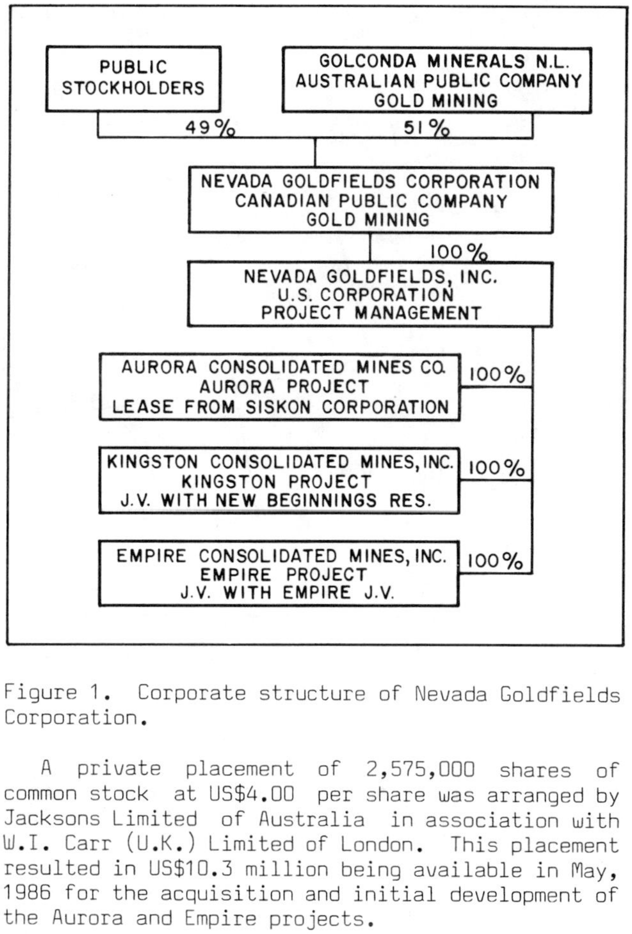

Figure 1. Corporate structure of Nevada Goldfields Corporation.

A private placement of 2,575,000 shares of common stock at US$4.00 per share was arranged by Jacksons Limited of Australia in association with W.I. Carr (U.K.) Limited of London. This placement resulted in US$10.3 million being available in May, 1986 for the acquisition and initial development of the Aurora and Empire projects.

While the financing was underway, project planning commenced on both the Aurora and Empire projects. The Aurora project was selected as the one to be initially developed because of its relatively advanced stage of development. The projected size of the project at 270 mtpd (300stpd) and proximity to Reno also helped make it a logical choice to establish the company's operations. A contract for plant design and engineering at Aurora was let and some of the mine and mill equipment sourced by April of 1986.

Also in April, 1986, a project was brought to the attention of Golconda which showed exceptional promise for the development of the long term future of the company. Examination of the property not only confirmed the good potential but also revealed several serious defects in the way the property had been managed during its recent production history in 1984-85. However, it was evident that these shortcomings could be rectified with better planning and additional capital expenditure to achieve good operating efficiencies and satisfactory profit margins.

Agreement was reached with the vendor in May and the property was transferred to Nevada Goldfields for operation in June, 1986. This acquisition gave Nevada three very promising projects for development over the following three years. The locations of these projects is shown in Figure 2.

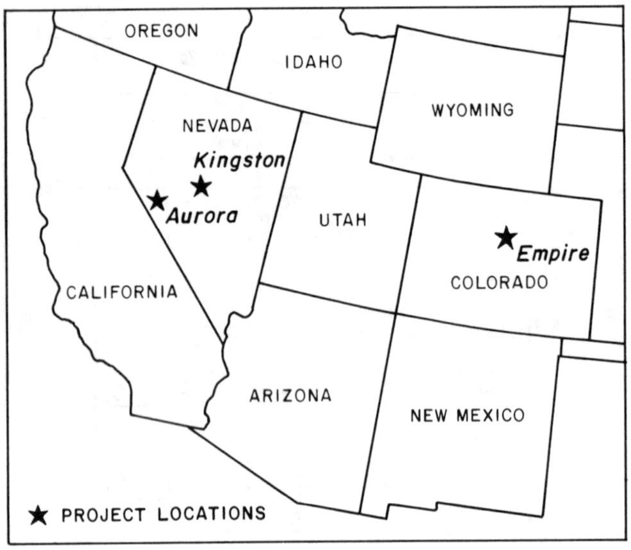

Figure 2. Location map of Nevada Goldfields' projects.

A US$12 million non-recourse project loan was arranged with the National Australia Bank for project construction and other corporate purposes. The company was listed on the Toronto Stock Exchange in October, 1986.

THE KINGSTON PROJECT

The Kingston Project consists of a new carbon-in-leach (CIL) mill and the old Victorine Mine, located in 1863 and patented in 1869. Figure 3 shows the location of the Kingston Project in central Nevada.

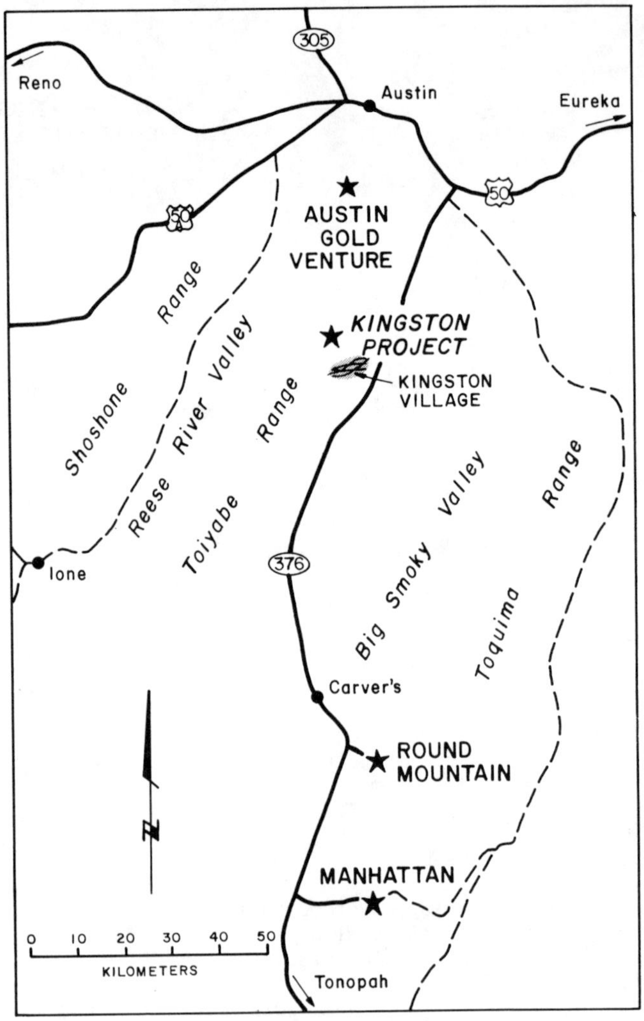

Figure 3. Kingston Project Location.

The early mills cound not handle the sulfide ore and the total recorded production of the Victorine Mine through 1979 was less than $100,000. The Victorine Mine was rediscovered and drilled in 1982 by New Beginnings Resources, Inc. An operating agreement with U.S. Energy was reached in 1983 under which U.S. Energy developed the mine and moved a small gravity/flotation mill from Colorado to the project. A CIP circuit and additional grinding capacity was added and although this mill was rated at 454 metric tpd (500 short tpd), it never reached sustained throughputs of more than 300 metric tpd (330 short tpd). Recoveries were poor and operating costs high, forcing a closure of the property in December, 1985. This allowed Nevada Goldfields to purchase the U.S. Energy interests in 1986.

After a detailed study of the project, both in-house and by outside consultants, Nevada Goldfields decided it was in their best interest to scrap the existing facility and build a carbon-in-leach process plant. Some equipment was able to be salvaged for the new plant although the amount was minimal. The new plant came up to the anticipated level of 640 mtpd (717 stpd) within 60 days of startup, confirming the correctness of the decision to build a new mill.

Geology

The Victorine gold deposit is located in the Bunker Hill and Summit mining district on the east side of the Toiyabe Range about 50 km (30 miles) south of Austin, Nevada on the east side of Victorine canyon which is a north trending branch off Kingston Canyon (See Figure 4.)

Rocks exposed in the area adjacent to the deposit are Paleozoic sediments consisting of lower Cambrian quartzite shales and limestones overlain by middle and upper Cambrian and Ordovician limestones and graphitic phyllites which were later cut by a thrust fault and intruded by Jurassic dioritic sills.

Both normal and thrust faulting exists on the property. In general the commercial gold mineralization occurs within a flat-lying quartz vein system located between the thrust fault and its upper thrust plate. The ore body consists of quartz-gold-sulphide lenses varying from 3 to 18 m (10 - 60 ft) in thickness, with each lens having several thousand square metres of width times length dimension. (See Figure 5). Usually a shaly limestone is found on the footwall and a graphitic phyllite shale on the hanging wall.

Finely disseminated gold occurs in the quartz, in fractures and along the margins of abundant pyrite. The sulphide assemblage consists of pyrite (often up to 20%), sphalerite, galena, chalcopyrite and tetrahedrite. Gold occurs throughout the deposit with assays ranging from less than 1 g/mt (0.03 oz/st) to over 300 g/mt (10.0 oz/st) with an average grade of 8g/mt (0.26 oz/st).

Mining

<u>Reserves.</u> The minable reserves were calculated by using surface diamond and rotary drill holes on 30 m (100 ft) centers and underground percussion drill holes on 10 m (30 ft) centers. Chip and muck samples from development were also used in the developed areas of the mine. The surface holes were assayed only in the ore zone on 1 m (3 ft) intervals whereas the underground holes were assayed full length on 1.2 m (4 ft) intervals. The resulting minable reserves were 0.95 million mt (1.06 million st) of ore averaging 8.99 g/mt (0.262

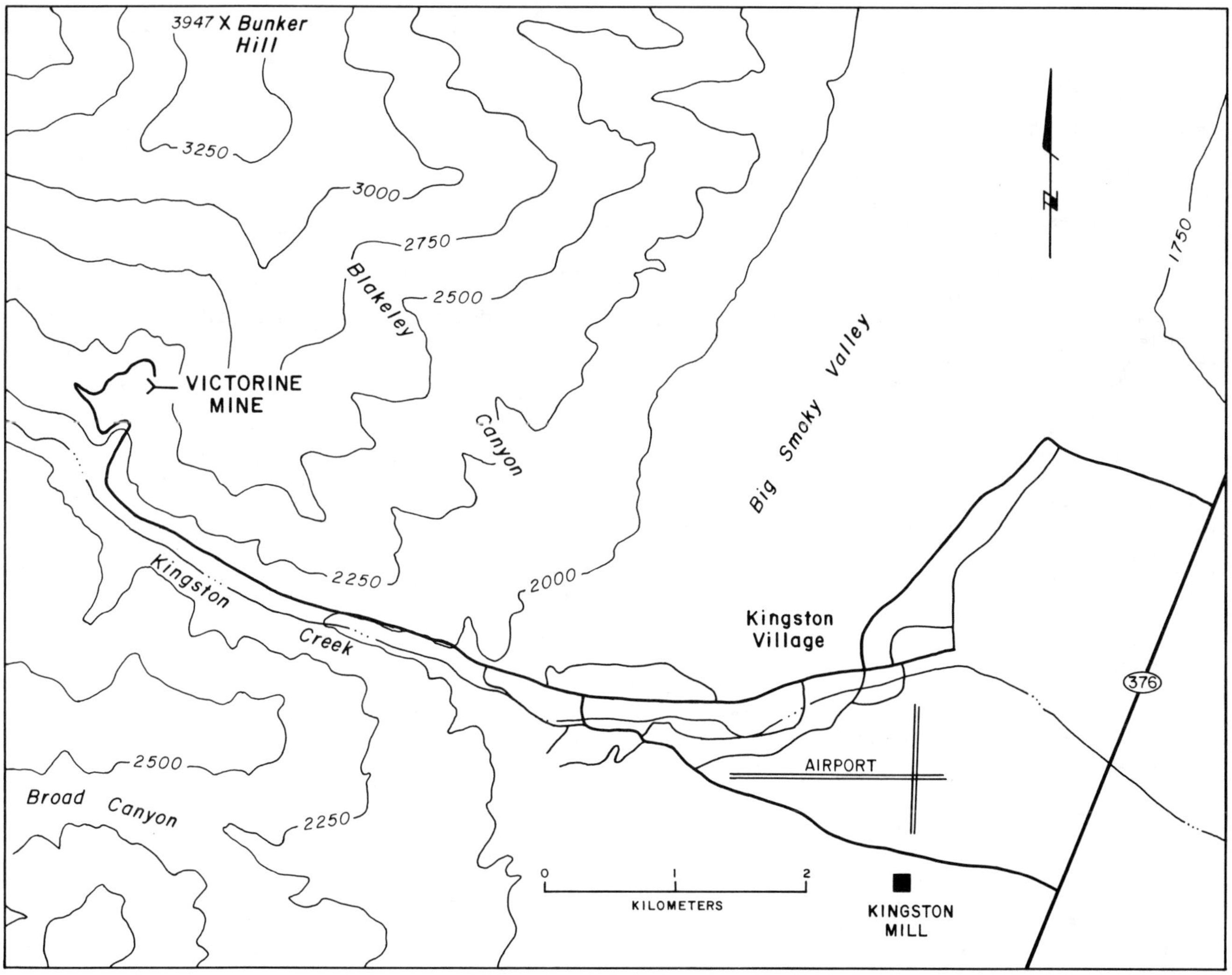

Figure 4. Kingston Project area map.

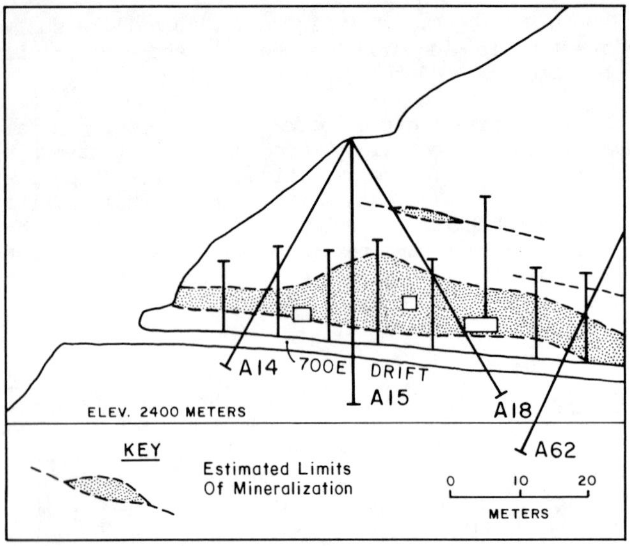

Figure 5. Generalized cross-section of the Victorine Mine.

oz/st) of gold, using a 3.0 g/mt (0.01 oz/st) cut-off grade.

Mine Production. An open cut was necessary to make room for a mine office, dry, shop and ancilliary facilities due to the lack of room on the steep slopes of Victorine Canyon. At the time of writing, the open cut is not finished but is scheduled for completion in July, 1987. About 50,000 mt of ore will be removed during the excavation of the open cut.

The mine operates two 8-hour shifts a day, 5 days a week with a total workforce of 34 including staff and geology. Underground production is scheduled to be 900 mt/day (1,000 st/day) from a random room and pillar method. The pillars will be removed in a retreat manner at the end of the project. The mine is accessed by a series of ramps within the ore zone, with rooms and pillars located to the sides of the access ways.

Production equipment consists of two 4 cu m (5 cu yd) LHD's, an electric-hydraulic drill jumbo, grader, prill truck, service truck and scissor lift for rockbolting. A 24 mt (26 st) haul truck and rockbolting jumbo are scheduled for next year.

To achieve maximum ore recovery, every round is sampled and fire assayed for gold and silver before it is moved to the surface. On the surface the ore is sorted according to grade: over 8 g/mt (0.26 oz/st) for immediate milling, 3-8 g/mt (0.1-0.26 oz/st) stockpiled for fill-in mill feed, 1-3 g/mt (0.03-0.1 oz/st) stockpiled for future recovery and less than 1 g/mt (0.03 oz/st) is wasted.

Ore Haulage. Ore is hauled from the mine to the mill a distance of 9.5 km (6 miles) over a combination of private, county and forest roads by a contract hauler. The grade of the road from the bottom of the Victorine Canyon to the mine averages 10% with grades approaching 16% for short intervals. A borehole ore pass is planned to shorten the haul and to eliminate the steep portion of the haul.

An adit will be driven from the bottom of Victorine Canyon under the mine bench a distance of 250 m (700 ft). A borehole will be pulled from the adit to the mine bench. A conveyor will then be installed in the adit from under the borehole to a truck loading facility on the surface. The primary crusher will be located on the mine bench and discharge directly into the borehole.

Scheduling. The mining sequence is outlined in a mine plan. The plan is a year-by-year schedule of tons, grade and location. Included are future development and exploration targets. In addition, monthly forecasts are made to allow for plan changes, shortfalls or averages.

Mill Construction

The general contractor for mill construction was The Industrial Company of Steamboat Springs, Colorado. Construction occurred in the winter from December, 1986 through March, 1987 and time elapsed from the first concrete pour to the first gold pour was 110 days. Total construction personnel peaked at 92.

Milling

The Kingston Project's ore processing plant uses conventional gold milling and recovery techniques. The plant is carbon-in-leach using Kambalda screens in the tanks. The original design was 500 dry mtpd (560 dry stpd); however, the used ball mill purchased had a greater capacity and allowed the plant to run at 640 dry mtpd (717 stpd).

The previous operators left an environmentally acceptable small tailings impoundment, which is being used until another partially built impoundment area, also from the previous operation, can be cleaned out, rebuilt and lined.

The mill employs 27 personnel including staff, operators and maintenance. The crushing plant and maintenance work 8 hour shifts 5 days/week. The mill operators are on a rotating shift basis using a Timken schedule. A diagrammatic flowsheet of the new mill is shown in Figure 6.

Crushing and Grinding. The crushing system is designed to have the primary crusher, a 0.64 m X 1.1 m (25 X 42 in.) jaw crusher, at the mine site; however, the jaw crusher is presently located at the mill until benching at the mine site is complete.

Stockpiled ore is fed to the jaw crusher with a 4 cu m (4.5 cu yd) front-end loader. The product from the crusher, nominally minus 63 mm (2.5 in.) is conveyed to the future truck dump hopper. An apron feeder under the truck dump hopper feeds a conveyor to a scalping screen and secondary crusher. The screen is 1 X 2 m (3 X 6 ft) with 20

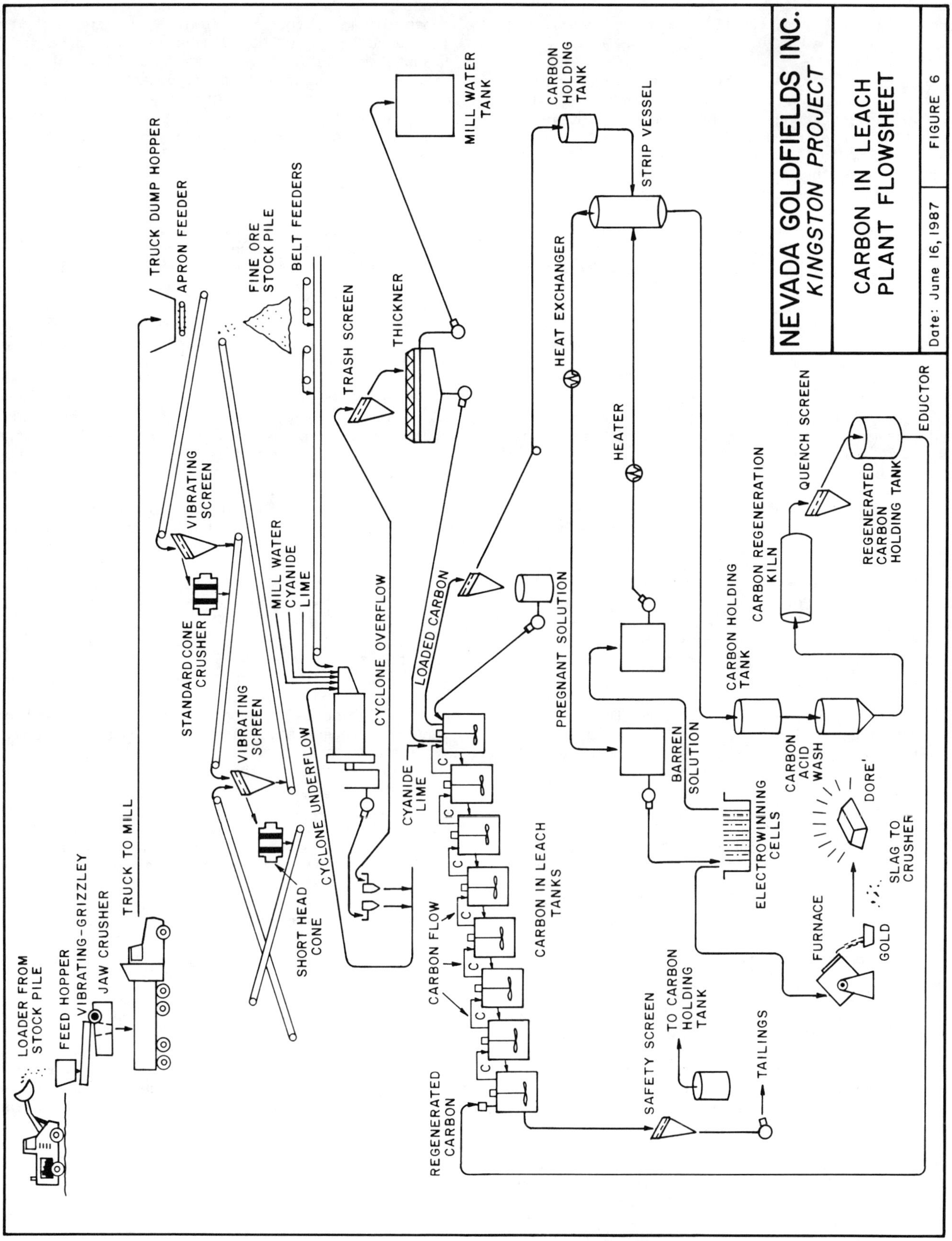
NEVADA GOLDFIELDS INC.
KINGSTON PROJECT
CARBON IN LEACH
PLANT FLOWSHEET
Date: June 16, 1987
FIGURE 6
LOADER FROM STOCK PILE
FEED HOPPER
VIBRATING-GRIZZLEY
JAW CRUSHER
TRUCK TO MILL
TRUCK DUMP HOPPER
APRON FEEDER
FINE ORE STOCK PILE
BELT FEEDERS
VIBRATING SCREEN
STANDARD CONE CRUSHER
VIBRATING SCREEN
SHORT HEAD CONE
MILL WATER
CYANIDE
LIME
CYCLONE UNDERFLOW
CYCLONE OVERFLOW
TRASH SCREEN
THICKNER
MILL WATER TANK
LOADED CARBON
CYANIDE
LIME
CARBON FLOW
CARBON IN LEACH TANKS
C
REGENERATED CARBON
SAFETY SCREEN
TO CARBON HOLDING TANK
TAILINGS
PREGNANT SOLUTION
BARREN SOLUTION
CARBON HOLDING TANK
HEAT EXCHANGER
HEATER
STRIP VESSEL
CARBON HOLDING TANK
ELECTROWINNING CELLS
CARBON ACID WASH
CARBON REGENERATION KILN
QUENCH SCREEN
REGENERATED CARBON HOLDING TANK
EDUCTOR
DORE'
SLAG TO CRUSHER
FURNACE
GOLD

mm (0.75 in.) openings used to scalp the product before the secondary crusher. The secondary crusher is a 1 m (3 ft) standard cone with a nominal discharge of 20 mm (0.75 in.).

Product from the secondary crushing system is conveyed to a 2 X 5.25 m (6 X 14 ft) screen with 13 mm (0.5 in.) openings. Undersize is conveyed to the fine ore stockpile, while the oversize discharges to a 1.25 m (4.25 ft) short head cone. Product from the cone is recycled back over the screen ahead of the tertiary crusher.

The crushing system has a baghouse dust collecting system covering all transfer and discharge points. A water spray system is used for back-up.

Fine ore from the 2 X 5.25 m (6 X 14 ft) screen is conveyed to a stacking tube and discharged to an open stockpile which covers two hoppers with variable speed belt feeders. The fine ore stockpile has 1,500 mt (1,680 st) live capacity and additional 5,000 mt (5,600 st) dead capacity. The dead portion of the stockpile can easily be pushed over the feed hoppers during periods of crusher maintenance if necessary.

The fine ore conveyor is set in a 3.6 m (12 ft) CMP tunnel under the fine ore stockpile. The fine ore is conveyed to a 3.1 X 3.9 m (10.5 X 13 ft) ball mill for grinding. A mixture of fresh and recycled process water is added to the ball mill to slurry the ore. Sufficient lime slurry is added to the ball mill to maintain a 10-12 pH level in the process circuit. Approximately 0.25 kg/mt (0.50 lb/st) of sodium cyanide is added to the grinding circuit.

Ball mill discharge is pumped to two 250 mm (10") cyclones. The cyclone underflow is recycled back to the ball mill for further grinding and the cyclone overflow is pumped to the gold recovery circuit.

Gold Recovery. The cyclone overflow flows to a 15 m diameter (50 ft) thickener, flocculant is added to the slurry at 0.023 kg/mt (0.05 lb/st) to aid in thickening. The thickener overflow is pumped to the mill water recycle tank. The thickener underflow is pumped to the carbon-in-leach tanks.

The leaching circuit consists of a series of eight tanks. Each tank is 6.1 m diameter X 6.1 m (20 ft diameter X 20 ft) high with a live capacity of 160 cu m (42,000 gal). Each tank is equipped with Kambalda-type screens, mechanical agitators and an air lift. Slurry flows by gravity through the CIL tank circuit. Compressed air is supplied to each tank to aid the dissolution reaction. Sodium cyanide is added at a 0.75 kg/mt (1.5 lb/st) rate to the No. 1 leach tank.

The Kambalda screens are horizontal carbon screens mounted in the CIL tank overflow launders to prevent carbon from moving tank to tank to tailings. Wipers are attached to the agitator shaft under the screen to prevent screen plugging.

Carbon is air lifted countercurrent to the slurry flow as often as once every 12 hours depending on carbon loading. Barren carbon is intoduced into the No. 8 tank and loaded carbon is transferred from No. 1 tank to the stripping circuit. The slurry from No. 8 tank is flowed over a safety screen to capture fine carbon which may have passed through the kambalda screen or splashed over the launder tops. The undersize from the safety screen is pumped to the tailing disposal impoundment.

Loaded carbon is washed, screened and transferred to a 1.2 m dia. X 4.9 m (4 ft dia. X 16 ft) high stainless steel strip vessel. Caustic cyanide solution is pumped from a holding tank, through heat exchangers, where it is heated to 138 degrees C (280 degrees F), upward through the stripping vessel where it desorbs the gold from the carbon. Heat for this process is provided by a dual burner, propane-fired hot water heater.

Pregnant solution from the stripping vessel flows through heat exchangers for cooling and is stored in a holding tank. It is then pumped through an electrowinning cell in the refinery where the gold is recovered from the solution onto steel wool cathodes. The barren solution resulting from the electrowinning is pumped back to the barren tank for recycling through the strip vessel.

Carbon is removed from the stripping vessel and checked for activity, if required it is acid washed and then transferred to a regeneration furnace where it is heated to 700 degrees C (1,292 degrees F) in a 0.75 m dia. X 6.1 m (2.5 ft dia. X 20 ft) long horizontal rotary kiln. Carbon from the kiln is water quenched, screened to remove fines and stored in a holding bin for reuse. Carbon which has a high activity after the strip vessel and not requiring regeneration is moved to the same tank.

In the refinery the steel wool cathodes are reduced with an acid wash. The resulting sludge is dried, fluxed and transferred to a propane-fired melting furnace. Dore' is poured as buttons and shipped to a commercial refiner. The resulting slag from the fluxes is returned to the crushing plant for reprocessing.

The plant is protected throughout with a series of sumps and retention walls, preventing the potential of contamination from circuit to circuit, of soils and of ground water.

Recovery during start-up approached 88%. Metallurgical testwork has shown that 93-94% is achievable. Minor system and routing changes performed during start-up along with operator training and procedural changes allows us to believe that recoveries in this range are attainable.

Mill Consumables. Consumable supplies used at the Kingston mill are:

Grinding balls 1.53 kg/mt (3.0 lb/st)

Lime 0.9 kg/mt (2.0 lb/st)

Sodium Cyanide 1.2 kg/mt (2.5 lb/st)

Flocculant 0.023 kg/mt (0.05 lb/st)

Caustic soda 0.3 kg/mt (0.74 lb/st)

Propane 5.76 l/mt (1.5 gal/st)

Diesel for power generation 2.67 l/mt (0.7 gal/st)

All supplies are delivered by truck to the plant. Diesel and propane are delivered in bulk. Other supplies are delivered in containers for yard and warehouse storage.

Maintenance

Maintenance is divided into three areas at the Kingston Project: mill, mine and surface. The mill and crusher maintenance is supervised by the mill foreman. The mine and surface maintenance is supervised by one maintenance foreman; however, the crews are separate. The surface crew is mobile to work at the mine or mill as needed. The remainder of their time is spent on surface mobile equipment. Seventeen personnel are employed in maintenance.

Community

The Kingston Project has a workforce of 80 employees. It is one of the leading contributors to southern Lander County's economy. Kingston Village has doubled in population as a result of the Kingston Project. The mine and mill facilities represent an $18.0 million investment, which has contributed directly to Nevada's economy.

Chapter 16

COST COMPETITIVITY DESIGN ELEMENTS AT PEGASUS'ZORTMAN-LANDUSKY GOLD OPERATIONS A CASE HISTORY

Michael L. Clark and Carson Rife

Pegasus Gold Corporation

INTRODUCTION

Zortman Mining Inc. is a wholly owned subsidiary of Pegasus Gold Corporation of Spokane, Washington. Pegasus has been producing gold and silver from the Zortman and Landusky mines since 1979. The Zortman and Landusky mines are two distinct open pit mines located approximately 6 kilometers apart in the Little Rockies Mining District of north central Montana. The two mines use large-tonnage, low-grade heap leaching techniques to efficiently process the gold and silver ore.

PROJECT DESCRIPTION

The Little Rockies rise abruptly 914 metres from the surrounding plains to about 1829 metres above sea level. The topography is quite steep, making efficient leach pad construction a major challenge. The climate is semi-arid with moderate summers and occasionally severe winters. Freezing temperatures usually arrive in late October or early November, influencing the effectiveness of leaching operations.

GEOLOGY AND MINERALOGY

The geology of the Little Rocky Mountains consists of steeply dipping sedimentary ridges forming an elliptical dome surrounding a core of intruding syenite porphyry. Within the syenite porphyry are well fractured and mineralized shear zones where the existing reserves and open pits are located. The ore is presently divided into oxidized, mixed, and sulfide. Only the oxides and mixed zones that are of ore grade are placed on the leach pads. The syenite porphyry host rock enables effective leaching of gold and silver to depths of 61 metres, having a major impact on leach pad construction costs.

MINING

All mining since 1979 has been by contractor. This enabled initial mining to commence without capital investment in equipment. Through competitive negotiations, Pegasus has been able to keep contract mining costs at levels very close to what would be experienced if Pegasus had purchased equipment initially. Contracts are negotiated yearly and are based on a unit price and haul distance for ore and waste. Haul distances have ranged from as little as 457 metres to as much as 3,050 metres. At present, Pegasus is developing a 27,000,000-tonne leach pad at the Landusky project that is adjacent to the mining area. This design feature will reduce the mining cost by 20 percent over the life of the mine. In 1987, over 5,900,000 tonnes of ore and 3,200,000 tonnes of waste will be mined at the Landusky project. Through 1986, 13,763,000 tonnes of ore have been mined at a waste-to-ore strip ratio of 0.60:1 at the Zortman project, and 13,005,000 tonnes of ore at a waste-to-ore strip ratio of 0.79:1 have been mined at the Landusky project. Current mining reserves are 32,769,000 tonnes of ore at a waste-to-ore strip ratio of 0.59:1. Through efficiencies of large-scale heap leaching, Pegasus has been able to lower the mining cutoff from 0.34 grams per tonne gold to 0.27 grams per tonne gold. This has reduced the mining cost per tonne of ore 6 percent through a reduction in strip ratio of 14 percent.

Beginning in 1984, all pits were designed based on computer-generated "floating cone" pits. The geologic models were developed by Pegasus staff with a series of economic models evaluated. The resulting computer-generated pits were used as a guideline in designing the mine plans, including haul roads for final calculation of tons and grade. Input to the floating cone models consisted of mining cost, processing cost, cutoff grade (0.34 grams gold per tonne), recovery (67 percent) and gold price.

Due to the gold being deposited primarily on the fractures of the syenite host rock, no crushing of ore is needed prior to leaching. This characteristic of the ore is a key in being able to profitably mine and process gold ore at an average grade of 0.69 grams of gold per tonne. Strict blasting control is also a key in success of the leaching operation. A system of splitting ore and waste in blast patterns has been devised to reduce dilution. Also, drill hole spacings are varied from 3m x 3m in breccia ore to 4m x 4m in syenite ore to 5m x 5m in waste. Drill holes

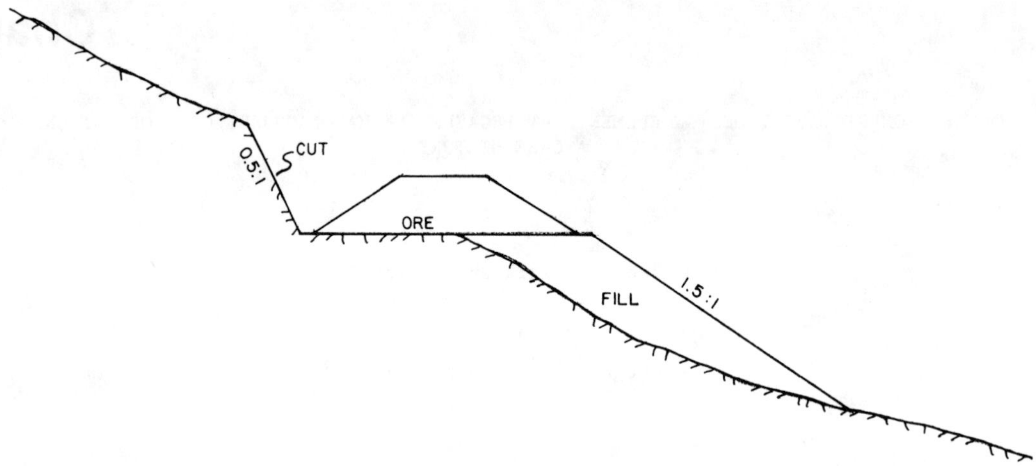

FIGURE 1. CUT & FILL LEACH PAD

are 16 cm diameter drilled 7.3m deep (1.2m sub-drill). Mining is done with 11.5m^3 front end loaders and combination of 77 and 45 tonne capacity trucks on 6.1 metre bench heights. Present production is 41,000 tonnes of ore and 23,000 tonnes of waste per day.

LEACH PAD DESIGN

Leach pad design is critical to the economics at the Zortman and Landusky projects. Depending upon the geometry of the ore heap, the cost of leach pad construction per tonne of ore can vary greatly. Beginning in 1979, leach pads were constructed using a cut-and-fill method for the leach pad base (see Figure 1). The cut-and-fill method was used until 1982. In 1982, a leach pad design was developed where waste material is utilized for construction of the leach pad. The area is cleared of all topsoil and subsoil, regraded to slopes not steeper than 2.5:1, and waste material placed in 1.5m lifts to construct the containment dike (see Figure 2). This method has been further refined and expended to the present.

The liner system used in 1979 through 1981 consisted of a 15 to 30 cm layer of compacted bentonitic clay shale placed on a graded surface. The leach pads were built such that the leach solutions would flow to the pregnant pond through a network of drain pipes and open ditches. In 1982, a 0.8-mm PVC (polyvinylchloride) liner was placed on top of the compacted bentonitic clay shale to alleviate any possibility of the bentonitic clay shale absorbing gold values. Mill tailings from the early underground mining is placed on top of the PVC liner to protect it from the ore as it is being placed by the haul trucks. All leach pads built since 1983 require pumping of leach solution from the leach pad to the pregnant pond. Table 1 shows a comparison between cost ($/tonne) and ore capacity (tonnes/m^2) for the years 1979-1987. Figure 3 shows the detail of the leach pad liner system.

Year	$/tonne	tonne/m^2
1979-1982	0.27	30.6
1983	0.24	30.1
1984	0.26	32.6
1985-1986	0.27	51.7
1987-1991	0.12	79.4

Table 1. Leach Pad Comparison

LEACHING

The ore is placed on the leach pad in 7.6m lifts with haul trucks. When placement of a section of ore is complete, it is ripped and cross ripped to a depth of 1.5m by dozer prior to spray line installation.

Design of the spray system incorporates vertical welded steel hub lines within the heap fed by a

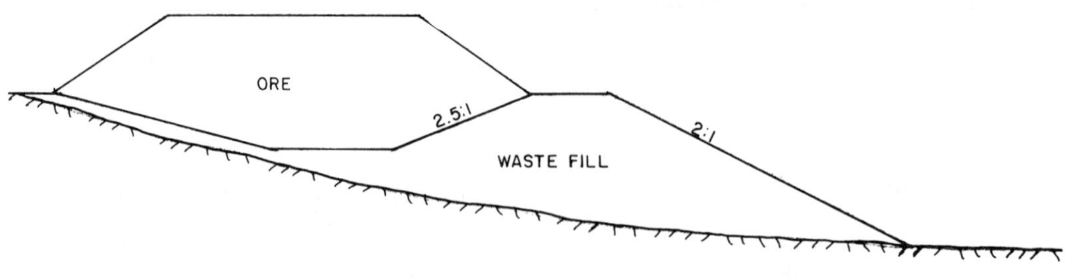

FIGURE 2. VALLEY FILL LEACH PAD

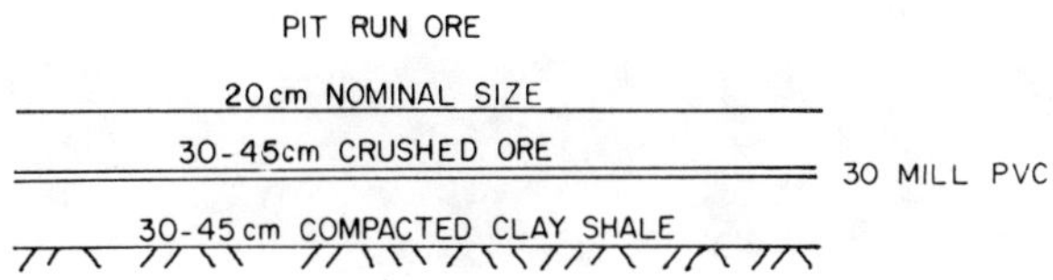

FIGURE 3. PAD LINER DETAIL

welded steel header pipe around the heap perimeter. A distribution hub is bolted to the top of each vertical pipe and 50 mm diameter PVC spray lines are installed in radial fashion from the central hub. As each lift of ore is placed on the heap, the hub is removed and a 6m extension is placed on the vertical line. The distribution hub and accompanying PVC pipe is put back into service when that section of ore placement is completed. This design allows for spraying new ore as quickly as possible, as well as providing rest cycles for ore in underlying lifts.

Sprinklers are placed on the heap in a 12m x 12m staggered arrangement to result in an application rate of 0.163 L/min/m^2. Wobbler-type sprinklers are used with a 5.6 mm diameter orifice.

Metal recoveries have historically been quite slow, taking up to five years to reach maximum recovery from a heap. A number of steps have been taken in 1986 and early 1987 to not only speed up metal recoveries, but improve overall recovery as well:

1) Well organized spray/rest cycles have been implemented on new ore, modifying the heap loading pattern to suit, as well as on the other heaps. Generally, older heaps require longer rest cycles and shorter spray cycles to enable further penetration into the rock fractures, with the oldest active heaps requiring only an occasional flush. It is common practice to use pregnant solution from older heaps to enhance the grade of barren solution being sprayed on new ore. This serves to maximize solution grades to the plants.

2) Better control over size reduction via blasting in the pit has enabled optimization of particle size for each particular rock type. Blast hole patterns are varied from rock type to rock type to insure optimum fragmentation. This practice is a result of a series of test blasts carried out in early 1987 in conjunction with laboratory column leach tests.

3) Overall plant capacity was increased with the addition of a carbon adsorption plant to supplement the existing Merrill-Crowe plants. The portable carbon adsorption plant, brought on line in April of 1987, is currently processing pregnant solution from older heaps.

4) Experience gained from winter leaching in 1986 is expected to lead to a design to permit longer leach seasons in ensuing years, shortening the recovery profile.

Maintaining consistent solution chemistry is extremely important to maximize throughput through the Merrill-Crowe plants, keep reagent consumptions at a minimum, and maintain high plant metal recoveries from solution. The current practice of controlling pH, free and total cyanide, alkalinity and calcium levels within very narrow ranges has accounted for considerable cost savings, particularly for cyanide and antiscalant. In addition tight reagent control has improved plant throughput rates more than 10%.

Chapter 17

MAINTENANCE AND INVENTORY FOR THE SMALL MINE

by

Mr. John W. Rushton
Rushton International
President
Salt Lake City, Utah

Maintenance and inventory are usually the last items considered in mine development. They need to be addressed properly and early. The micro-computer brings sophistication within the reach of even the smallest mines. Maintenance and inventory programs, if properly installed, will reduce costs 15 to 25% and allow management time to address the mining and metallurgical problems at start up. Shops and warehouses are rarely designed properly. Large engineering firms spend 5 to 20 times more than is necessary for first class facilities. There is an optimum design, size, and layout. Take the time necessary to assure you get the right people, programs, and facilities in that order.

Maintenance, inventory, and inventory control must be addressed early if a small mine is to have a chance at optimizing costs during the first two years of operation. This is hard work and is usually put on the back burner until the mine is in some degree of trouble. Hiring an experienced maintenance superintendent is important but it is not the complete solution. Maintenance systems, design of facilities, and use of computers are not skills that maintenance people usually have. In addition, the work of ordering spare parts and working with erection people is a full time job and any maintenance people on site during construction are quickly recruited. The end result is that few mines start up with an adequate Maintenance or Inventory System in place. The months, and in some cases years, that the mine attempts to operate without a system are both painful and expensive.

Designs of successful integrated maintenance and inventory programs are virtually 100% supported with some outside assistance. Our system is the approach that we are most familiar with and this paper will isolate how Rushton International would go about setting up systems and designing support facilities.

Micro-computers are the only way for a small mine to go. Mini-computers cost more originally and operating costs are many times higher than even the most sophisticated micro system. Micro-computer software is developed further and recent advances in hardware makes medium priced units equal to most mini's in RAW COMPUTER POWER. Mines in general, and especially small mines, have no need for main frame power.

MAINTENANCE

A good maintenance program has:

1. Equipment Master List
    a. Equipment Detail Reports
2. Work Order and Backlog Control Program
3. Component Control Program
4. Tire Program
5. Wear Materials Program
6. PM Control Program
7. Maintenance History Program
8. Performance Program

We also track miscellaneous items such as MSHA training and inspections, timekeeping, etc.

Rather than going through the program step-by-step, it makes more sense to go through a typical Maintenance Day and show how the programs are used. The first thing maintenance supervision needs to know is "how we are doing." We usually print the "PM Trigger Report" (Fig. 1) and the "Previous Day Performance Report" (Fig. 2) early in the morning. If a unit is overdue for PM, action can be taken immediately. If one type of equipment is performing poorly, priorities can be adjusted immediately.

It is important to note that all the representations of computer reports have been trimmed to match a 48 character format. They are not complete but the information shown is accurate.

EQUIP NO.	INT	DATE LAST PM	HRS FRM LAST PM	HRS TO PM	(%)
LOADERS MINE					
187	125 HRS	02/19/87	253	-128*	202
169	125 HRS	03/10/87	123	2	98
150	125 HRS	12/09/86	89	36	71
212	125 HRS	03/25/87	35	90	28
DOZERS					
171	250 HRS	01/28/87	204	45	81
200	250 HRS	04/10/87	96	153	38
211	250 HRS	04/13/87	41	209	16

* Note that loader #187 is 128 hours overdue for PM. This is a common on an initial PM Trigger Report

FIG. 1 PM TRIGGER REPORT

Display range : All records selected.
2nd requirement : All records in display range.

EQ NO.	DATE	HOURS OPRT	HOURS RPR	HOURS IDLE	M.A.	UT.
HAUL TRUCKS						
211	04/17/87	21.0	0.0	3.0	100	88
209	04/17/87	19.0	2.0	3.0	90	88
158	04/17/87	21.0	0.0	3.0	100	88
149	04/17/87	15.0	0.0	9.0	100	63
142	04/17/87	20.3	0.7	3.0	97	88
137	04/17/87	15.0	6.0	3.0	71	88
133	04/17/87	13.0	2.5	8.5	84	60
118	04/17/87	18.7	2.3	3.0	89	88
HAUL TRUCKS		143.0	13.5	35.5	91	80

FIG. 2 DAILY PERFORMANCE

The Planner controls the Maintenance System and will usually start the day by reviewing the backlog for equipment that is scheduled for PM or in the shops for major repairs. (Fig.3) We normally supplement the computer by displaying the backlog on a magnetic board. This visual aid makes it easy for anyone to quickly see the backlog by equipment.

The planner will usually start the day by reviewing the backlog for equipment that is scheduled for PM or in the shops for major repairs. (Fig.3)

#133 EQUIPMENT

Display range : All open work orders
2nd requirement : All records in display range.

WORK ORDER	EQUIP NUMBER	RPR CODE	ORIG	PRI-ORITY	DATE OPENED	AGE	(%)
00001	133	MSU	RH	3	03/05/87	104	253
REPLACE TRAILING ARM PINS AND BUSHINGS.							
00012	133	MEL	RH	3	03/10/87	99	471
REPAIR HORN AND BACKUP ALARM AND LIGHTS							
00013	133	MEL	RH	2	03/10/87	99	1650
REPAIR BACKUP ALARM AND TACHOMETER							
10029	133	MEN	RH	2	04/02/87	76	1900
REPAIR ENGINE OIL LEAK: BY-PASS FLTR HOSES LEAK							

FIG. 3 BACKLOG BY EQUIPMENT

This allows maximum use of available downtime by allowing the planner to schedule additional work on units that are in for major repair or PM. After communicating pertinent information to the foreman or general foreman, he immediately starts to prepare a plan for the next days work. He again uses the Work Order, Backlog and Priority Systems provided to make up a tentative schedule for review by production personnel.

During the day, the Planner can usually find time to monitor the one line Repair History file by Equipment Number and Repair code. This is a good way to evaluate trends indicating problem areas such as the example of a Brake System shown in Fig. 4.

Display range : Equipment number 133
2nd requirement : All records in display range.

RPR HIST NUMBER	EQU NO.	DATE COMPLETED	OPER. HOURS	RPR CODE	SUPV
647	133	03/16/87	36098	MAS	RH
REPAIRED AIR LEAK: UNIT DOWN 2 HRS.					
96	133	10/31/86	35286	MBR	RH
REPLACED REAR BRAKE SHOES,WHEELBRGS,CUPS,SEALS					
101	133	04/05/86	34861	MBR	RH
REBUILT REAR BRAKES					
86	133	02/14/86	32556	MBR	RH
REBUILT FRONT BRAKE SHOES					

FIG. 4 REPAIR HISTORY

This file differs from the Closed Work Order File in that it contains only those jobs that are considered significant by Management. The Closed Work Order File accumulates so much information that it is difficult to efficiently utilize and generally is not.

After the initial schedule is completed, a planner has to do his daily maintenance of the Work Order and Backlog Files. New work is entered either by creating a Work Order or using a Job Master on work that has been done before. (Fig. 5)

```
W/O NO. 10040        STATUS: OPEN        PRIORITY:  2
EQUIP NO. 133        ORIGINATOR:  ET
REPAIR CODE: MBR - BRAKES
JOB MASTER NO:        1004

DESCRIPTION OF WORK:
REPLACE FRONT BRAKES COMPLETE, USE HEAVY DUTY
BRAKE PADS.

CREW: MECH                      DATE OPENED:  06/19/87
NO.OF PEOPLE: 2                 DATE REQ:     06/20/87
JOB DURATION (HRS): 16          DATE COMP:    06/22/87

DAYS TO COMPLETE      2 DAYS       PLANNER           RH
AGE    (   0 %)       0 DAYS       SUPERVISOR        RK
DAYS REMAINING        2 DAYS       RPR HISTORY        Y

**----------------------------------------------------
:[Esc]        DESCRIPTION OF WORK              PARTS  :
:NEW W/O NO.        MANPOWER:                  TOOLS  :
:EDIT THIS PAGE                    CREATE JOB MASTER  :
------------------------------------------------------
```

FIG. 5 JOB MASTER

** The window at the bottom of the form is the same window that is used to create a Work Order. With the Work Order the option to expand on the "Description of Work" allows an unlimited amount of text to be used on large or complex jobs. If you expect to do a job again, a Job Master can be created by moving the cursor to "Create Job Master" and hitting a return. A Job Master is stored by Equipment Number and Repair Code and is used as a template to make repetitive work orders quick and painless.

The planner then has time to check on high cost items such as Components and Tires. (Fig. 6 and Fig. 7)

2nd requirement : Component name ENGINE

EQUIP NUMBR	DATE INSTALLED	(HOURS) LST REB	BDGT LIFE	(HOURS) REMAINING	INTERVAL PERCENT
151	05/07/86	20792	8500	-12292	245
208	01/01/73	18808	8500	-10308	221
118	05/01/86	16825	8500	-8325	198
108	04/29/86	15685	8500	-7185	185
137	11/22/85	14121	8500	-5621	166
133	12/01/86	10522	8500	-2022	123
171	11/26/86	10190	10000	-190	101
209	05/01/81	8559	8500	-59	100
150	05/21/82	8648	10000	1352	86
210	03/01/82	7104	8500	1396	83
211	09/16/83	8312	10000	1688	83
187	02/21/87	7609	10000	2391	76
149	03/08/85	6908	10000	3092	69

FIG. 6 COMPONENT PROJECTIONS

The example of a component projection shown in Fig.6 might indicate that the mine had a large percentage of high life engines and it may be wise to change out some of those with the highest hours prior to failure.

EQUIP #133

POSI-TION	BRAND NUMBER	SIZE	MFR	TYPE	DATE INST
LF	ZPY59A4A	21.00R-35	MICHLN	XHD	07/16/86
LRI	FPB53A0A	21.00R-35	MICHLN	XHD	06/26/86
LRO	FPB55A8A	21.00R-35	MICHLN	XHD	06/02/86
RF	UVB58A6A	21.00R-35	MICHLN	XHD	05/17/86
RRI	CPB56T0A	21.00R-35	MICHLN	XHD	12/28/85

FIG. 7 ACTIVE TIRE STATUS BY EQUIPMENT

Tires are very difficult to track without a specialized tire program. One of the more important reports is the Tire Cost by Manufacturer. It is not unusual to have a 100% or more tire cost differential, when comparing one manufacturer against another at a particular property.

In the afternoon the planner has to meet with production, finalize the next days schedule, and spend some time in the field. Being a GOOD PLANNER is hard work. Every mine of any size at all can and should have a planner. The real small mine may add some additional clerical duties to fill up time, but we DO NOT recommend that you combine Supervisory and Planning duties. All of our Maintenance Systems can and have been done manually with good success. The computer is an inexpensive tool and we highly recommend its use. The use of a computer is optional, but a system is an absolute if a mine is to have any chance of being first class.

INVENTORY

Inventory and warehousing are integral parts of the successful maintenance program. Most systems are designed for ease of accounting and little thought is given to utilizing the system as a tool for Maintenance and Production Supervision. Rather than trying to outline a total system design we will try to show how a our system differs from the average "off the shelf program." Any system requires a unique stock number for each type of part. Ours, like most systems, can adapt to virtually any type of number that anyone can dream up. For operations that do not have a stock code, we recommend a system that uses the first two digits to describe a general class of parts, that may or may not be related to a specific Vendor or Manufacturer. The rest of the code is a unique number for that part and is used exclusively for internal control. A sample listing based on the Internal Stock Code might be as shown in Fig. 8.

44-1439 06 3 1/4 ID Rotating Seal

MFG PT#	EQUIP#	REPAIR CODE	VENDOR CODE
34567G	H-23	BR	14
8765TR	H-24		08

Fig. 8 SAMPLE INVENTORY LISTING

An important feature of the program is that all information about the part that can have more than one option is on the vertical axis allowing an unlimited number of entries. This listing is simplified and the program contains more information about the part, including items such as location, purchasing status, stocking information, cost information and a longer more detailed description. Virtually any field can easily be used for quick look up or accounting purposes. It is normal for mines to start out by using the Manufacturers Part Number as the primary look up tool and gradually start using the Internal Stock Code, Equipment Number and Repair Code as they get more familiar with the system. The system allows a mechanic that knows only that he is working on the brakes of a haulage truck to enter the Equipment Number and Repair Code to get a list of only the items stocked for brakes on that particular group of equipment. A properly installed system can replace a whole library of parts books that usually does not have the particular book you need at that time anyway.

A good way to start the system is to enter the parts as they are ordered and use the computer to generate bin labels as the parts are received and stocked in the warehouse. This is a good visual check to insure that a part has been properly accounted for. Warehousing is hard work and the computer as a tool is, in our opinion, indispensable. Properly organized warehouses do not have to be very big to be efficient. Any size warehouse will be filled to capacity. In the opinion of warehouse management, all warehouses are too small! How much warehouse and how much stock can you afford?

FACILITIES

Mines, historically, have gone from one extreme to the other in providing Shop and Warehouse Facilities. Within the same mining area you might find a major mining company repairing equipment in the dirt while their neighbor down the road has shops that look like they were built for the U.S. Government on a Cost Plus Contract. Neither concept contributes to high Equipment Availability or productivity. Shops need to be properly designed and the right size. The warehouse should be an integral part of Maintenance shop, since the largest percentage of warehouse activity is maintenance related.

Wash Bay

GIVE YOUR WASH BAY TOP PRIORITY. You will be better off working on clean equipment outside than filling an elaborate shop with dirty equipment. Wash Bays should be fairly inexpensive but must be designed properly. Some points to remember are:

1- It should be high enough to raise the bed of a truck.

2- You need about 12 feet on each side of your largest unit.

3- Flush trenches should be provided to remove dirt as it is brought in. Sumps and wide trenches are always a mistake. A properly designed flush trench is 6 to 12 inches wide with the ability to flush the trench with 200 to 300 GPM of water as needed.

4- A Recirculating Water System is important from both an environmental and economic point of view.

PM Shop

Next in order of importance is the PM Shop. The "Assembly Line Method of PM" works every time. Conventional PM programs fail at the same rate. The time to establish the flow of your PM Line is during the design of the facility. We like the Wash Bay to be independent, but Service and Mechanical PM should flow through two drive through bays. These bays do not need an overhead crane but a bulk lube system is extremely important. Eighty percent of the maintenance work at the mine will be done in these two bays and tooling should be first class. The Supervisory and Planning offices should be located near the PM Line.

Warehouse and Major Repair

These buildings are also important. Proper sizing takes time and study. Mines generally need one major repair bay for every 15 units of major equipment. An overhead crane can be justified, but it is not an absolute must if you have adequate mobile cranes available. Designs for these facilities are fairly conventional and even the major engineering companies will do a reasonable job if you can afford them.

General

Some general thoughts on facility design are:

1- Most shop designs use about twice as much concrete as is necessary. Design for minimum thickness, if you do get a few hairline cracks, it is not that big of a problem.

2- Design for standard modular buildings. We have a case where the construction cost of a Wash Bay went from $750,000 to $200,000 with the use of a standard building and thinner concrete.

3- Make aprons a minimum of 20 feet wide.

4- All shops should be washed down daily. This requires properly spaced flush trenches and a good high pressure water distribution system.

CLOSING COMMENTS

You cannot do much about the geology of an ore deposit. Its a Mine or it isn't. There are a lot of mines out there that can be developed if management will properly address the items that can be controlled.

The most cost significant items, for the prospective mine or the mine that is already in operation, are Maintenance and Inventory Systems. These systems touch every aspect of an operation, and ultimately will influence every part of it, positively or negatively. You must control the result. It will materially control the bottom line. It is common to see an acceptance of emergency breakdowns and emergency parts supply, as if nothing can be done about it. That is simply not true.

Take a close look at your current or projected operations from a maintenance and inventory standpoint. Do the following:

1. Decide if improvement is necessary.

2. A good decision is to improve maintenance and inventory systems. This is a very important step and a major advantage for smaller companies. The message must come from the top and be communicated to every department.

3. Select and implement a program which addresses your situation. No two mines are alike and your people already have ideas which can improve performance.

4. Monitor results. Find out if you are getting improvement and make changes to optimize results.

The key point is to do something about maintenance and inventory control. The justification is there. We have found that an 15% increase in availability and a 15% operating cost decrease is achievable for most operations.

Surprisingly, we usually see a 20% increase in production. These results can make a big difference to the small operation.

Chapter 18

SMALL MINES DEVELOPMENT IN CHINA'S GOLD MINING INDUSTRY

Xie Longshui

Changsha Institute of Mining Research, China

INTRODUCTION

The gold production is from the state-owned gold mines, the local gold mines and the by-product of metal mines. Since 1975, the small gold mines have developed in rapid strides. Currently, there are about 3000 small gold mines in China. Most of them are operated by the local, and the production is about 50% of China's gold output. In the small gold mines,the normalized mining equipments are employed at most of mines, but some mines still use such backward tunnelling and mining tools as pick,shovel and hand hammer etc.. The mining techniques have developed from the primitive to the advanced, especially the methods of open stope, shrinkage and filling have been widely used. Such new mining techniques as bolting, vibrating draw, non-electric ignition and diamond drilling etc. are also introduced in the small gold mines. In recent years, the development of small gold mines is very fast under the guidance of new mining policy which is ' rapid and great development of large-sized mines, helping small mines to develop, and quick mining for the existing mines; and encouraging the collective and the individual to operate a mine.'

GEOLOGICAL FEATURES

The gold veins are dominant, which are divided into two types, ie. the mesothermal deposit with altered rock in a crushed zone and the hypothermal deposit with quartz veins. The orebody distribution is not centralized, and the size of deposit and single minable orebody is small respectively. Most of deposit dip at 30° to 50°, characterized by expansion and shrinkage, ramification and compound, thinning out and reappearance, separation and composite at the parallel veins etc.. The ore grade is variable, and the factor of grade change is above 100%. The waste zones with lean or submarginal ore usually exist in deposits. As for the gold deposits with altered rock, the ore boundary is indistinct. Most of gold deposits, particularly the gold deposits with altered rock, are intruded by the late structural rock, thus the faults and joints are developed,and the rock and ore are crushed,especially the host rock of roof wall is very crushed. It is obvious that the conditions of both formation of deposit and mining are considerably complicated.

HISTORY

With a long mining history and rich experiences, the local gold sectors are small, but they play an important role in China's gold mining industry.

Dated from Xia and Shang Dynasty, the gold goods was made. In the Tang Dynasty, the development of gold mining industry was very fast.A Chinese poem , written by Liu Yuxi -- one of famous poets in the Tang Dynasty, says:

> As sun is shining over Dengzhou land,
> The fog on river slowly goes away,
> The panning girls together meet at bay,
> The beauty's jewels and stamps of king
> are gold from river sand.

This poem shows the local gold mining was flourishing at that time. In 1888, the local gold mining sectors were all the rage at the moment, and the annual output of gold was as high as 21.5 tons.

Since the founding of New China, the development of local gold mining was subjected to several backsets, but certain advances are made on the whole, especially since 1970. Round about 1970,the simple, backward and indigenous mining was dominant in the small gold mines; for instance, the hand-hammer drilling, the hand-operated winch hoisting and the hand haulage etc.were main methods of gold mining . Up until the late 1970s,great changes took place in stoping, haulage and so on. Currently,some advanced technology and large-sized equipments are adopted in the local gold mines.

Generally, the local small gold mining sectors play an important role in development of China's gold mining industry, and are important and attractive sectors.

PRESENT DEVELOPMENT

The gold mineral resource is abundant and widespread in China. Lots of small and rich gold deposits are suitable for the local small-scale mining.

In recent years, the development of local small gold mines has been much better under the guidance of the correct policy of gold mining. The total output of gold from all small mines in 1984 was six times of that in 1975, averaging 20.5% of increase with years.

The small gold mines, operated by the local, mainly distribute over some provinces in the north of China; for example, Shandong, Hebei, Henan, Nei Mongol,Liaoning, Jilin and Heilongjiang etc. which are the leading regions of gold mining. The output of small gold mines in the above provinces was about 84% of the total production of local small gold mines in China.

According to incomplete statistics, China has about 70 small mines of vein gold which the ore production of mining and processing is over 25 tpd and 20 gold dredges. The stable capacity of mining is so provided for the local small gold mines.

Currently, the local gold mines have some difficulties existed in mining technology and production control etc.. The engineers and technicians,who are engaged in mining,processing,geology etc., are still short in most of local small gold mines.The mining equipments used in some mines are old and backward. Built in 1970s, some mines are not good in the producing technology, and even now the improvement is not made necessarily and reasonably.In the mining operation,some mines employed such way as ' killing the hen to get the eggs', which is the Chinese idiom, and chasing for production and benefits. The continuation of mining operation, therefore, was affected.

DEVELOPMENT EXPERIENCES AT TWO MINES

The common experiences obtained from Honghuagou and Tongguan gold mines are that they take effective measures according to local conditions,change and improve the mining method to increase ore output and lower the dilution and loss of ore.

Honghuagou Mine

It is situated in Nei Mongol Autonomous Region, with 40 Km from the southwest of Chifeng city. An actual capacity of mining and processing is 70-80 tpd of ore in this mine.

Geology The deposit is the gold-quartz vein,which is 0.4-1.2 m width, averaging about 1 m. The vein dips at 65° to 80°, and its extent is about 200 m down dip and 250 m along strike. The ore grade varies greatly, averaging 20-30 gpt; the rich ore centralizes partially. The host rock consists of hornblende anorthosite gneiss,and its stability is not so good, especially the altered sericite zone contacting with the orebody is more crushed and could scale off.

Shrinkage to Filling The shrinkage method was employed in 1965 to 1971. It has the advantage of simple process, high efficiency and output,and low working load etc., but the exposed trouble is as follows:

-- the ore dilution in stoping was higher,especially the roof wall spalled off and the second dilution was much higher in a prodigious amount of draw, due to the unstable host rock, developed joints and fissures, and crushed roof and foot wall;

-- the orebody is the lens with varied sizes and quartz vein, dipping on strike. The partial is rich ore, the partial is lean ore or rock. The shrinkage method is only used for the combined stoping, rather than for the selective mining. The ore dilution, therefore, was increased and the ore grade of stope mucking was reduced;

-- the loss of ore was considerable.In the large drawing, the roof wall scaled off and the big rock held down the ore and blocked up the draw cones, so the ore in stopes could not be drawn out and the fine ore was remained owing to the unsmoothed foot wall. What a pity these fine ore with a high grade was lost. In the other hand, the loss of ore occurred in the pillar recovering,generally was up to 50%.

In view of above problems, the shrinkage stoping was changed into the filling method in 1972.

Fig. 1 shows that the stope of filling is 30-40 m high and about 40 m long. The pilot raise,ie.the original exploratory raise, was arranged at the two ends of stope, sizing 1.5 m x 2.5 m. The filling raise was sunk in the centre of stope, with a dimension of 1.8 m x 1.8 m . The sill pillar is 2-4 m high. The ore pass was spaced at about 10 m. With the steel concrete (body steel is 14-16 mm diam, fine steel being 6-8 mm diam and wire mesh sizing 250 mm x 250 mm), the man-made sill is 0.5 m thick.

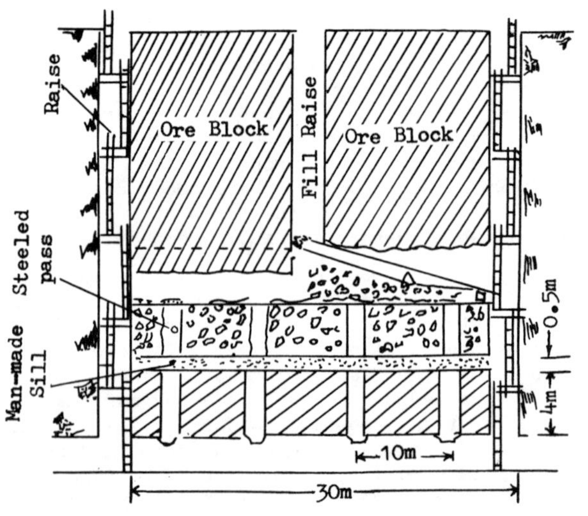

Fig. 1 The Schematic Diagram of Filling Stope

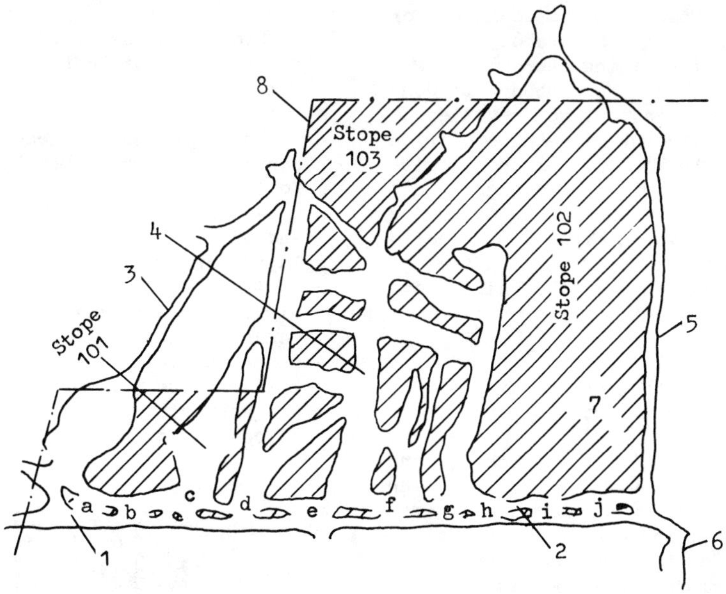

Fig. 2 Sectional View Showing Small Face Advancing And Retreat Mining: a - j. draw cone; 1. haulage level; 2. stope drift; 3. No 1 inclined drift; 4. No 2 inclined drift; 5. No 3 inclined drift; 6. cross-cut; 7. barren zone; 8. ore boundary

The practical operation has shown that it is effective and rational for the filling method to mine Honghuagou thin gold vein with steep inclination, unstable host rock, high and irregular grade of ore. The loss and dilution of ore were reduced from 30% to 3-5% and from 40% to 25% respectively. The grade of mined-out ore was increased from 7.8 gpt to 14.7 gpt.

Experiences in Application of Filling Method

-- The different filling methods were used on the basis of vein thickness. The dry-placed filling was adopted when the vein was more than 0.5 m thick. The fill was the rock from driving and the gangue in stopes. The cutting wall and filling method was employed for the vein less than 0.5 m thick. In this way, the ore dilution was decreased significantly.

-- Compared with the shrinkage stoping, the cutting wall and filling can make the grade of mined-out ore improve greatly and be used to mine the droppers complicated in shape.

-- As a result of replacement of the sill pillars by the man-made sill which the steel concrete was 0.5 m thick, the recovery of sill pillars was increased from about 60% to more than 95% .

-- The timber lining pass was replaced by the steeled pass, which was made of steel plates with a size of 4-6 mm thick. It is convenient for the steeled pass to be installed, and adapted for the change of vein dip. In addition to decreasing timber consumption, the loss of fine ore in the pass with timber lining can be avoided.

-- The straw pad or the used belt of conveyor was put on the fill to space the fine ore from the fill, with the result that the gold content in the fill was reduced from about 2 gpt to 0.1-0.3 gpt.

-- The continuous and safe operation in the mine was guaranteed, owing to that it is not necessary to handle the mined-out area.

Tongguan Mine

It is located at Tongguan county in Shensi province, with about 50 tpd of ore. This mine has obtained a good experience in mining the flat and thin gold vein. Here is introduced it briefly.

Geology The gold-quartz vein is dominant. The orebody dips flatly to the northwest at 3° to 10°, averaging 0.6-0.8 m thick. The strike is the northeast. The gold-bearing ore is pyrite, chalcopyrite and galenite. The gold grade varies greatly, averaging 7-15 gpt. The roof and foot walls are the gneiss with stronger migmatization and f=8-12. The roof wall adjacent to the ground was fully oxidized. The joints and fissures were developed.The stability of roof was getting better.

Experiences in Mining Method Initially, the full face stoping without support was used in Tongguan mine. However, there were so many problems in the practical operation that the original method could not be gone on.

According to the original design of mining, Level 1060 was divided into three stopes,e.g. Stope 102 and Stope 103. When Stope 101 was advanced for 13 m by the full face stoping, the orebody thinned out. In the other hand, the mining operation was transferred to Cone c because the orebody at Cone a - b was located downward and not yet revealed.It was no way for the full face stoping to be carried out, due to falling of the roof. Under this circumstance, the full face stoping was changed into the room-and-pillar method by small face advancing and retreat mining, as shown in Fig.2 . The shortcomings of this method found in the mining are:

-- low production efficiency at the initial operation;
-- poor ventilation;
-- the working face was limited;
-- no space for storing the rock picking;
-- no way to improve the dilution of ore.

The overburden atop Stope 102 was thicker, and the roof gradually trended towards stability. The full face stoping by temporary supporting, therefore, was begun to try along No 2 inclined drift. It was not successful for this method to be used because the problems existed in supporting were not solved and the roof collapse was much worse.In this case, the room-and-pillar method by small face advancing and retreat mining was employed instead of the full face stoping by temporary support illustrated in Fig. 3. However, the overlong distance of scraping and lower production efficiency were presented in mining operation. To assure the ore supply, the two sides were firstly cut when the access had not reached the end of room. As a result, the host wall was overcut owing to the thin vein, the dilution of ore was high and the recovery of pillars was more difficult.In addition to these, the roof and the side wall were fallen.

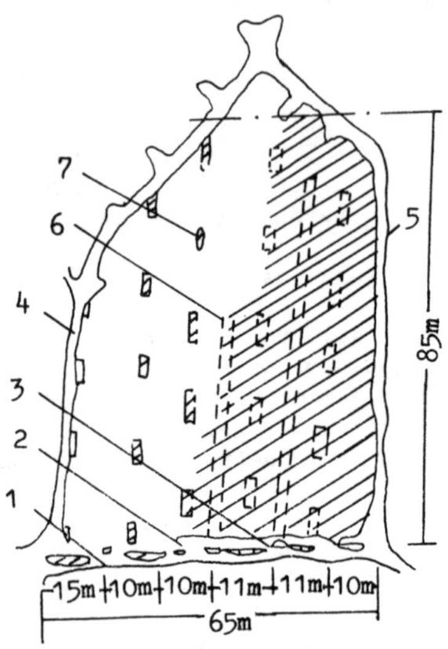

Fig. 3 Sectional View Showing Small Face Advancing And Retreat Mining: 1. haulage level; 2. stope drift; 3. draw cone; 4. No 2 inclined drift; 5. No 3 inclined drift; 6. access with a small dimension; 7. intervening pillar.

Stope 103 was at the top of Stope 101, without the independent bottom infrastructure, as shown in Fig. 2.

To solve the problems above mentioned, the technical measures taken by Tongguan mine are as follows:

-- the scraper drift was excavated at Stope101, and one set of scraper was added to form the relayed scraping with Stope 103;
-- the remnant of Stope 102 beyond the scraper distance was redivided into stope. In general, the scraper distance was 40-50 m;
-- the full face stoping was used, by which the pillars were not left or less recovered, the face was supported preliminarily with timbers and the slice was advanced along strike, as shown in Fig.4.

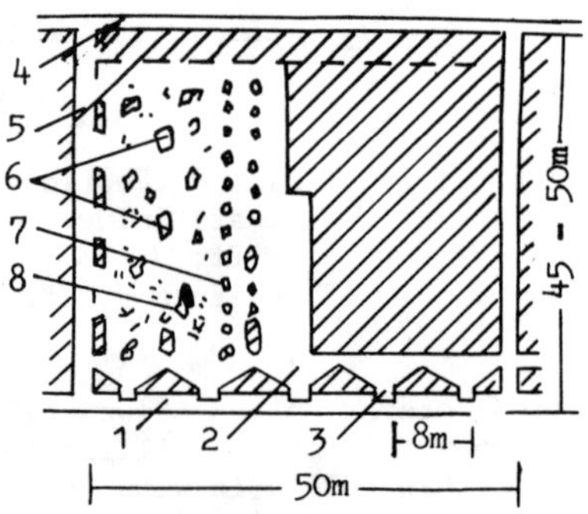

Fig. 4 Typical Diagram Showing Full Face Stoping: 1. haulage level; 2. stope drift 3. draw cone; 4. return tunnel; 5. inclined drift; 6. pillar in stope; 7. temporary support; 8. rock picking.

-- the minable limit was determined on site. It is very important for the mining operation, especially in the early stage of mining;
-- the mining width was controlled. When the vein width was less than 1.6 m, the mining width was fixed to 1.6 m; When the vein width was more than 1.6 m, the vein was completely mined;
-- the rock in stope was picked out by hand to lower the dilution of ore.

It has proved, in the mining practice, that the improved full-face stoping has many advantages,for examples:

-- favourable for ventilation and convenient for mining operation;
-- higher drilling effect. The working efficiency for one set of drill, on an average, was up to more than 40 tons per shift, increasing by a factor of 19/25 as compared with the previous 22.9 tons per shift;
-- easy to control the width of mining;
-- large space enough to store the rock. It was significant for the dilution of ore to be reduced, as given in Table 1;

Table No 1

Stope	Dilution, %		Loss, %	
	RPM *1	FFS *2	RPM	FFS
101, 103	55	28.6	43.9	
102	65.2	43.3	34.7	6.5

Notes: *1 --- room and pillar method

*2 --- full face stoping

-- good economic results. The annual output of ore was 16500 tons estimated at 50 tpd.

The rock mixing into ore could be reduced by 3300 tons per year if the dilution of ore, on an average, was lowered by 20%.

The loss of ore could be decreased by 5133 tons per year if lowered 28%, corresponding to 16426 grams of gold estimated at 4 grams of gold per ton and 80% of milling recovery.

The mining practice has shown that it is ideal for the full face stoping, by temporary support of face and advancing of slice along strike, to be used for mining the flat lode with less change of both dip and strike on condition that the host wall is not stable or less stable. To use this method not only improves the production efficiency but lowers the dilution and loss of ore significantly. In addition, a good economic result can be obtained.

FUTURE DEVELOPMENT

The future development of China's small gold mines is optimistic in view of the following key points.

-- The gold mineral resource is abundant and widespread in China. The gold deposits are small and scattered, and very suitable to develop the small-scale mining.

-- China has a long history and rich experience in the small-scale gold mining, which provide a good condition for the great development of small gold mines.

-- Since the founding of New China, lots of experts and technicists engaged in the gold industry have been cultivated, playing an important and positive role in promoting the gold industry of China.

-- Currently, the gold production has been laid stress on from the centre to the local in China. Under the guidance of mining policy of ' rapid and great development of large-sized mines, helping small mines to develop, and quick mining for the existing mines ', a rush for gold mining has risen in China.

-- With the development of economic construction of China, the manufactories producing the mining equipments are making a positive effort to develop the equipments suitable for the small gold mines. The mechanized mining in the mines will be greatly improved.

-- With the reform of scientific research system in China, many institutes of mining design and research have transferred to provide the technical service for or do the cooperative business operation with the small mines. It is believed that the development of small gold mines will be more flourishing.

CONCLUSION

The mining methods used for the gold deposits in China vary with the different formation. It is important for the precious metal mines to select and use the mining methods with low dilution and loss of ore. This is a main experience in the gold mining in China. Author hopes this paper would be a little help to the gold mining in the world.

Chapter 19

REFRESH YOUR EQUIPMENT REPLACEMENT DECISION-MAKING SKILLS

Roger S. Dewey

Minerals Engineer and Economist
Arvada, Colorado

1.1. INTRODUCTION

Despite indications of a general economic recovery in the U.S. during the mid-1980's, mine operators have not generally shared in the benefits of these reported better times. Budgets have not rebounded, borrowing is tougher, parent companies' pockets are not as deep as in previous years, and margins are not any better -- if there is any left over at all.

Managing capital equipment budgets is always a challenge, and the last five years of stretched equipment lives make these decisions more tenuous. Delays and avoidance of replacements due to tough times inevitably makes those decisions increasingly difficult and usually more expensive.

The purposes of this paper are to present a brief refresher outline of the fundamentals of equipment replacement analysis, to suggest a number of papers where analysis methods are explained in greater detail, and to recommend the use of a widespread microcomputer software capability (1-2-3) as the easiest path to a much-needed solution for a very difficult, and usually under-supported management decision problem.

2. REPLACEMENT DECISION POLICY

Investments in equipment replacements must compete for capital with all other investment alternatives. Capital for replacements often represents a greater commitment than do the new projects which a company might handle in a given year. But, sadly, formal policies for standardized analysis of replacements rarely exist, even in major companies. Yet this policy can be very significant in determining the competitiveness of a firm.

Whereas many replacement decisions are based on cost minimization, economists and financial analysts agree that wealth maximization or return on investment is the superior criterion. Though the difference may seem slight, agreement on a firm's financial objective is vital. Replacement policy should be straightforward, have a logical basis, and a proper level of sophistication.

There are several reasons to develop a policy for replacement decision-making: [1]

1 - equipment's optimal economic life is nearly always shorter than its physical life;

2 - mine life often exceeds equipment life, so management must face these decisions routinely

3 - replacement capital costs, on an undiscounted basis, can run 25 to 75% of original capital

4 - technological changes may result in a newer machine or system which can operate more efficiently, more safely, or less expensively

5 - estimates should be made consistently in the fundamental areas of life expectancy, availability, utilization, price escalation, taxes

6 - long-term and short-term budgeting without decent data and analytical capability is nearly impossible, or at least ineffective

Policy development assumptions which must be considered include: the planning horizon, cost patterns, interest rates, availability of capital, replacement vs. expansion, technology advances.

---

[1] Gentry, D.W., 1985, course notes, Mining Engineering 440, Colorado School of Mines.

Under normal economic circumstances, some 30 to 40% of new equipment sales are intended to replace old equipment [2]. During the recent recession in the minerals industry, this percentage has probably adjusted somewhat, but the poor sales records of most manufacturers during the 1980's would lead to the conclusion that a large number of pieces of equipment were not replaced on 'normal' schedules.

Those equipment units that are still active, beyond their normal replacement time, may have been rebuilt, or they may still be struggling along without either solution to the inevitable demise of any piece of capital equipment.

Operators may have been forced to consider rebuilding some equipment rather than replacing units whose cost-to-performance ratio may have declined to 'uneconomic' levels. And many operators have probably postponed rebuilding as insufficient operating budgets precluded all but the most necessary maintenance.

As reinvestment decisions are postponed, they usually become more difficult. The possible management inattention or under-explained non-replacement that accompanies postponements can sometimes lead to decreased care and commitment to maintenance of a particular unit. Rationalization of replacement decisions often results in decreased attention to good record-keeping, causing the necessary cost and performance data to become harder to find. All of these rather subjective occurrences are undeniable, and are also nearly unmeasurable, making the replacement analysis less valid or at least more difficult to prepare.

Replacement decisions are important on a macroeconomic level because they represent at least $2 billion of annual reinvestment decisions in the mining industry. And because they are often the most flexible expenditures in a mine manager's grasp, they are viewed as maybe the most important decisions in an operating budgeting context (or microeconomic sense).

Therefore, it is quite likely that most mining companies are facing an unusually large number of replacement decisions during the next few years. It is also quite likely that these decisions will often be based on less-than-optimal information, analysis techniques, and financial criteria. This paper will suggest several potential improvements in these tasks.

----

[2] Frost & Sullivan, 1979, Mining Equipment Market Survey and Forecast, Frost & Sullivan, Inc.

3. CONCEPTS OF EQUIPMENT REPLACEMENT

George Terborgh in 1949 published an explanation of the gradual change in competitiveness of a machine caused by the 'inferiority gradient'. [3] Figure 1 illustrates these two components:

obsolescence, reflecting the difference between the cost per unit of work or the earning rate of the challenger vs. the incumbent when it was new, this (external) component is called ALPHA

deterioration, the amount by which the cost per unit of work or the earning rate of the incumbent has fallen below or risen above its own level when new, this (internal) component is called BETA

Certainly the advent of modular drive trains and the increased usage of hydraulic power could be considered examples of the technological change which would cause an Incumbent machine to suffer from Alpha, the obsolescence component. Examples of Beta, the deterioration component speak for themselves, unfortunately.

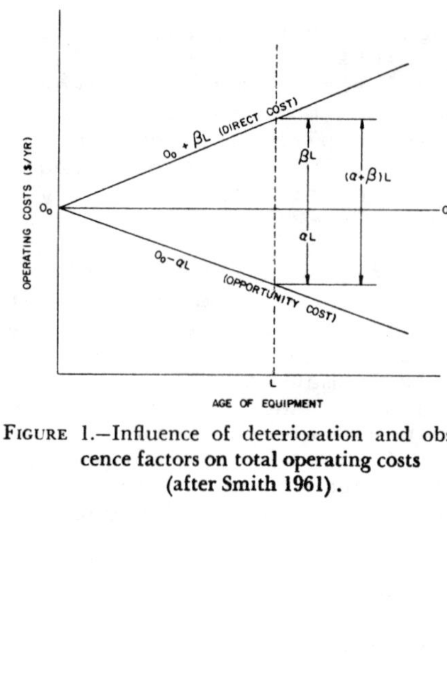

FIGURE 1.—Influence of deterioration and obsolescence factors on total operating costs (after Smith 1961).

----

[3] Gentry, D.W., and T.B.Johnson, An investment policy for equipment replacement, Colorado School of Mines Mineral Industries Bulletin, vol.17, no.1, January 1974.

3.1. Engineering Economics Approach

Thirty to forty years ago the engineering economics approach to investment analysis was popularized in industry and in college courses. It is essentially a cost minimization technique in which the best (lowest cost) challenger is identified, then it is compared to the incumbent Use of Capital Recovery Factor or discounting was a tremendous advance over any previous method.

Traditional analysis of equipment replacement involved the calculation of total average annual equipment costs, a combination of four elements:

the capital investment,

the return or interest on that investment,

the operating cost, and

the obsolescence and deterioration costs.

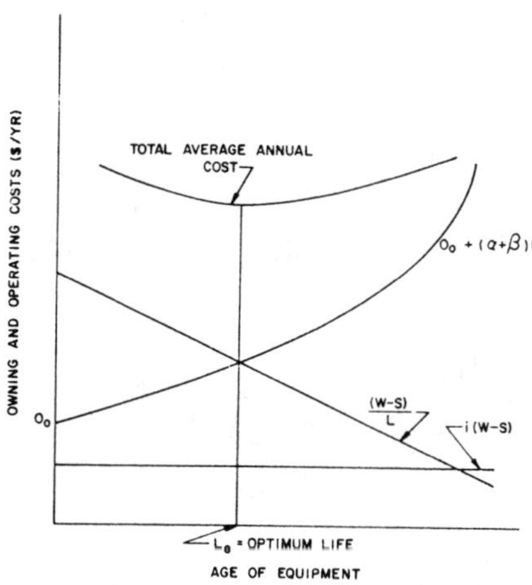

FIGURE 2.–Components of total average annual equipment cost.

The concept of optimal equipment life, that age at which the total average annual cost is minimum, is used to compare an Incumbent with a Challenger. Replacement is made whenever the Challenger has a realizable minimum average annual cost sometime in its future life lower than the future minimum average annual cost of the existing or Incumbent machine. Gentry and Johnson [3] clearly summarize this technique.

Another advance in replacement anal-ysis came in the 1960's when the marginal cost of keeping the Incumbent one more year was compared to the average annual cost of the newer machine. This was a refinement of the cost comparison method.

3.2. Investment Evaluation Approach

In the 1970's an investment approach which addressed the wealth maximization criteria adopted by most companies was popularized by many writers, including Stermole [4], Rathbun [5], Kearns [6], and Gentry & Johnson [3].

Recent texts have not emphasized the equipment replacement problem, leaving the development of much of the analysis up to the individual, perhaps because replacement is not really a unique set of the investment evaluation world. However, this writer believes that replacement has been and still is somewhat of a step-child in most firms, and that increased attention to the special needs of the operating manager should be provided by the financial and engineering support staffs because of the importance of these decisions in annual budget-making and the ongoing competitiveness of a mining operation.

Replacing equipment units whenever a higher discounted cash flow-rate of return (DCF-ROI) was anticipated with a Challenger became vogue during the 1970's as capital budgeting and wealth-growth objectives were reinforced by most organizations.

Benefits measurement, accounting of costs as well as performance, and determination of corporate cost of capital were added to the knowledge domain of the engineer-manager. Under capital budgeting objectives, the ranking of alternative investments within a firm required consistent criteria and analytical procedures, a painful exercise for most mining professionals when working within the conglomerates and diversified industries which largely dominated the mining industry.

----

[4] Stermole, F.J., 1984, Economic Evaluation and Investment Decision Methods, fifth edition, Investment Evaluation Corporation, Golden, CO.

[5] Rathbun, G.A., 1970, Equipment Replacement --- Evaluation by Return on Investment, chapter in CIM Special Volume 12: Decision Making in the Mineral Industry.

[6] Kearns, J.L., etal, 1979, Taking the Guesswork out of Equipment Replacement Decisions, SME Preprint 79-6, New Orleans February Meeting.

3.3. Management Science Contributions

Somewhat beyond the scope of this paper, but worthwhile for those pursuing the latest thinking in the area of replacement investment analysis, are numerous papers coming out with applications of mathematical programming - linear, integer, dynamic, etc. and simulation or decision analysis.

Papers from the biannual APCOM (Applications of Computers in Mining), The Institute for Management Science, and the Operations Research Society of America are beginning to establish another generation of decision-making technology with great implications for heavy process industries.

Though not yet widely popular because of both their relative young age and their relatively sophisticated logic or procedures, these methods are now being accepted and applied, particularly with the availability of desktop computing power which makes them more controllable and inexpensive.

One example of an OR technique applied to the equipment replacement problem is Sherer and Gentry [7], who developed a network approach employing integer programming, a method of constrained optimization.

This computerized model is formulated to use historic cost and production data to predict equipment cost vs. age for the existing and potential replacement units and to predict the optimum replacement age or time.

Another part of the model determines the best strategy for accomplishing estimated production goals at a minimum discounted cost.

Additional applications of management science to decision-making in the mining industry include the use of simulation techniques to assess various uncertain outcomes. Widely used in the petroleum industry, mining geology and quality control areas, these techniques hold promise for use in financial analysis of mining projects and equipment-buying whenever probabilistic predictions would be more realistic than the usual deterministic methods.

Decision-tree illustration of alter -natives and known-unknown situations, with Expected Value or Rollback estimations of cost or return forecasts can add tremendously to discussions and evaluations.

[8] Stewart, R.E., 1980, Proper Equipment Replacement -- The Key to Replacement Economy, _CIM Bulletin_, vol.73, no.823, November.

4. CHOICES -- NEW VS. REBUILT VS. USED

Buying new equipment is a wonderful option except for one factor -- price. A well-financed buyer can choose the exact machine and options he wants. He can have maximum confidence in the performance and reliability of the equipment. He knows that the manufacturer will warrant his equipment, and that he has a fully depreciable item. Buyers in the current market can usually get personalized financing through the dealer or manufacturer, so this choice is made even easier.

But if the price of new equipment is still too steep, the option of rebuilding is not a bad deal if it's well understood. Similarly, buying used equipment can be a preferred option, but one which carries increased risks, of course.

An equipment owner should be armed with knowledge of at least the seven major options available in the realm of replacement equipment: [9]

- -- buy new equipment
- -- buy OEM-rebuilt equipment _updated_ to today's safety and performance specifications
- -- buy OEM-rebuilt equipment _restored_ to its original condition
- -- have your existing equipment OEM-rebuit and _updated_
- -- have your existing equipment OEM-rebuilt to its _original_ condition

---

[7] Sherer, H.E. and D.W. Gentry, 1982, An Operations Research Approach to the Equipment Replacement Problem, 1982 APCOM Proceedings.

[9] Richards, Rod, (Lee-Norse Co.) Buy new, used or rebuilt equipment, _Coal Mining & Processing magazine_, September 1983.

-- buy equipment rebuilt by an independent facility to its original condition with required safety updates (or rebuild it yourself)

-- buy straight 'as is' used equipment

Each choice offers trade-offs between costand risks of performance and reliability. The variables that determine your choice usually depend on a company's own financial situation -- largely economic questions involving accounting methods, credit, tax situation, and cash flow.

5. INFORMATION REQUIREMENTS

As the decision models get more sophisticated and more realistic, they will become increasingly input-sensitive. The demand for better data is obvious.

Accounting and engineering record-keeping to aid the decision maker are of increasing importance but fortunately the capability to collect and to analyze such data is already in place in almost every office. Relevant, current equipment costs and production-performance data on a unit basis have become fairly standard practice in most modern mining firms.

Mining and contractor organizations have developed similar cost accounting systems, though the exact 'code of accounts' may vary. Historical cost data needed for replacement analysis are the following:

purchase price
overhead costs
fuel-power costs
operating labor costs
tire costs
other consumables costs
repair parts costs
repair labor costs

Basic performance data requirements include:

hours operated and downtime per time period
units of work produced per time period
expected future benefits from new machine
expected utilization of old and new machines

Financial information is also needed:

salvage value of existing machine
tax loss on sale of existing machine
salvage value of new machine
projections of interest rates and debt/equity ratios, desired discount rates

Equipment costing typically done for mine accounting reporting is often insufficient or not well-correlated with performance data. Another obstacle is the tendency for maintenance costs to be recorded by fleets rather than by individual units. This lazy practice cannot be allowed unless replacements are decided on a fleet-wide basis, which is rather unlikely.

Ordinarily, companies already collect and accumulate ample equipment cost data to provide the basis for replacement analysis, but access to or manipulation of the data within the accounting system is sometimes difficult. Sharing data is becoming easier, however, with the spreading usage of microcomputers between which floppy disks of selected cost data can easily be hand-transported.

Mine-level sharing of these data is fairly easy to arrange, particularly when a committed management team wants the results of this analysis.

Adjusting these historical cost data to reflect price trends should be automatically incorporated into the computer database. Raw or uncorrected costs, coupled with updating factors based on specific U.S. Bureau of Labor Statstics or real-evidence multipliers, will allow evaluation in constant dollars. Published indices are probably the easiest to research and to use, but actual data may furnish a more applicable index in some instances.

Assembling all of this data to provide the foundation for this fairly sophisticated analysis may appear to be an awesome task, but once the procedure is in place and the database is compiled, staff and management comfort should rise. Being able to provide financial verification of the wisdom of making equipment replacements will be a worthwhile investment in staff attention.

6. INVESTMENT PROFITABILITY CRITERIA, NPV vs. ROI-IRR

Whenever there are two or more mutually exclusive projects, NPV should be used because IRR can be misleading. There is a theoretical reason, but a simple example is easier to understand:

Suppose you have a choice between earning 100% for one year on an investment of $1, or 10% on an investment of $100. If you made this decision based on IRR, you would go with the first option and your earnings would be $1 for the year.

NPV is a better criterion; it tells you to choose the second option, which leads to earnings of $10.

When a single investment project is considered by itself, IRR and NPV usually produce the same answer, but even in a 'go-no go' decision, the IRR can conflict with the NPV if there are both positive and negative cashflows during the life of the project. For instance, overhauls during the life of a machine could cause negative cashflow in a mid-life year. This causes the IRR calculation to have more than one solution, one of which is not correct.

Some companies use a variation of the IRR called a profitability index or a benefit/cost ratio. This calculation has the same disadvantages as the IRR, signaling the wrong selection when there are negative cashflows during the life of the equipment item.

The only sure way to stay out of trouble in ranking capital investments is to rely on the NPV.

7. 1-2-3 TO ANALYZE EQUIPMENT INVESTMENTS

Once the time-consuming problems of gathering historical cost information and estimating future costs are done, computerized spreadsheets can make short work of evaluating the comparative returns.

Following is a primer on setting up a 1-2-3 spreadsheet for equipment replacement analysis, liberally borrowed from Mr. Gardner's excellent article in LOTUS magazine. [10]

Beginning with the usual titles to identify the worksheet, project or mine, equipment type, date, analyst, etc., perhaps in the first four rows, proceed to identify the two competing alternatives in cells A5 and G5.

Set up column headings: "CASHFLOWS" in cell B7 and "YEAR" in cell E7. Subheadings "INVEST", "OPSAVE", and "C.F.SUM" go into cells A8, B8, and C8. "C.F.CUMUL" goes into cell D8. For the Challenger, duplicate all these headings from the range A7..E8 into cell G7 (appearing in range G7..K8). Column F is blank for visual separation between the cases.

Insert cashflows, beginning in row 10, negative for investments in columns A and G, and positive for cost savings in columns B and H, for the Incumbent and the Challenger, respectively. Continue in annual steps down as many rows as your analysis will run.

Enter a summing formula "A10+B10" into cell C10, then copy it into the range C11..Cn, where n is the number of years the analysis will run. Enter the formula "@SUM(C$10..Cn)" in cell D10, and copy it into the range D11..Dn. These formulas accumulate the total cashflow from the initial outlay to the current year. In cells E10..En, enter values for years 0 through n. These five columns, A..E, are for the Incumbent. Duplicate the range of column D formulas and column E values into columns I and J (cell I10) for the Challenger analysis.

Enter the label "DISCOUNT RATE" in cell A17. Enter the value of your selected discount rate in decimal form (.2 for 20%) in cells C17 and I17, with format for percent, two decimal places. This is the minimum return desired on any new or old investment, and should reflect both the cost of capital and the risks involved in owning.

Enter the label "NPV" in cell A19, and the formula "+A10+@NPV(C17,C11..Cn)" in cell C19. Format cell C19 with two decimal places. The initial investment is outside the @NPV expression because it is made immediately; the discount rate has no effect on it, only on the payback amounts.

The @NPV value calculated here is the net present value of the investment rather than the total; it is the amount for which the payback values associated with the equipment would yield a break-even proposition. The formula in cell A19 adds the negative initial investment to the positive but discounted annual savings to get the NPV. Anytime NPV is greater than zero, the investment is a bargain. When evaluating alternatives, one looks for the greater(est) NPV as the most attractive path to increasing wealth.

----

[10] Gardner, E.S., Making Capital Investment Decisions, article in LOTUS magazine, May 1987.

EQUIPMENT REPLACEMENT ANALYSIS USING NPV

anymine anyunit anytime anybody anyboss

Dozer A (rebuild existing unit)

INVEST	OPSAVE	C.F.SUM	C.F.CUMUL	YEAR
($120,000)		($120,000)	($120,000)	0
	$84,000	$84,000	($36,000)	1
	$60,000	$60,000	$24,000	2
	$36,000	$36,000	$60,000	3
	$12,000	$12,000	$72,000	4

DISCOUNT RATE 15.00%

NET PRESENT VALUE $28,944

Dozer B (purchase new unit)

INVEST	OPSAVE	C.F.SUM	C.F.CUMUL	YEAR
($540,000)		($540,000)	($540,000)	0
	$188,000	$188,000	($352,000)	1
	$164,000	$164,000	($188,000)	2
	$140,000	$140,000	($48,000)	3
	$116,000	$116,000	$68,000	4
($54,000)	$164,000	$110,000	$178,000	5
	$140,000	$140,000	$318,000	6

DISCOUNT RATE 15.00%

NET PRESENT VALUE $21,077

4

Low Cost Processing For Precious Metals Recovery

T. M. Plouf
MinProc USA, Ltd.
Englewood, CO

Chapter 20

STRIPPING FOR FUN AND PROFIT

MR. MICHAEL C. COSTELLO

Chief Chemist
Minproc Engineers,Perth, Australia

MR. JOHN H. MCCOWAN

Technical Director
Minproc Engineers, Perth, Australia

DR. MARK E. WHITE

Manager Metallurgy,
Minproc Engineers, Perth, Australia

ABSTRACT

Minproc Engineers has carried out the design construction and commissioning of thirty two process plants in the period 1978-87. Twenty eight are gold silver recovery plants, and of these twenty two were gold alone and six were gold and silver. In most plants carbon was used for gold adsorption. Desorption has been carried out using the USBM Homestake method, the Anglo American Research Laboratories method, and the Micron Methanol method. Electrowinning has utilised steel wool, stainless steel wool, and aluminium cathodes. Electrorefining using polished stainless steel cathodes is practised. This paper summarises the experience gained in these operations.

DEFINITIONS

In the text of this paper the elution procedures will be referred to in a shortened form as follows: Zadra, Anglo, and Micron. Zadra will refer to the process invented by him and published by the United States Bureau of Mines (Zadra 1950). Anglo will refer to the process developed by Davidson and Duncanson at the Anglo American Research Laboratories in South Africa (Davidson 1977). Micron will refer to the process developed by Griffin & Costello & Self at Micron Research Laboratory Western Australia (Griffin 1982).

ZADRA SYSTEM DESIGNS AND OPERATING EXPERIENCE

Water Conservation and Planning for Gold Room Operation in Arid Areas

The total elution procedures include both the acid wash and rinse, the presoak and elution, and the carbon cool, water demands. In the period after the use of the atmospheric Zadra system it seemed that the Anglo process with an elution cycle time of less than eight hours and the added benefit of presenting extremely clean acid washed carbon to the regeneration kiln, was a better choice than the pressure Zadra System so that the latter system went out of favour and was replaced for a period by the Anglo and Micron systems. The Anglo process in its normal direct form however consumes a relatively large volume of potable water per tonne of carbon so that in arid areas where potable water was at a premium a return to the pressure Zadra system was made. The following table indicates the water useage of the various forms of the elution systems and highlights the low water consumption of the Zadra System particularly if acid wash and rinse is placed after the kiln and carried out in process water. These water demands are from operating plants where every effort has been made to recycle water particularly carbon transport water. In the case where acid wash has been located after the kiln there is a tradeoff situation between the very considerable savings in water useage and the potential damage to the regeneration kiln by presenting carbon contaminated with inorganic salts to the kiln.

WATER USEAGE VARIOUS STRIPPING SYSTEMS

ELUTION METHOD	PROCESS VARIATIONS	WATER USEAGE M^3/t BARREN REUSED	ELUATE NOT REUSED
ANGLO	NORMAL DIRECT PROCEDURE	21.0	26.8
ANGLO	SPLIT ELUTION TECHNIQUE	14.2	20.0
ZADRA	NORMAL PROCEDURE ELUATE USED ONCE	14.0	19.8
ZADRA	NORMAL PROCUDURE ELUATE USED TWICE	NA	16.9
MICRON	NORMAL PROCEDURE	NA	11.8
ZADRA	PROCESS WATER ACID RINSE ONE CYCLE	3.5	9.3
ZADRA	PROCESS WATER ACID RINSE TWO CYCLES	NA	6.4

The Temora Goldmine, Temora, New South Wales, Silver Effects

At the time of designing this system it was known from metallurgical testwork that silver loadings on the carbon would be high and desorbed loadings of 2450 g/t Au and 3800 g/t Ag were allowed for. It was also known that silver would elute before gold and carry the current in the electrowinning cell initially to the exclusion of gold. The danger exists in such a mixed system that if either the cell or the rectifier are undersized then gold will recycle back to the elution column and prohibitively long cycle times will result. The cell was therefore generously designed with 18 parallel plate cathodes and 1500 amps supplied by 2 x 750 amp rectifiers.

Graph one shows the actual system performance. Note that silver actually carried the current to the exclusion of gold until simultaneous deposition occurred at 140 g/m^3 gold and 100 g/m^3 silver, and that a three hour delay in the stripping of gold occurred:

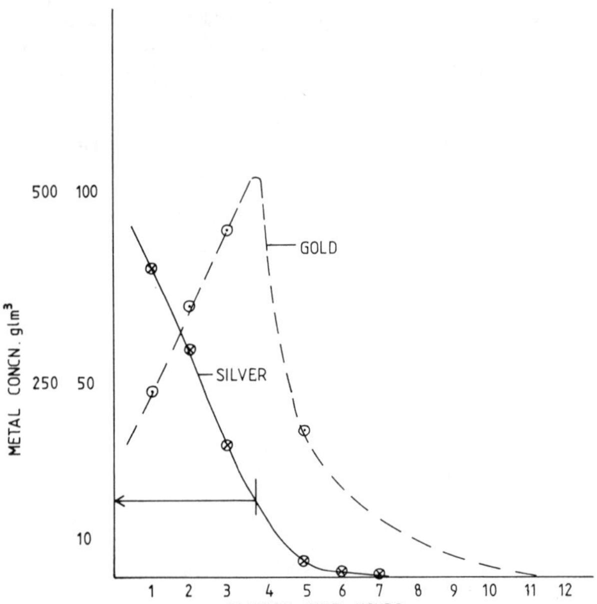

GRAPH ONE GOLD/ SILVER IN ELUATE VS. TIME TEMORA

CHOICE OF CELL FORM, RELATIONSHIP WITH ELECTROREFINING The original Zadra cylindrical cells have proven themselves perfectly adequate to the task of winning the gold in numerous atmospheric systems throughout the world. In the 1 percent w/v sodium hydroxide electrolyte used in this system high cell efficiencies are achievable. In later designs however, we wished to combine electrowinning with electrorefining and therefore used the conventional packed bed parallel plate cell using knitted or woven stainless steel with cathodes modelled on the Ortiz system and reported by (Arnold 1986).

INTEGRAL COLUMN AND PRESSURE ELECTROWIN ZADRA SYSTEM The idea of electrowinning at pressure in a closed system was discussed by Baxter and McCauley who reported on the design and operation of a pilot unit operated at Hickey Mines at Blayney in New South Wales (Baxter 1984). This system has now completed in excess of 1000 cycles and is claimed to have the lowest water, power and chemical consumptions of all the systems. The performance has been so impressive that the decision has been made to scale up the 330 kg pilot plant to a full scale production plant for the Scheahan Grants project in New South Wales. Stripping times of less than six hours to barren carbon levels of less than 10 g/t are being achieved in the pilot unit.

AARL SYSTEMS DESIGNS AND OPERATING EXPERIENCE

The Harbour Lights Gold Mine, Leonora, Western Australia.

This unit used conventional AARL technology available at that time. The unit was a three tonne rubber lined column with a six to one height to diameter ratio. Due to the inability of the rubber linings to withstand temperatures in excess of 110 deg C this design limit was imposed on the system. Carbon transport into and out of the column was by hydraulic transport and visual observation ports were built into the top and bottom of the columns to allow operator observation of the full and empty condition. A separate presoak solution tank was used with separate presoak and potable water positive displacement progressive cavity pumps. All process operations were automatic using air actuated valves controlled by the central plant PLC.

The individual process functions of acid wash, acid rinse, presoak and elution were all carried out in the same single column. The 45 m^3 batch of pregnant eluate was stripped of its gold by electrowinning in two nine cathode parallel plate packed bed cells running in parallel. Operational experience showed that only one cell was adequate so that later operations used one cell only. Gold was separated from the steel wool by dissolution in concentrated hydrochloric acid followed by filtration drying and smelting. The typical elution profile applicable to this unit is given as graph two for cycle 300.

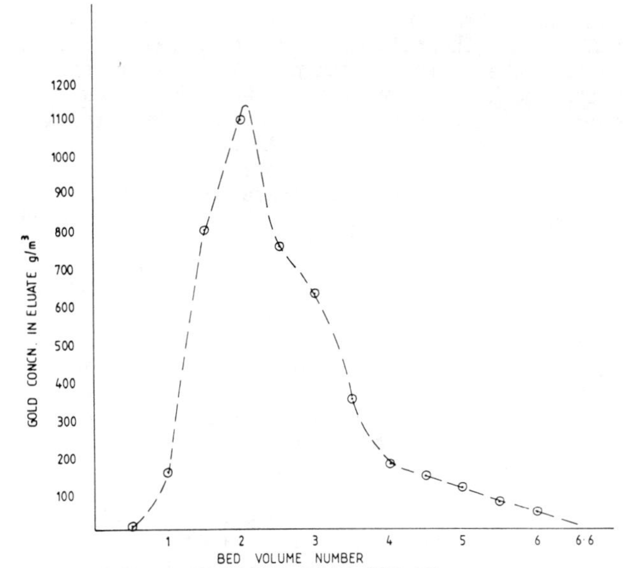

GRAPH TWO HARBOUR LIGHTS ELUTION CYCLE 300
CARBON HEAD 3210 g/T CARBON TAIL 45 g/T

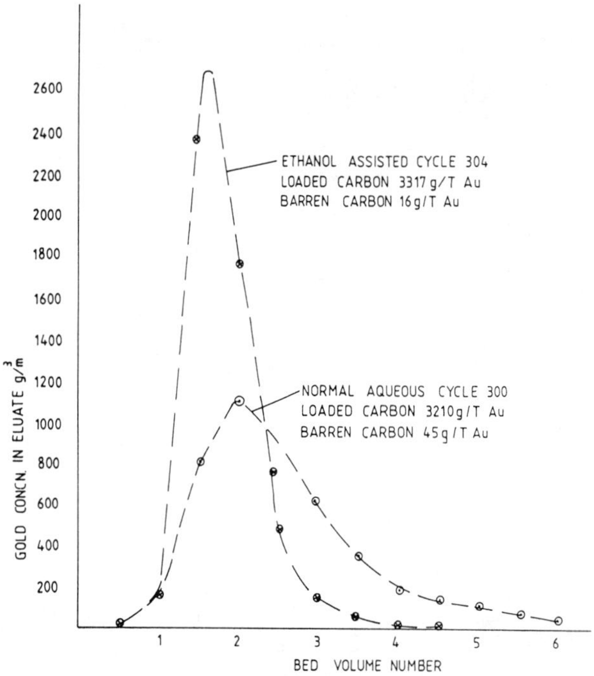

GRAPH FOUR. HARBOUR LIGHTS ELUTION CYCLES 300-304
COMPARISON OF AQUEOUS AND ETHANOL ASSISTED STRIPS

<u>The Effect of Alcohol to Enhance Elution Performance in the AARL System</u> This effect is well known and has been used in the Duval modification of the Zadra system for some time. We wanted to learn the effect of alcohol addition to the AARL system and carried out controlled additions of ethanol as industrial methylated spirits to the presoak solution and to the first bed volume of eluate. The effect was significant and resulted in faster elution using less water and to a lower barren carbon level. The data is presented as graph number three and a comparison with the non alcohol assisted strip is given as graph number four.

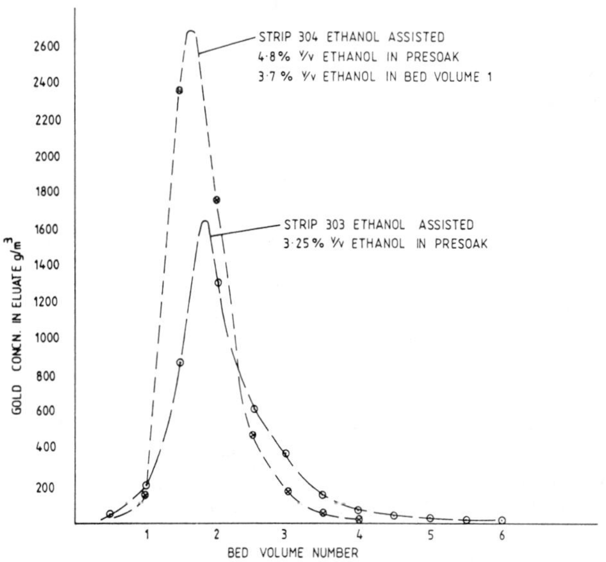

GRAPH THREE HARBOUR LIGHTS ELUTION CYCLES 303-304 ETHANOL ASSISTED

ELECTROWINNING CELL PERFORMANCES IN THE AARL SYSTEM Design of cells for use in this system have followed the principles given by (Paul 1983). In the original Harbour Lights design even though the data indicated that a single nine cathode cell would service a three tonne column we conservatively supplied two cells operating in parallel. The data indicating the performance of the two cell system is given in the following Table Number Two.

TABLE TWO
HARBOUR LIGHTS, TWO AARL CELLS IN PARALLEL

TIME hrs	CELL IN Aug/m³	CELL OUT Aug/m³	FRACTION PER PASS
0	245	75	0.69
1	104	21	0.80
2	64	15	0.77
3	42	8	0.81
4	30	4	0.86
5	20	3	0.85
6	15	2	0.86
7	10	1	0.90
8	7	<1	–

The allowable time for electrowinning in an AARL system cycling once per day is sixteen hours so that one cell and one rectifier is therefore only used with the performance indicated in Graph Number Five.

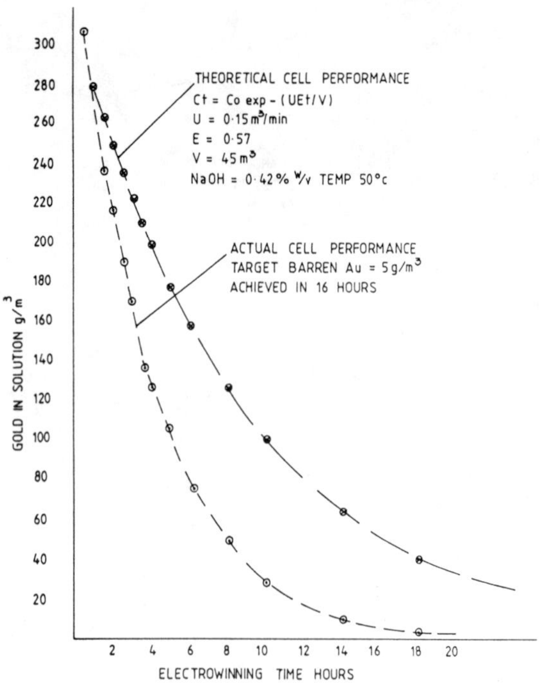

GRAPH FIVE. SINGLE CELL ELECTROWINNING COMPARISON OF ACTUAL AND THEORETICAL PLATING RATE

The caustic concentration in the eluate effects the electrowinning cell efficiency. The caustic from the presoak is not sufficient to serve this purpose so that a separate caustic injection is made into the last bed volume of eluate after the colummn in order to raise the caustic concentration to a target 0.5 per cent w/v NaOH solution. Under the conditions shown in Graph Five, a mean extraction per cell pass of 0.57 with an overall current efficiency of 12.5 per cent plated 9500 g of gold in the target 16 hours.

EFFECT OF WATER QUALITY ON THE ANGLO AMERICAN STRIPPING SYSTEM The combination of raising the temperature to 110 deg C or greater and raising the pH to 12 plus causes carbonate and hydroxide scaling to occur in the AARL heating system heat exchangers. At relatively low total solids and hardness levels of 450 g/m^3 at the Mt. Percy and Jubilee Mines (predominantly sodium chloride) no water pretreatment is used and intermittent acid cleaning of the plate heat exchangers is practised. With 600 g/m^3 dissolved solids at the Red Dome installation all composed of calcium bicarbonate 19 kg of scale was being produced per cycle causing heater failure and blockage of column inlet screens. Base exchange softening of the water solved this problem. At the Harbour Lights and Granites installations with water dissolved solids exceeding 2000g/m^3 reverse osmosis is used to prepare the feed water to the AARL system.

THE TWO COLUMN AARL SYSTEM MT. PERCY, WESTERN AUSTRALIA The Kalgoorlie mining area is over 600 km from the source of its potable water in the Perth foothills. Water is a very expensive commodity in this area, so that the traditional design of the single column AARL system was modified to a twin column system. The logic is that if the acid wash is separated from the elution then the absence of rubber lining imposes no limits on the elution temperature and a sharper elution profile should be possible. Additionally, the change was made in that the presoak solution tank and pump were eliminated and presoaking was achieved by using the eluate in closed cycle to recirculate caustic and cyanide which were mixed in the column itself. This system allowed presoak temperatures to be raised to 100 deg C and elution temperatures to be raised to 115-120 deg C. The result was a sharper elution profile with lower barren carbons and the indication that water useage could be reduced. Additionally, the eluate potable water pumps were driven by two pole motors allowing operation at two or four bed volumes per hour so that a single unit acts as eluate pump and transport water pump for hydraulic carbon movement. The transport water was filtered and recycled to the same potable water tank so no nett useage of water occurred in the hydraulic tranpsort of carbon from the acid wash column to the elution column or from the elution column to the reactiviation kiln feed hopper. The performance of this modified AARL system is given in Graph Number Six.

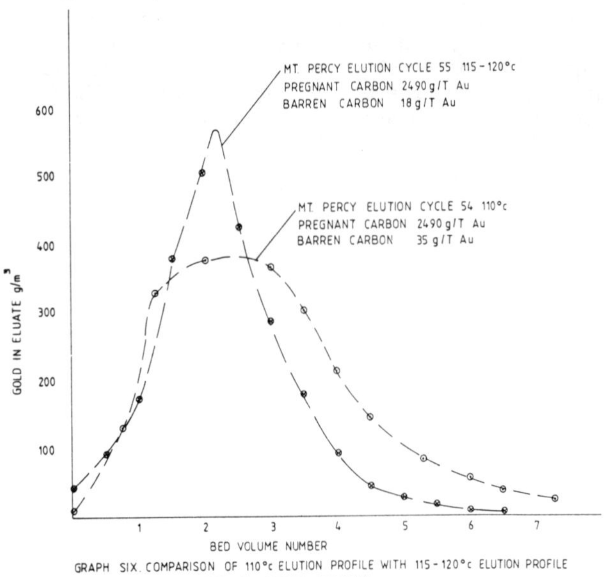

GRAPH SIX. COMPARISON OF 110°c ELUTION PROFILE WITH 115-120°c ELUTION PROFILE MT. PERCY MINE, KALGOORLIE, WESTERN AUSTRALIA

The Granites Gold Mine in the Tanami Desert Area of the Northern Territory of Australia used the same system as that described above with the addition of a reverse osmosis feed water treatment system.

TWO COLUMN AARL SYSTEM WITH ELECTROWIN ONTO KNITTED STAINLESS STEEL WOOL AND ELECTROREFINE ONTO PLAIN STAINLESS PLATE The Red Dome AARL system was designed after technical visits were made to the Ortiz operation in America in New Mexico. The operation of this system has been reported (Arnold 1986). The AARL system at Harbour Lights was controlled by the plant PLC, however, all subsequent systems were controlled by a simple sequencer. It became clear that the system reliability was so high that the gold room operator had little to do apart from pushing the sequence start button. Most of his time was used in the preparation and stripping of cathodes and the aciding of residual steel wool, filtering drying and smelting. The addition of electrorefining to the AARL system had the potential to eliminate the need for a special gold room operator and to reduce actual gold room operations hours to aproximately four man hours per week. This decision was made and after many months of development by both the contractor and the client the system has achieved its design purpose. The Red Dome AARL/Electrorefine system was influential in choosing the same system for the Mt. Leyshan project and Hampton Jubilee Project. The former is now commissioned and operating satisfactorily and the latter is currently undergoing commissioning trials. The new system lends itself to a design where the gold room sequencer and electrowinning and electrorefining controls are all located in the plant control cabin and the cells are viewed through a window. The actual gold room high security area is only entered by two people for half a day per week during which time the plate scraping and smelting operations are carried out. The system is considered at this time to be the most labour and capital efficient method of final bullion production and has been included in the flow sheets for the Colosseum, Copperstone and Ridgeway Projects in the USA and for the Telfer and Karonie Projects in Australia, all of which are still in the design and construction stage of development.

THE EFFECT OF HIGH LEVELS OF CYANIDE SOLUBLE SILVER As mentioned under the section dealing with design of the Temora Zadra system, orebodies with high cyanide soluble silver cause problems at all stages of the design. Firstly, silver elution efficiency is low leaving high silver on carbon Tails. Secondly, the silver electrochemical equivalent weight is low leading to the need for large electrowinning and electrorefining cells and rectifiers. Thirdly plate quality for co-deposited gold and silver is poor leading to sludging and plate detachment and the need to deal with large masses and volumes of sludge. At the Mt. Leyshon installation where the gold to silver ratio is 2:1 satisfactory performance is achieved. At the Temora site where the gold silver ratio is 0.77:1 plating difficulties are occurring. On the basis of this experience the decision has been made to avoid electrorefining for both the Pajingo and Bottle Creek projects which are currently undergoing flowsheet design.

THE SPLIT ANGLO AMERICAN SYSTEM The AARL system has shown such high reliability and ease of use that it is generally favoured above the Zadra and Micron systems in Australia. The water consumption for the direct system at 26.8m^3/t is high and as mentioned previously this can be a serious problem in the remote desert mining areas. To economise on water it is possible to use the split elution technique and experience has been gained by others (Blanks 1987) (Dunstan 1987) at the Bluebird and Westonia mines in Western Australia. The experience is mixed. In some installations poor elution efficiency and high carbon tails have occured when the second half of the eluate is recycled to feed the first half of the next elution cycle, while in others completely successful performance is occuring particularly when the feed water quality is good and softening of the feed is practised. On the basis of this experience the decision has been taken to include the split Anglo technique in the flowsheet for the Karonie project which is situated on the Nullabor Plain east of Kalgoorlie in Western Australia.

MICRON METHANOL SYSTEMS DESIGNS AND OPERATING EXPERIENCE

Development of the Micron System

This system was developed with the aim of producing a low cost system which should have the advantages of a short desorption cycle time and leave a product of relatively high carbon activity. The initial concept was reported by (Griffin 1982). Development has continued from pilot plant stage to early production designs using gravitational condensate return systems to units using pumped and spray distributed condensate return. Five years of operational experience is now available and Minproc has installed eight units ranging in size from 350 kg to 1000 kg.

Operational Experience with the Micron System The initial units installed at the Rogex-Youanmi desorption facility and at the Black Cat and Sons of Gwalia mines all used high height to diameter ratio columns with gravity distribution of returned condensate. All of these units proved unreliable in service and were all withdrawn from service and replaced with Zadra or Anglo systems. The second generation designs introduced by Minproc at the Mt. Bonnie plant used a level controlled condensate surge tank and a pumped spray condensate return system. This unit was completely successful returning barren carbon assays during commissioning of less than 20 g/t Au in cycle times of less than twelve hours. Scale up from this size i.e., 400 kg to the next generation of units at Tower Hill, Porphry and Cracow of 900 kg caused further problems. Basically, two major problems have emerged (a) controlling condensate droplet size to the optimum between adequate distribution and spray entrainment and (b) controlling superheating leading to bumping and foam boil up. The optimum droplet size has now been fixed and is achieved using a single cone distribution spray. Foam boilup when it occurs, for instance in the Cracow

Tailings retreatment operation has been controlled by the use of antifoam. The extent to which this has been successful is indicated by the relative success of the third generation design installed at the Galtee More Plant at Mt. Magnet in Western Australia. This unit is a 900 kg unit and has been in service for 15 months. Data was monitored for the first 290 desorption cycles where carbon heads were in the range 2500-3500g/t Au with a mean carbon heads of 3000 g/t Au and carbon tails in the range 50-120 g/t Au with a mean carbon tail of 100 g/t Au. Plant solution tails were 0.01-0.02 g/m^3 for the first 190 cycles and slowly rose to 0.02-0.03 g/m^3 during the next 100 cycles.

NEED FOR REGENERATION KILN WITH MICRON SYSTEM All of the plants which have had Micron systems installed and which have not had accompanying regeneration kilns have in the long term had to install a kiln. The Galtee More figures given above are a good indication of what can be achieved when treating a normal orebody ie., in the absence of special foulants such as flotation reagents. A project which has limited up front capital available can expect to install a Micron system for the lowest capital cost without a regeneration kiln but with acid wash, and operate successfully for at least six months without the need for reactiviation. Once cash flow has been established retrofit of a kiln can then be made. The system has the further advantage in that the moisture content of the steamed out carbon at approximately 30 per cent w/w is suitable as feed to the latest generation of vertical tube kilns without the need for the predryer and continuous kiln operation can occur.

ELECTROLYTIC CELLS FOR USE IN MICRON SYSTEMS The high grade (1.5-2 x carbon head grade) and low volume (0.8 x carbon volume) of the micron pregnant eluate typically 5000 g/m^3 Au in 0.75m^3 lends itself to plating on plain alfoil surfaces at high current efficiencies. Cell barrens howerver, vary with the cathode material used, for plain aluminium a 10 g/m^3 tail is typical which reduces to 2-3 g/m^3 if shredded alfoil is used. Expanded alfoil yields a 0.7 g/m^3 tail while steel wool is capable of 0.25 g/m^3. Choice of cathode material is governed by the method of disposal of barren electrolyte. If plain or shredded aluminium is used then return of barren eluate to the carbon adsorption circuit is necessary. If expanded aluminium or steel wool is used rejection of barren eluate is recommended. The loss in revenue due to acceleration of carbon activity deterioration will almost certainly be greater than that due to loss of the gold in the barren eluate. For a 250,000 tpa plant using shredded aluminium the gold loss in money terms due to discharging barren eluate to tails is $A9,550/annum or $A900/annum if steel wool is used. This must be offset against a loss of A$70,000/annum per 0.01 g/m^3 rise in solutions tails loss for an $A gold price of A$500/ounce. Clearly maintenance of carbon activity and discharge of contaminated barren eluates to waste is better economics than potentially causing faster carbon deterioration, and expanded aluminium cathodes or steel wool are necessary to achieve this condition.

PROCESS ECONOMICS

COMPARISON OF ZADRA ANGLO AND MICRON SYSTEMS

Plant operation in all three systems has reached the stage where no specialised gold room operator is needed so that for all systems the manpower component of the proces control is the same. For the Anglo and Zadra systems as mentioned earlier process control is automatic by programmable sequencer and operator involvement is minimal. For the Micron system two automatic systems have been built (Tower Hill and Porphry) but they have proven themselves unreliable in service so that a return to manual control was made. The manual functions can however, still be included in the duties of routine plant control. A complete and detailed breakdown of the cost components of each of the systems has been made and is presented in the Appendices (1 and 2). Clearly energy costs, water costs and chemical costs vary from site to site. The costs used are those applicable to the Kalgoorlie area of Western Australia at May, 1987 prices using LPG gas fired heating units and quote costs in Australian Dollars.

Capital costs taken from as built units (Hampton Jubilee Anglo plant, Gidgee Zadra Plant, Galtee More Micron Plant) indicate minimal cost difference between the systems. Royalty payments on Zadra plants are nil, Anglo approximately $A20,000/unit and Micron approximately $A35,000 per unit.

SUMMARY AND CONCLUSIONS

The design and operating experience obtained during the construction and operation of twenty-eight precious metal recovery plants has lead to the conclusion that the Zadra system is the cheapest system to install and operate and is economical in the use of potable water especially if acid rinse with process water is used.

The Anglo System is dearer to operate than the Zadra System and a high consumer of water so that both these facts lead to a 26 per cent increase in costs over and above the Zadra system. Against these disadvantages the Anglo system presents very clean acid washed and rinsed carbon to the kiln ensuring good reactivation and long kiln life. Further due to the very short elution cycle time of 7.5 hours carbon treatment rate can be doubled or tripled merely by the addition of extra elution tanks and electrowinning cell capacity. The Micron system has the highest operationg cost but the lowest water consumption and can function without the need for a regeneration kiln for 6-12 months. It is chosen in arid areas for minimal upfront capital cost.

APPENDIX ONE
COMPARATIVE COSTS OF ELUTION SYSTEMS
(1) ONE TONNE ZADRA SYSTEM WITH ACID WASH AND RINSE BEFORE COLUMN

OPERATION	WATER		ENERGY		ACID		CAUSTIC		CYANIDE	
	AMT m3	COST A$	AMT KWH	COST A$	AMT kg32%	COST $A	AMT KG	COST A$	AMT KG	COST A$
Carbon Fill	0.91	0.79								
Acid Inject	1.25	1.09			127.9	44.55				
Acid Rinse	9.09	7.9								
Solution										
Heat	2.8	2.44	468.2	16.76			75	60	10	19.05
Water										
Displace	1.36	1.19	44.34	1.59						
Stripping			395.8	14.17						
Soln.										
Displace										
Carbon										
Cool	1.36	1.19								
Carbon										
Trans.										
TOTAL	16.78	14.61	908.3	32.5	127.9	44.55	75	60	10	19.05

OVERALL TOTAL COST $1.71/TONNE

APPENDIX ONE
COMPARATIVE COSTS OF ELUTION SYSTEMS
(2) ONE TONNE ZADRA SYSTEM WITH ACID WASH AFTER KILN

OPERATION	WATER		ENERGY		ACID		CAUSTIC		CYANIDE	
	AMT m3	COST A$	AMT KWH	COST A$	AMT kg32%	COST $A	AMT KG	COST A$	AMT KG	COST A$
Carbon										
Fill	0.91	0.79								
Soln Heat	2.8	2.44	468.2	16.76			75	60	10	19.05
Water										
Displace	1.36	1.19	44.34	1.59						
Stripping			395.8	14.17						
Soln										
Displace										
Carbon										
Cool	1.36	1.19								
Carbon										
Trans										
Acid										
Wash					127.9	44.55				
TOTAL	6.44	5.61	908.34	32.52	127.9	44.55	75	60	10	19.05

OVERALL TOTAL COST $1.62/TONNE

APPENDIX ONE
COMPARATIVE COSTS OF ELUTION SYSTEMS
(3) ONE TONNE ANGLO SYSTEM NORMAL SINGLE COLUMN SYSTEM

OPERATION	WATER		ENERGY		ACID		CAUSTIC		CYANIDE	
	AMT m3	COST A$	AMT KWH	COST A$	AMT kg32%	COST $A	AMT KG	COST A$	AMT KG	COST A$
ACID INJECT	1.25	1.09	–	–	127.9	44.55				
ACID RINSE	9.09	7.91	528	18.93	–	–				
PRESOAK INJECT	1.15	1.00	79.2	2.83			27.3	15.71	27.3	54.56
PRESOAK SOAK										
ELUTION	13.64	11.87	792	28.4						
CARBON COOL	1.36	1.19								
CAUSTIC INJ.							47.7	27.48		
TOTAL	26.5	23.06	1399.2	50.16	127.9	44.55	75	43.19	27.3	54.56

OVERALL TOTAL COST $216/TONNE

APPENDIX ONE
COMPARATIVE COSTS OF ELUTION SYSTEMS
(4) ONE TONNE MICRON SYSTEM

OPERATION	WATER		ENERGY		ACID		CAUSTIC		CYANIDE		METHANOL	
	AMT m3	COST A$	AMT KWH	COST A$	AMT kg32%	COST $A	AMT KG	COST A$	AMT KG	COST A$	AMT L	COST A$
ACID WASH	1.25	1.09			127.9	44.55						
ACID RINSE	9.09	7.91										
PRESOAK SYSTEM	0.76	0.66	28.8	1.03			7.6	5.88	75.7	144.6		
HEAT STRIPPING			846.7	30.31								
METHANOL RECOVERY			235.7	8.43							165	107.25
FURTHER RECOV & STEAMOUT	0.65	0.57	570.1	20.41								
TOTAL	11.75	10.23	1681.3	60.18	127.9	44.55	7.6	5.88	75.7	144.6	165	107.25

OVERALL TOTAL COST $373/TONNE

APPENDIX TWO
COMPARATIVE COSTS OF ELECTROWINNING CELL OPERATIONAL METHODS AND MELTING TECHNIQUES

(1) Normal nine cathode anglo cell with acid digest and smelt costs are for 5 x 1 tonne strips per cell operation.

ITEM	AMOUNT KG	UNIT COST $A	TOTAL COST $A
Steel wool	4.5	5.4	24.30
Acid	34.8	0.35	12.18
Borax	7.5	2.90	21.75
Soda Ash	1.5	0.78	1.17
Nitre	1.5	4.00	6.00
Silica	4.5	0.90	4.05
Gas	371	0.26	9.67
A100 Pot	15 smelts per pot		12.94
TOTAL			92.06
COST PER CYCLE			$18.41

Labour Component Acid and Filter	4 hours
Make up new Cathodes	2 hours
Strip Loaded Cathodes	0.5 hours
Smelt Supervision	6.0 hours
TOTAL 5 STRIPS	12.5hours
Cost $16.30/hour	$203.75
Cost Per Cycle	$40.75
Overall Total Cost Per Cycle	$59.16

(2) Nine cathode parallel plate cell Zadra system with direct smelting of cathodes costs are for 5 x 1 tonne operational cycles.

ITEM	AMOUNT KG	UNIT COST A$	TOTAL COST A$
Steel wool	7.02	5.40	37.91
Silica	11.96	0.90	10.76
Nitric	13.73	4.00	54.92
Boric Acid	7.18	4.20	30.16
Gas	22.21	0.26	5.77
A100 pot	5 smelts per pot		38.82
TOTAL			178.34
COST PER CYCLE			$35.67

LABOUR COMPONENT

Strip Loaded Cathodes	0.5 hours
Make Up New Cathodes	2.0 hours
Smelt Supervision	6.0 hours
TOTAL 5 strips	8.5 hours
Cost at $16.30/hour	$138.55
Cost per Cycle	$27.71
Overall Total Cost/Cycle	$63.38

(3) Electrorefining onto plain stainless steel basis is 400 electrorefining cycles per batch woven stainless and 5 strips per cycle.

ITEM	AMOUNT KG	UNIT COST $A	TOTAL COST $A
Stainless Wool	1/400	3000	7.50
Cyanide	42	1.905	80.01
Borax	7.5	2.90	21.75
Soda Ash	1.5	0.78	1.17
Silica	4.5	0.90	4.05
A100 Pot	20 smelts/pot		9.71
Gas	371	0.26	9.67
TOTAL			133.86
COST PER CYCLE			$26.77

LABOUR COMPONENT

Strip Cathode	3 hours
Transfer and Wash	1 hour
Smelt Supervision	4 hour
TOTAL 5 Strips	8 hours
Cost $16.30 1 hour	$130.4
Cost per cycle	$26.08
Overall Total Cost/Cycle	$52.85

REFERENCE LIST FOR PAPER - STRIPPING FOR FUN AND PROFIT

1. Arnold Jr and Pennstrom WJ 1986 "The Gold Electrowinning/Replate Circuits at Gold Fields, Ortiz and Mesquite Operations" Society of Mining Engineers Preprint Number 86-309.

2. Baxter K and McCauley J 1984 "Integral Pressure Strip Electrowin Circuit for Gold Recovery from Activated Carbon" The Australiasian Institute of Mining and Metallurgy Regional Conference 9-11 October, 1984 Paper 9.

3. Davidson RJ and Duncanson 1977 "Desorption of Gold from Activated Carbon with Deionised Water". Journal of the South African Institute of Mining and Metallurgy Volume 77 Number 12 Page 254.

4. Griffin AF and Costello M. C. 1982 "Carbon Desorption Process Development at Micron Research (WA) The Australasian Institute of Mining and Metallurgy Regional Conference Kalgoorlie July, 1982 Page 140-151.

5. Paul R. Filmer J and Nicol M 1983 "The Recovery of Gold from Concentrated Aurocyanide Solutions" Proceedings of the Third International Symposium on Hydrometallurgy Atlanta Georgia page 690-704.

6. Zadra JB 1950 "A Process for Recovery of Gold from Activated Carbon by leaching and Electrolysis" US Bureau of Mines Report of Investigations 4672.

7. Blanks Dr. R and Dunstan A. "Personal Communication.

Chapter 21

NEW INNOVATIONS IN GOLD RECOVERY COLUMN DESIGN FOR IN-SITU STRIPPING OF CARBON

Mr. Nicholas S. Lynn

N. S. Lynn & Associates

Introduction

Borne out of the need for reductions in capital expenditure and increased production, several new and innovative carbon systems have been developed for small precious metal heap leach operations throughout the United States. Perhaps the most important aspect of these systems is the carbon column. Though diverse in design, these columns can be categorized into three basic types. The first is known as the gravity flow type or Cortez type. As the name indicates, solution flows via gravity, throughout 4 to 6 columns which are arranged in a normally stepped configuration which is conventionally upflow in design. The second type, known as the vertical column, is an arrangement whereby flow is established through pressurization in an upward manner, throughout 4 to 6 columns in a vertically stacked configuration. This arrangement is typically intended for use with small portable heap leach operations and employs a semi-continuous mode thus preventing gravity from occurring. The third type and newest development in carbon columns is an arrangement whereby the solution flows throughout 4 to 6 columns placed in a side by side configuration and pressurized by a single feed pump (Figure 1). A unique aspect to the third type, is that the adsorb and desorb cycles are accomplished in the same column, due to a fixed carbon charge. An in-depth analysis is warranted due to numerous advantages of this arrangement, and is therefore the subject of this paper.

Advantages and Variations

One advantage to the third type of carbon column is the fact that initial capital expenditures and operating manpower requirements are reduced. Another advantage is that the system can be made fully portable up to a 0.0315 m^3/s (500 gpm) capacity. Systems of this nature are operational at several sites. Above this flow rate, the system's portability is one which can be maintained, but disadvantages (i.e. larger diameter carbon column and greater weights) far outweigh the advantages.

There are several variations in each of these systems which allow for unique operation. For example, carbon columns made highly portable, can be moved from unitized heap leach operations to central stripping facilities. Although this method has been used in Colorado, transportation of these units has been known to be quite cumbersome. If distances exceed 5 km (3.2 miles) then special front-end loaders are required. This limits heap locations to within 5 km (3.2 miles) of the central processing facility. Another example transports the carbon charge from the heap leach locations to a centralized processing facility; this method is quite effective. For the past 5 years one mine has employed this technique of transporting carbon in 907 kg (1 ton lots), in DOT containers, towed by company pick-ups. This operation uses a vertical column with a 907 kg (1 ton) carbon charge. A third example uses static carbon columns with fixed carbon charges.

Carbon Adsorption System

The carbon adsorption system (Figure 1) is fed by a single preg solution feed pump. This feed pump pushes the preg solution through all four to six fixed charged carbon columns, which are on line. The sequencing of the carbon columns is achieved using valves. Expansion of the carbon charge is dependent upon the carbon selected. The carbon typically used is a 6 x 12 mesh size, with a nominal flow rate of 0.00135 - 0.00138 m^3/s (22-25 gpm) per square foot of carbon, based on a 40% bed expansion. With 8 x 16 mesh size, the nominal flow rate decreases to approximately 0.00114 - 0.00126 m^3/s (18-20 gpm) per square foot of carbon based on 40% bed expansion. Carbon is retained in each tank through an upper carbon screen which is located at the outlet nozzle.

The flow is equally distributed throughout the system by a unique three-phase flow distributor. The flow distributor, up to 0.0315 m^3/s (500 gpm), utilizes a side entrance. This unique side entrance is why the solution flow is equally distributed throughout the entire carbon column in both the adsorption and desorption phases. Solution slows through this side entry at a velocity of less than 2 mps (6 fps) and passes through a flow diversion device which reduces the velocity to less than 0.030 mps (1 fps). The solution then passes through a flow distributor which then produces a pressure drop of 3% of the entire pressure drop across the carbon column. This allows for a pressure drop across the entire system to be approximately 0.010 Pa (15 psig) or 0.002 Pa (3 psig) of loss per tank.

Flow distribution through the strip cycle is such that pressure drop across the distributor system, is approximately 3% of the entire pressure drop across the piping system. This pressure drop allows for a flow distribution across the carbon bed with a variance of plus or minus 7%, thus allowing for both adsorption and desorption (stripping) within the same unit.

The carbon system is drained via a drain system back to the preg pond primarily because solution will be of the highest grade.

Carbon Strip Cycle

The carbon strip cycle for the in-situ strip columns is highly dependent upon chemical composition, pressure, and temperature of the strip solution. The conventional in-situ strip utilizes a solution heater, whereby the temperature difference of the incoming and outgoing solutions of the column is 27.7^{o} C. (50° F.).

Chemical composition of the glycol strip solution is 20% ethylene glycol, 1% sodium hydroxide and 0.25% sodium cyanide. The effective flow rate through the columns was about 3 bed volume per hour The overall strip solution shows a direct function with the temperature and the solution flow rate. The data shows that a spike occurs approximately 3 hours after the initial strip solution cycle takes place, with an exponential decay. An integration of this volume, versus solution assay, shows that the overall strip solution can be calculated from the input and output.

An overall metallurgical balance based upon solution into the carbon columns, stripping solution from the carbon columns to the electrowinning cell, and the electrowinning solution to recovery of Dore' from the furnace, should be capable of being monitored to a plus or minus 5% on a gross scale. With the flow meters calibrated accurately, this metallurgical balance is easily held within a plus or minus 1% range. This plus or minus 1% range is very accurate for accounting purposes, although the flow meters must be calibrated. Figure 3 shows the carbon adsorbed gold, carbon stripped, and the carbon electrowinning results. There is an approximate lag time of 9 days throughout the entire circuit and this nine days must be recovered through the overall analysis of the unit.

The overall strip solution can be enhanced by the addition of a heat reclaim heat exchanger. This allows for the solution to heat up in approximately one hour and reduces the heat burden and the overall evaporation from the circuit. This allows the solution in the electrowinning circuit to be cooled to a temperature of 66.6^{o} C. (120° F.). This reduces the overall solution losses from 5% to less than 2% on an overall basis. These solution losses are figured on a monthly basis and are very difficult to determine because of the entrained solution losses on the strip cycle, and the overall effects of the operating system.

The overall stripping cycle has been shown to be very effective through the entire operation of this circuit. Stripping analysis shows that the carbon can be taken from an overall loading for new carbon of 15.43 kg/mt (450 oz./st), to less than 102.9 g/mt (3 oz./st) in a 24 hour strip, or less than 171.5 g/mt (5 oz./st) in a 16 hour strip.

Carbon Performance

The following case study of a new carbon system illustrates carbon performance from startup through 10 cycles. After 5 cycles of usage the carbon is semi-activated. After 10 cycles of usage, the carbon is fully activated and the barren solution analysis increases from non-detect to 0.137 g/mt (0.004 oz./st). At this point in time, a sample of the carbon was analyzed, and found the calcium carbonate to be in excess of 8%. Calcium carbonate levels were then washed from the carbon by immersing the carbon, for approximately 2 hours, in a quiescent system of one normal HCl. The carbon was then rinsed with a very dilute caustic wash and put back on line. THe barren solution analysis revealed a drop to approximately 0.069 g/mt (0.002 oz./st).

After this many cycles, the specific gravity of the dried carbon had decreased from about 75 to 64 grams per 100 ml's. When this new carbon was added after 20 cycles, substantially higher loadings of gold was realized.

At this point in time the gold/silver ratio in the system was approximately 80% gold, 20% silver (silver analysis was not being done in the analytical lab). These results showed that throughout the entire circuit, the initial metallurgical balances were within plus or minus 3%.

Once the carbon contains an excess of 6% calcium carbonate, the overall stripping efficiency decreases. The high peaks that were initially encountered were reduced to the level shown in figure 2. This shows that the calcium carbonate level, in actual practice, affects reactivity of the kinetics of carbon, and therefore produces an overall reduction to the stripping efficiency. The carbon profile (Figure 2) shows performance of new carbon and carbon after 10 cycles. It has been assumed that this inefficiency is caused by calcium carbonate and calcium sulfate molecules blocking the carbon's pores.

<u>System Modification</u>

On the newer carbon systems, carbon retention screens have been installed to reduce the carbon loss throughout the circuit, to less than 1% on an annuallized basis. Prior to this installation, significant quantities of carbon were lost from the tail column, due to upset conditions. When screens were installed in the circuit, approximately 453.6 kg (1000 lbs.) of carbon was recovered over an operating period of 4 months. These screens have significantly lowered carbon usage.

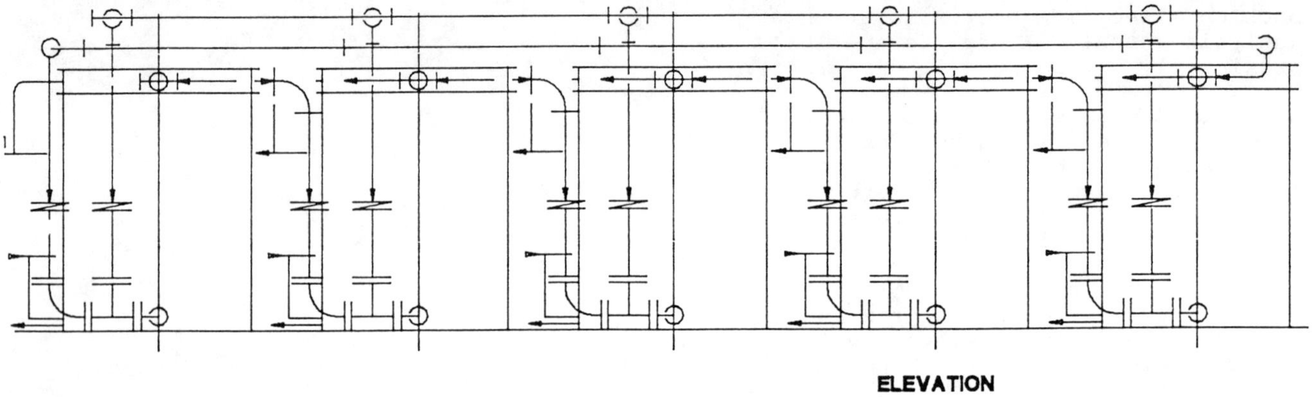

SIDE VIEW
500 GPM PLANT

(Figure 1)

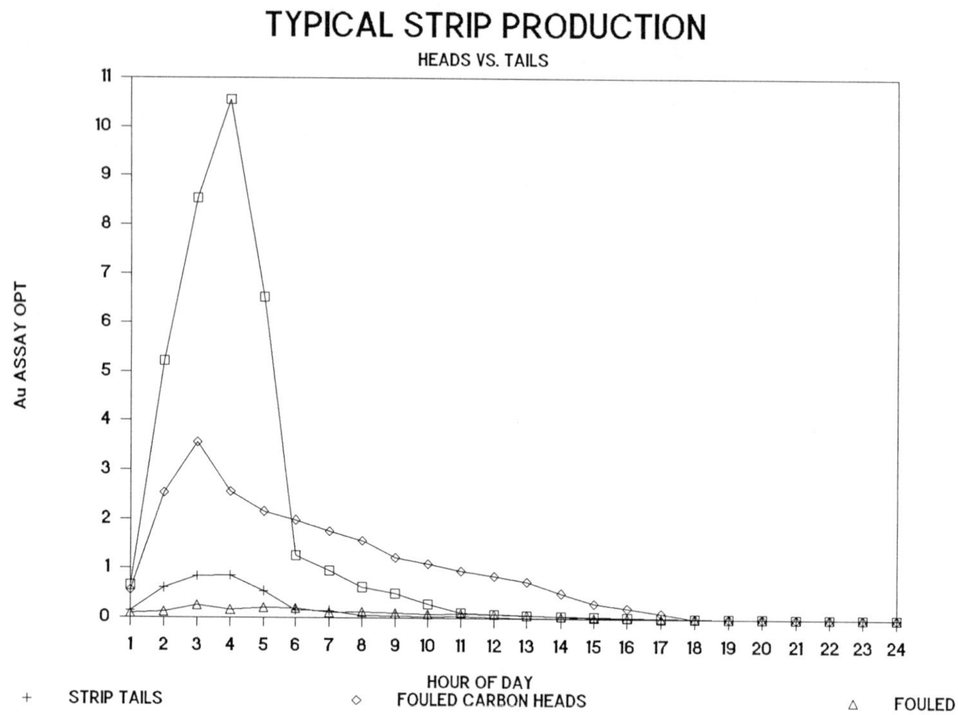

(Figure 2)

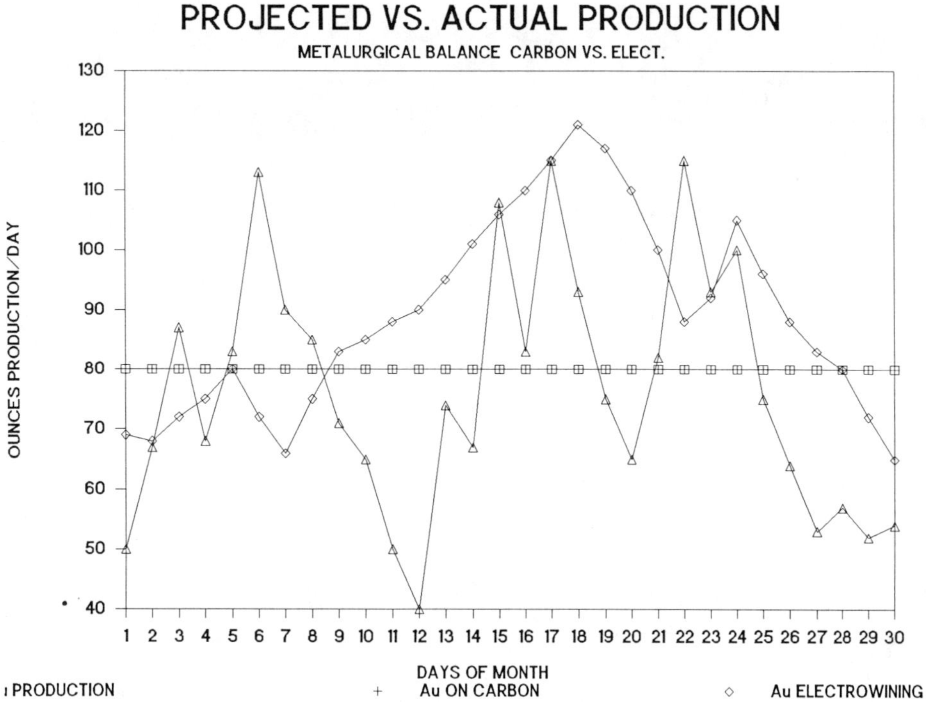
PROJECTED VS. ACTUAL PRODUCTION
METALURGICAL BALANCE CARBON VS. ELECT.
OUNCES PRODUCTION/DAY
130
120
110
100
90
80
70
60
50
40
1 2 3 4 5 6 7 8 9 10 11 12 13 14 15 16 17 18 19 20 21 22 23 24 25 26 27 28 29 30
DAYS OF MONTH
PRODUCTION
Au ON CARBON
Au ELECTROWINING

(Figure 3)

Chapter 22

COMPUTER SIMULATIONS FOR OPTIMAL GOLD PROCESS DESIGN

by

Mr. John T. Bartlett
Pincock, Allen & Holt, Inc.
Manager, Process Systems
Lakewood, Colorado

ABSTRACT

During the last decade, the increasing price of gold has renewed interest in the various metallurgical techniques for recovering precious metals. Much of the recent technology advance has been aimed at low grade deposits resulting in more complex processes and the necessity of tighter economic control. These factors along with more stringent environmental constraints have made it necessary to optimize each process to minimize energy and raw material consumption and reduce waste discharge. In most cases, detailed process analysis is required to detect inefficiencies and provide alternative solutions for process improvement. Also during this same period, the computer has become widely accepted and computerized mathematical modeling has been applied to an increasingly broader range of problems. Microcomputer based process modeling and simulation systems are routinely used to analyze and improve comminution, hydrometallurgical, and pyrometallurgical processes for the recovery of precious metals.

INTRODUCTION

Process modeling using computers has become more common place with the availability of lower cost computer hardware, an increase in the number of computer trained engineers, and the advent of process simulation systems. Early computer models were often developed for specific processes and run on large main frame computers. These models were expensive to develop because they often required formulation and testing of new algorithms, were programmed in tedious computer languages, were run on non-interactive main frame computers, and were created by engineers with little prior experience. Now, there are process simulation packages available which can be run on desktop personal computers. These systems reduce the time and cost of creating a process model. Application of such programs have proven to be successful in increasing the efficiency of plant operations by demonstrating the effects of process alternatives and expediting the decision making process. It should be considered that even though some models may not be able to give exact quantitative results, qualitative results may be invaluable in making decisions concerning alternative operations.

When should process optimization take place? The following scenario may be all too common. A new project is initiated by the exploration department and data passed to the metallurgist to develop a process. We cannot optimize now, we do not have enough process information. The feasability stage is started with two or three alternative processes to be evaluated. We select the alternative with the best return on investment. The project economics show this one to be a bonanza. No need to optimize now, build it. The project goes into the design and construction phase. Cannot optimize now, changes would interfer with the constuction schedule. We are in the plant startup phase, no time to optimize, too many fires to put out. The operation is running smoothly, we now have a choice:

1) We are making enough money and do not need to optimize our process.
2) We are losing money and cannot afford to optimize our process.
3) We cannot afford not to optimize our operation.

A gold mine and mill, processing 1000 tons of ore per day, operating 350 days per year, with a head grade of 0.1 ounces per ton, recovering 85 percent of the gold, and breaking even at a $350 per ounce gold price, would profit by $104,000 per year for each 1 percent reduction in operating cost or 1 percent increase in gold recovery.

When should process optimization take place? Simple: Before you spend money.

PROCESS SIMULATION

Process simulation is a problem solving technique whereby mathematical equations are used to describe the mechanisms of processing operations. When these equations are rigorously solved, they will accurately predict the outcome of the process.

Process simulation can be used:

1) to evaluate pilot plant data.
2) to evaluate alternative flowsheets.
3) to perform feasibility studies.
4) to scaleup design calculations.
5) to train plant operators.
6) to analyze actual plant performance.
7) to study process control schemes.

Some of the advantages and benefits of process simulation are:

1) It is less costly in time and money to simulate a proposed process than to operate a pilot plant. Simulation reduces the test program by permitting identification of key parameters early in the development stage.
2) Hazardous and high cost processes can be experimented with at a lower risk to the environment and finances.
3) It requires the engineer to develop a detailed understanding of the process using consistant design criteria.
4) It provides a structured framework for evaluating alternative processes.

One must keep in mind, process simulation is not a replacement for, but a supplement to sound engineering practices and judgement. A process engineer must have complete control over and therefore responsibility for the accuracy and applicability of the calculations.

ECONOMICS

Process optimization is tied directly to economic factors. Although some optimizations may be based upon maximum precious metal recovery, minimum enviromental impact, or some other criteria, economics usually dictate the objective function. Therefore, any process modeling must account for capital and operating cost variations. The author is not aware of any modeling system for the mining industry which can simulate the process in sufficient detail to optimize both capital and operating cost. Capital costs can be approximated by some computerized estimating systems, but these require manual data input and are not currently suited to online optimization. The same is true for indirect operating costs. Operating costs that are calculated directly from process and equipment parameters are suitable for inclusion in the optimization objective function in current process models. This is accomplished by including a program module that calculates and totals such items as ore and reagent costs, fuel and energy costs, wear material and maintenance costs, and emission and tailing treatment costs. Credits can be included for salable or otherwise usable products. The process model can then be exercised with varying operating parameters and flowsheet configurations in order to optimize the operation. As future databases are developed and computer hardware and software improve, more elaborate optimizations will become routine procedures.

MODEL BUILDING

The model can range from a simple "black box" to a complex detailed dynamic model of the process. The "black box" model reveals little about the internals of the process, but can be used to estimate the effect of major parameters on process behavior and economics. The dynamic model can be used to design and evaluate process operation and control strategies for process optimization. Experience has shown that the most useful and cost effective method of process modeling is to start with development of the model early in the feasibility stage and to modify and refine the model as more detailed data are made available and more specific results are required.

Early work on modeling by Brittan in 1973 as well as some more recent efforts by Lloyd in 1978 and Splaine in 1982 dealt with analyzing past plant performance data to develop predictive equations. In these efforts multiple regression techniques were utilized to develop equations of the form:

$$\text{Gold Recovery} = A + BX + CY + \ldots DZ$$

This type of analysis has lead to individual process plant performance improvement through a better understanding of the process. However, with this technique, the models are site specific and cannot be transferred to other operations without repeating the tedious regression analysis. These models are specific to the current process configuration and will not predict alterations to the process flowsheet or operations outside of current practices. Also, these models are usually linear in nature, whereas the actual influence of many variables, such as concentration and temperature, are non-linear. A more accurate, and more difficult, approach is to develop models which are based upon first principles. These models can then be used for predictive purposes in gold recovery circuits at other locations with only minimal laboratory testing and parameterization. This approach was utilized in developing the current comminution models which have become widely accepted for simulation of grinding mills.

The following text covers the current status of development of predictive models for various gold recovery processes. In keeping with the scope of this paper, great detail in the mathematics will be excluded but references are provided.

Unit Operations

Unit operation modules are the primary building blocks for process model construction. A generalized gold processing flowsheet is illustrated in Figure 1. Any given precious metals plant would not incorporate all of the unit operations noted. This flowsheet shows the interrelationship of the various steps and serves as an outline for the discussion to follow. Each major flowsheet operation should be represented by a complete mathematical model of that particular unit operation.

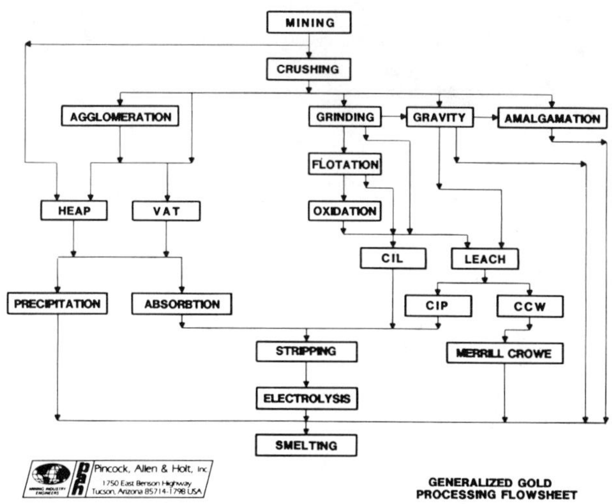

FIGURE 1. Generalized gold processing flowsheet showing major unit operations.

Comminution

Comminution circuits have been the focal point for much of the initial effort of computer modeling in the mineral processing field. This is largely due to the commonality of such circuits to many minerals but is also due in part to the physical nature of the problem as opposed to the more abstract chemical phenomenon associated with other unit operations. Three texts which deal with comminution are by Lowrison in 1974, Lynch in 1977, and Mular in 1982. Predictive modeling is relatively advanced in the crushing, rod and ball milling, and classification operations whereas autogenous milling simulation is now more art than science. Mineral liberation and SAG mill modeling are the current focal point in comminution research.

Agitated Leaching

A predictive model for agitated leaching circuits requires combining a leaching kinetics model and a fluid flow model. Laboratory testing is used to ascertain, for a specific ore, the influence of time, particle size, cyanide concentration, pH, aeration, and other factors upon the rate of extraction. The fluid flow model is used to correct the batch laboratory test data to actual operations by accounting for any deviation from plug flow and ideal mixing that may be experienced in actual plant operation. The model for a series of continuous stirred tank reactors has been solved rigorously for operations with simplistic flow schemes. However, a CIL circuit where carbon is transferred countercurrently on a discontinuous basis results in an extremely complex mathematical model. A population balance model, such as those proposed by Himmeblau in 1977 and Sepulveda in 1978, plus a simple probability model can be used to determine the detrimental effect on the residence time distribution of such a flow scheme. The preliminary results of such modeling indicate that the effect of carbon transfer is minor (See Figure 2 for the computer output of such a model) and would most likely be overshadowed by short circuiting or other flow pattern distortions. This is the type of process question that can be answered by use of process models.

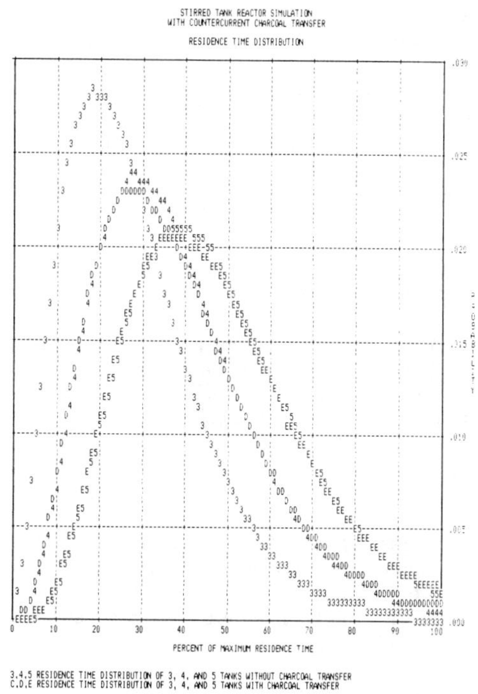

FIGURE 2. CIL circuit residence time distribution curves with and without carbon transfer.

The chemical aspects of the leaching model are well documented with regard to the influence upon extraction of individual parameters such as temperature, cyanide concentration, etc. Numerous investigations have been made of gold cyanidation chemistry, Kudryk in 1954, Kakovshii in 1960, Cathro in 1963, Hallet in 1981, Akhtar in 1982, and Taylor in 1983. However, no unifying model has been published. Leaching tests have shown that the rate of gold extraction is known to increase with a reduction in particle size, increased agitation, higher oxygen content, higher temperature, and higher ferric ion concentration; exhibit a maximum at optimum cyanide and pH levels; and decrease with higher copper, ferrous and sulfide ion concentrations. Several of these variables such as aeration, agitation, cyanide level and pH are known to interact thus resulting in a much more complex rate expression. The author is developing an overall model, encompassing the above parameter, by analyzing a large number of laboratory tests results. While all of the effects are well known, the difficulty lies in developing the overall rate expression.

Carbon Adsorption

The successful modeling of gold adsorption from cyanide solution onto activated carbon depends upon adequately defining the mechanism of adsorption as it affects the equilibrium or ultimate gold loading and the kinetics or loading rate. Davidson in 1974 and 1982, McDougall in 1980, and Muir in 1982 reported investigating equilibrium carbon loadings as a function of several influencing factors. Fleming and Nicol further refined this work in 1984. Based upon currently accepted mechanisms and the above investigations, the carbon equilibrium gold loading capacity is known to increase with carbon surface area (S), carbon activity (A), and ionic strength (I) of K, Na, Ca2; and decrease with pH, temperature (T), free cyanide concentration (CN), contained organics (O), and competing metallic ions (M)

The kinetics of adsorption have been investigated by several researchers including Pitt, et al, in 1978 and 1979, Hussey, et al, in 1978 and 1979, Davidson in 1982, Hahn in 1982, and Fleming in 1984. The rate of adsorption of gold on carbon generally increases with carbon concentration (W), carbon activity (A), agitation and mixing (Mx), temperature (T), and ionic strength (I) of K, Na, Ca2; and decreases with increasing carbon particle size (R), increasing pH, and cyanide concentration (CN). Rate expressions based upon diffusion phenomenon have also been proposed as the controlling rate equation. The general equilibrium and rate expression for adsorption should therefore contain at least the following terms.

$$Auc^* = f(A,CN,I,M,O,pH,S,T,Aus)$$
$$dAu/dt = f(Aus,Auc^*,Auc,W) \times f(D,Mx,R) \times f(T)$$

Where: Aus = solution gold concentration.
Auc = carbon gold loading.
Auc* = carbon equilibrium gold loading.
t = time since last carbon transfer.

The author has developed a technique for determining these specific equations by use of the differential method of data analysis incorporating the methods of isolation, initial rates, and least squares. The concentration-time curves are smoothed using manual or multiple-spline regression techniques. The slopes at various points on the curve are then regressed against all of the test parameters. Normalization of all variables allow for the blending of different experimental data sets. One example of the application of this technique is presented in Figure 3. The data, taken from USBMRI-8368, was fit to an equation of the form:

$$dAuc/dt = K\ (1-Auc/Auc^*)a\ (Auc)b\ (Aus)$$

This equation with one proportionality constant, K, and three exponents a, b, and c was backsolved and the results plotted. This demonstrates that a relatively simple technique can be used to generate a complex and accurate predictive model.

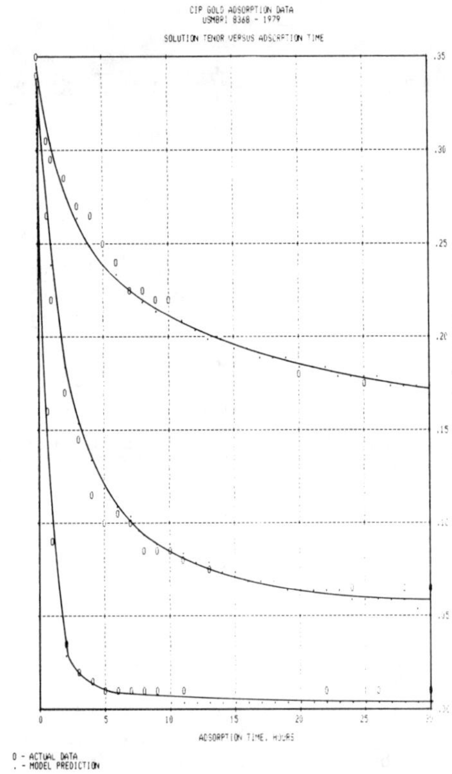

FIGURE 3. CIP adsorption test data showing actual data versus plot of prediction from single equation model.

The overall objective of developing the adsorption model is to facilitate modeling of complete CIP and CIL plants. However, a working model for CIP and CIL plants must also include a fluid flow model, a leaching model and complete mass and energy balance capabilities. An example of computer output from a mass balancing program by the author in 1979 and 1987 is included to demonstrate the amount of detail which can be generated by a complete modeling system. Figure 4 is the flowsheet for a gold cyanidation circuit that was modeled, optimized, and subsequently revamped. Figures 5 and 6 are samples of stream by stream flow rates and gold assays output from the generalized process modeling system used in the optimization of the process illustrated in Figure 4.

Heap Leaching

While much has been written regarding the proper methods for building and operating heaps, little has been published regarding the techniques for modeling the operation of heaps. That which has been written, has been simplifications of copper leach dump models. The copper leach models serve as an excellent frame work for building a precious metal heap leach modelas reported by various investigators, Bartlett in 1974, Roman in 1974, O'Brien in 1982, and McKee in 1984. The complexity of modeling precious metal heap

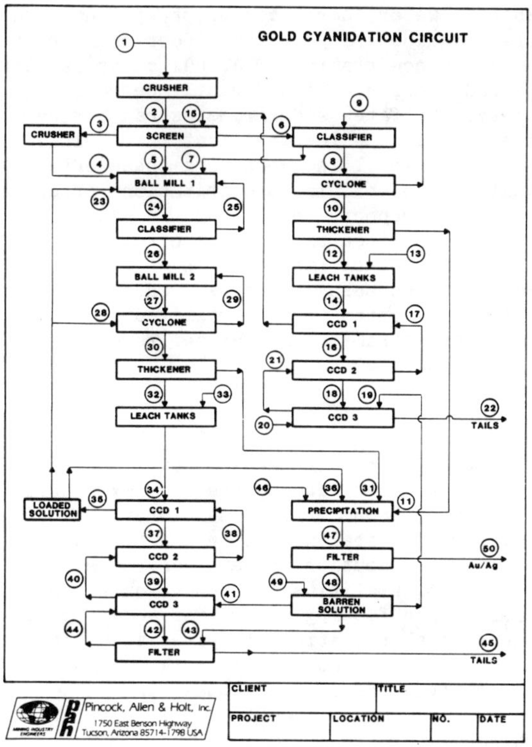

FIGURE 4. Gold cyanidation circuit with agitation leach, countercurrent washing and zinc precipitation.

METSIM
GOLD CYANIDATION CIRCUIT

DESIGN FLOW RATES

NO.	STREAM	TIME	ST/H-S	ST/H-A	ST/H-T	GPM
1	GOLD ORE	100.000	229.649	31.316	260.965	477.88
2	CRUSH PROD	100.000	229.649	31.316	260.965	477.88
3	SCRN OSIZE	100.000	149.272	28.461	177.733	340.23
4	SEC CRS PROD	100.000	149.272	28.461	177.733	340.23
5	SCRN MIDS	100.000	34.447	9.487	43.935	89.89
6	SCRN USIZE	100.000	45.930	151.794	197.724	662.08
7	CLASS OSIZE	100.000	13.347	2.355	15.703	29.68
8	CLASS USIZE	100.000	65.165	163.403	228.568	736.89
9	CYCL UFLW	100.000	32.583	13.964	46.547	104.47
10	CYCL OFLW	100.000	32.583	149.439	182.022	632.42
11	THICK OFLW	100.000	.000	88.928	88.928	346.55
12	THICK UFLW	100.000	32.583	60.511	93.093	285.86
13	CYANIDE LIME	100.000	.000	3.660	3.660	13.87
14	SLM LEACH DS	100.000	32.583	64.171	96.753	299.72
15	SLM 1 CCD OF	100.000	.000	158.427	158.427	614.23
16	SLM 1 CCD UF	100.000	32.583	60.511	93.093	284.70
17	SLM 2 CCD OF	100.000	.000	154.767	154.767	599.21
18	SLM 2 CCD UF	100.000	32.583	60.511	93.093	284.35
19	BARREN WASH	100.000	.000	89.602	89.602	351.66
20	WASH WATER	100.000	.000	65.165	65.165	247.05
21	SLM 3 CCD UF	100.000	.000	154.767	154.767	598.86
22	SLIME TAILS	100.000	32.583	60.511	93.093	284.20
23	MILL SOLN	100.000	.000	114.534	114.534	449.69
24	BMILL DSCH	100.000	591.199	253.371	844.570	1901.78
25	CLASS OSIZE	100.000	394.132	98.533	492.665	991.86
26	CLASS USIZE	100.000	197.066	154.838	351.904	909.92
27	2 BMILL DSCH	100.000	591.198	323.752	914.950	2178.94
28	CYCL SOLN	100.000	.000	211.161	211.161	829.07
29	CYCL UFLW	100.000	394.132	168.914	563.046	1268.70
30	CYCL OFLW	100.000	197.066	365.999	563.065	1739.79
31	SAND THK OF	100.000	.000	.019	.019	.07
32	SAND THK UF	100.000	197.066	365.980	563.046	1739.71
33	CYANIDE LIME	100.000	.000	5.576	5.576	21.01
34	SND LEACH DS	100.000	197.066	371.556	568.622	1760.69
35	SND 1 CCD OF	100.000	.000	997.788	997.788	3917.57
36	LOADED SOLN	100.000	.000	672.093	672.093	2638.81
37	SND 1 CCD UF	100.000	197.066	213.488	410.554	1140.92
38	SND 2 CCD OF	100.000	.000	839.719	839.719	3297.81
39	SND 2 CCD UF	100.000	197.066	213.488	410.554	1141.16
40	SND 3 CCD OF	100.000	.000	839.719	839.719	3298.04
41	BARREN WASH	100.000	.000	560.543	560.543	2199.94
42	SND 3 CCD UF	100.000	197.066	213.488	410.554	1141.22
43	BARREN WASH	100.000	.000	131.377	131.377	515.61
44	FILTRATE	100.000	.000	279.177	279.177	1096.39
45	SAND TAILS	100.000	197.066	65.689	262.754	560.71
46	ZINC DUST	100.000	.001	.000	.001	.00
47	AU PRECIP	100.000	.001	761.040	761.042	2985.42
48	BARREN FILT	100.000	.000	761.039	761.039	2985.42
49	NEW WATER	100.000	.000	20.483	20.483	81.81
50	AU CAKE	100.000	.001	.001	.002	.00

FIGURE 5. Simulation model output showing all stream flowrates in tons per hour and gallons per minute.

METSIM
GOLD CYANIDATION CIRCUIT

GRAMS OF GOLD PER METRIC TON

NO.	STREAM	G/MT-S	G/MT-A	G/MT-T
1	GOLD ORE	3.831	.00000	3.371
2	CRUSH PROD	3.831	.00000	3.371
3	SCRN OSIZE	3.831	.44470	3.289
4	SEC CRS PROD	3.831	.44470	3.289
5	SCRN MIDS	3.831	.44470	3.100
6	SCRN USIZE	3.831	.44470	1.231
7	CLASS OSIZE	3.831	.44470	3.323
8	CLASS USIZE	3.831	.44470	1.410
9	CYCL UFLW	3.831	.44470	2.815
10	CYCL OFLW	3.831	.44470	1.051
11	THICK OFLW	.000	.89572	.896
12	THICK UFLW	1.762	.89572	1.199
13	CYANIDE LIME	.000	.00000	.000
14	SLM LEACH DS	.811	1.32779	1.154
15	SLM 1 CCD OF	.000	.53261	.533
16	SLM 1 CCD UF	.811	.57237	.656
17	SLM 2 CCD OF	.000	.21845	.218
18	SLM 2 CCD UF	.811	.23261	.435
19	BARREN WASH	.000	.05159	.052
20	WASH WATER	.000	.00000	.000
21	SLM 3 CCD UF	.000	.08561	.086
22	SLIME TAILS	.811	.09002	.342
23	MILL SOLN	.000	1.08113	1.081
24	BMILL DSCH	2.719	2.33075	2.602
25	CLASS OSIZE	2.719	2.33075	2.641
26	CLASS USIZE	2.719	2.33075	2.548
27	2 BMILL DSCH	1.787	2.78297	2.140
28	CYCL SOLN	.000	1.08113	1.081
29	CYCL UFLW	1.787	2.11112	1.884
30	CYCL OFLW	1.787	2.11112	1.998
31	SAND THK OF	.000	2.11112	2.111
32	SAND THK UF	1.787	2.11112	1.998
33	CYANIDE LIME	.000	.00000	.000
34	SND LEACH DS	.381	2.82544	1.978
35	SND 1 CCD OF	.000	1.08113	1.081
36	LOADED SOLN	.000	1.08113	1.081
37	SND 1 CCD UF	.381	1.16835	.790
38	SND 2 CCD OF	.000	.33149	.331
39	SND 2 CCD UF	.381	.36497	.373
40	SND 3 CCD OF	.000	.12724	.127
41	BARREN WASH	.000	.05159	.052
42	SND 3 CCD UF	.381	.13437	.253
43	BARREN WASH	.000	.05159	.052
44	FILTRATE	.000	.10283	.103
45	SAND TAILS	.381	.10283	.311
46	ZINC DUST	.000	.00000	.000
47	AU PRECIP	99.999	.05297	1.059
48	BARREN FILT	.000	.05297	.053
49	NEW WATER	.000	.00000	.000
50	AU CAKE	99.999	.05297	99.999

FIGURE 6. Simulation model output showing gold assays for solids, liquids, and pulp.

leaching is not due to the chemistry involved, but to the lack of knowledge regarding solution flow patterns, ion diffusion and parameters for characterizing the ore. At the minimum, a model for heap leaching must incompass the following:

1) Fluid flow model for solution flow through the heap.
2) A model for estimating the degree to which the ore is wetted.
3) A diffusion model from wetted ore surfaces into the porous rock.
4) A material balance on all reactants.
5) An inventory model to monitor performance of individual sections of the heap.
6) An oxygenation model for cyanide solution passing through a heap.
7) A transport model for recovering pregnant solution.

The flow model can be based either upon first principles of fluid dynamics or by a weighted probability distribution based upon an assumed distribution of permeabilities. The first principle approach utilizes the mathematics of unsaturated flow through porous media. For heap leaching, the flow model becomes complex due to the number of flow sites or initial solution distribution points, but the model does allow for the prediction of saturation curves as a function of flow. The unsaturated flow model requires that the rock properties be characterized as to the maximum attainable saturation, unreducible saturation, distribution of pore sizes and a measure of air rate head pressure. The second approach is to model the heap using small increments with varying permeabilities. After assuming an initial flow distribution atop the heap, the flow through each unit volume is a function of permeability. Areas where compaction has occurred can be assigned lower permeabilities, while areas where larger ore fractions may accumulate can be assigned higher permeabilities. A model for agglomeration of the heap feed, such as the one by Milligan in 1984, can be incorporated in the program to evaluate the influence of improved permeability. The important parameters are the affect of cement, cure time and moisture content upon permeability.

Given that the flow or saturation in a given volume of heap can be estimated, the effect upon leaching must be determined. Leaching will commence when diffusion has begun from a wetted surface. The time to wet a surface may be less than to reach a given degree of saturation as predicted in the unsaturated flow model. Likewise, if a flow rate is predicted for given volume unit, only a percentage of this flow will wet the ore with the rest shortcircuiting through the ore. While the wetting parameter cannot be determined analytically, laboratory testing can give an adequate measure of the degree of wetting versus flow rate and particle size distribution. The above diffusion model allows the effect of particle size distribution to enter into the kinetics of extraction.

Of prime importance in the heap leach model is the ability to follow the change in concentration of various elements in the heap. This requires that detailed material balances for each reactant be performed in each unit volume of the heap. The material balance program can calculate, as a function of time, precious metal inventory in the heap, the concentration of ions in solution and the analysis of leached ore in various sections of the heap. The last major step is to provide for renewal of oxygen in the cyanide solution by diffusion of air through the dump. This refinement is not critical for small dumps, but can be influential in examining the performance of large, permanent dumps.

Zinc Precipitation

The zinc precipitation or cementation process has existed since 1890 and in its most common form, Merrill-Crowe, since 1932. Given that the process is so widely used and extensively studied, little thought is given to modeling the process. Extensive data does exist to allow for development of the model which could aid in the operation of existing plants or for the process development of new plants. In the U.S., the process is still used extensively for recovering silver from heap leach operation. The important parameter affecting the reaction rate are the oxygen content of the solution, temperature, cyanide concentration, alkalinity and trace ion concentration as reported by Leblane in 1942, von Hahn in 1967, and Adamson in 1982. Over the normal ranges of cyanide concentration and pH, the affects of these parameters on gold recovery is nominal, however there does exist optimum levels of free cyanide and pH. The effect of trace ions, such as a concentration of 0.0001 moles per liter of sulfide ion, can significantly decrease recovery. Sulfates and chromates can reduce recovery to a much lesser extent than sulfide ion. The use of a model incorporating these effects can be utilized to optimize precipitation performance. Currently as reported by van Zyl in 1983, precipitation circuits can be controlled by feedback control on the addition of zinc to hold a barren solution assay. With a predictive model, feed forward control can be utilized incorporating more process parameter for higher recovery or lower operating costs.

Flotation

Flotation is one of the most difficult unit operations to model due to the complex nature of the flotation phenomena. Most flotation models are either empirical models or kinetic models involving distribution of rate constants for particle floatabilities and retention times for particles in a flowing pulp. For modeling of precious metal flotation circuits, most data is acquired from laboratory test work and is analyzed using multiple regression techniques.

Roasting

The modeling of roasting has been primarily done on concentrates other than precious metals. While some research has been performed on roasting of pyrite and arsenopyrite concentrates, little modeling has been performed on the data. For the most part, correlation have been made between effects of the roasting parameters and precious metals recovery during subsequent cyanidation. Due to the lack of understanding of how roasting affects leaching kinetics, linear regression models are the current tool being used in the analysis of precious metal roasting data.

Chemical Oxidation

The use of chemical oxidation in place of roasting as a pre-treatment is becoming commonplace. In particular, the use of oxygen autoclaves has gained acceptance as a means of treating non-cyanidable concentrate. Little modeling has been performed on the chemical oxidation processes for precious metals. However, there is little reason to doubt that techniques applied to modeling other autoclave technologies cannot be bought to bear on the processing of precious metal ores or concentrates.

Model Input And Output

The required input data include a flowsheet description, process operating conditions, unit operation parameters, and compositions and flowrates of feed streams. This input is generally derived from process design criteria, theoretical and empirical data from technical references, and laboratory, pilot plant, and full scale plant operations. Economic evaluation of direct operating costs will require input of fuel, power, water, and other utility unit costs as well as raw material, wear material, and reagent unit costs.

Process model output generally consists of stream data including mass, volumetric, and energy flowrates, specific gravities, temperatures, screen analyses, concentrations, and assays. If appropriate, a heat balance around each unit operation is output along with a summary of the heats of reactions and heat losses. The economic portion of the output will contain direct operating costs for consumables and gross income from salable products.

APPLICATIONS

Process simulation has been used on comminution processes to evaluate the effect of feed rate, rock hardness, feed size distribution, operating time, crusher and mill type, and other operating parameters on product size distribution and operating costs. Precious metal process models have included, gravity concentration, cyanidation, flotation, carbon-in-pulp, carbon-in-leach, smelting, and refining circuits. The author has been involved in the computer simulation and modeling of fourteen gold processing plants, five of which are cited as follows:

1) A CIP model was used to determine the required number of adsorption stages, optimum carbon inventories, and overall system hydraulics.

2) A non-steady state heat balance model was used to establish the most economic schedule, firing rate, and charging rate for a gold precipitate smelting furnace.

3) A complete flowsheet model, see Figure 4, included crushing, grinding, classification, sand and slime leaching circuits, counter current decantation, filtration, and precipitation of gold using the Merrill-Crowe process, facilitated the optimization and subsequent modification of the CCD circuit of an existing gold mill, minimizing cyanide consumption and maximizing gold recovery.

4) A comminution model for a small gold mill was used to optimize a grinding ciruit with a rod mill and three different sized ball mills. The model demonstrated that the ball mill feed split was critical. Precise control of the split and a higher ball charge increased the throughput by sixteen percent.

5) A non-steady mass balance model simulated a vat leaching process for gold cyanidation using agglomerated ore. The goal of the optimization was to reduced the stockpiling area to a minimum, maximized gold recovery, and minimize the flow of pregnant liquor. This was accomplished by using a staged countercurrent washing circuit on the eight vats for each of the wash liquors.

CONCLUSION

As noted in the preceding text, numerous modeling techniques are currently available to the process engineer for evaluating precious metal projects. The usage of computers for process evaluation, design and control is available at all levels of engineering activity. Significant benefits have been realized from the application of computer modeling techniques to actual plant operations. Each engineer and operator should assess the potential benefit of the application of computers for process optimization and subsequent economic enhancement to thier current or future precious metal properties in order to take advantage of this money saving tool.

REFERENCES

Adamson, R. J., "The Chemistry of the Extraction of Gold", Gold Metal. S. Africa, pp 327-333.

Akhtar, S., "Cyanidation of Gold/Silver Ore", International Gold/Silver Conference II, Reno, Nevada, May, 1982, Y. S. Kim, Chapter 8.

Auck, Y. T. and Wadsworth, M. E., "Physical and Chemical Factors in Copper Dump Leaching", International Symposium on Hydrometallurgy, Evans, Shoemaker - Editors, AIME, pp 645-699.

Bartlett, J. T., "Metallurgical Process Simulator", 18th Conference of Metallurgist, CIMM, Sudbury, Canada, August 1979.

Bartlett, J. T., "Process Simulation and Optimization Using Metsim", Annual Meeting, AIME, Denver, Colorado, February, 1987.

Bartlett, R. W., "Application of Diffusion Models in Estimating Heap Leach Extraction", SME Preprint 74-AS-351, SME Fall Meeting, 1974.

Brittan, M. I. and van Vuren, E. J. J., "Computer Analysis, Modeling and Optimization of Gold Recovery Plants of the Anglo-American Group", Journal South African Institute of Mining and Metallurgy, February 1973, pp 211-222.

Brittan, M. I. and MacLeod, D. S., "Improvement of Gold Recoveries at Western Deep Levels by Mathematical Modeling and Plant Testwork", Trans Institute of Mining Metallurgy, Volume 83, 1974, pp C133-C142.

Cathles, L. M. and Schlitt, W. J., "A Model of Dump Leaching Process that Incorporation Oxygen Balance", Leaching and Recovering Copper from As-Mined Materials, W. J. Schlitt, Editor, SME, AIME, pp 9-28.

Cathro, K. J., "The Effect of Oxygen in the Cyanide Process for Gold Recovery", Proc. Aust. Inst. Min. Metall., No. 207, 1963, pp 180-205.

Cho, E. H. and Pitt, C. H., "Kinetics and Thermodynamics of Silver Cyanide Adsorption on Activated Charcoal", Met. Trans. B, Amer. Soc. for Metals and Metall. Soc. of AIME, Vol. 10B, June 1979, pp. 185-189.

Davidson, R. J., "The Mechanism of Gold Adsorption on Activated Charcoal", J. of S. Afr. Inst. of Min. and Met., 1974, pp. 67-76.

Davidson, R. J., Douglas, W. D., and Tumilty, J. A., "The Selection of Granular Activated Carbon for Use in a Carbon-In-Pulp Operation", Carbon-In-Pulp Seminar, The Aus.I.M.M. Perth and Kalgoorlie Branches and Murdoch Univ., July 1982, pp. 199-218.

Davidson, R. J., Douglas, W. D. and Tumilty, J. A., "Aspects of Laboratory and Pilot plant Evaluation of CIP with Relation to Gold Recovery", XIV Intnl. Min. Proc. Cong., CIM, Canada, Oct. 17-23, 1982, pp. II-6.1-6-19.

Dixon, Steve, Cho, Eung H., and Pitt, C. H., "The Interaction Between Gold Cyanide, Silver Cyanide, and High Surface Area Charcoal", The AIChE., Univ. of Utah, 1978, pp. 75-83.

Fleming, C. A. and Nicol, M. J., "The Absorption of Gold Cyanide onto Activated Carbon. III. Factors Influencing the Rate of Loading and Equilibrium Capacity", J. S. Afr. Inst. Min. Metall., Vol. 84, No. 4, 1984, pp. 85-93.

Hallet, C. J.; et al, "Oxygen Mass Transfer in Pachuca Tanks", Extraction Metallurgy 1981, Inst. of Min. and Met. - London, pp 308-320.

Heinen, H. J. and McClelland, G. E., "Enhancing Percolation Rates in Heap Leaching of Gold-Silver Ores", Department of Investigation - 8388, Bureau of Mines, Report of Interior.

Himmeblau, D. M. and Bischoff, K. B., "Population Balanced Model, Chapter 4", Process Analysis and Simulation, Wiley and Sons, N. Y., N.Y.

Hussey, S. J., Salsibury, H. B., and Potter, G. M., "Carbon-In-Pulp Silver Adsorption from Cyanide Leach Slurries of a Silver Ore", Bureau of Mines Report of Investigations 8268, U.S. Dept. of the Interior, pp. 1-22.

Hussey, S. J., Salisbury, H. B., and Potter, G. M., "Carbon-In-Pulp Gold Adsorption from Cyanide Leach Slurries", USBMRI 8363, U.S. Dept. of Interior, pp. 1-22.

Kakovskii, I. A. and Kholmanskikj, Yu, "Kinetics of Cyanidation of Copper and Gold", Invest. Akad. Navk, S.S.S.P., Otpeg Tekh Nauk, No. 5, 1960, pp 207-218.

Kudryk, V. and Kellogg, H. H., "Mechanism and Rate Controlling Factors is the Dissolution of Gold in Cyanide Solution", Journal of Metals, Vol. 6, 1954, pp 541-548.

Leblane, R., "Precipitation of Gold from Cyanide Solution by Zinc Dust", Canadian Metallurgical Quarterly, Vol. 63, No. 1, 1942, pp 213-19, 297-306 and 371-379.

Lloyd, P. J. D., "Maximizing the Recovery of Gold from Witwatersrand Ores", Minerals Sci. Engr., Volume 10, No. 3, July, 1978, pp 208-221.

Lowrison, G. C., "Crushing and Grinding", Butterworths Monographs in Chemistry and Chemical Engineering, 1974.

Lynch, A. J., "Mineral Crushing and Grinding Circuits". Elsevier Scientific Publishing Company,Amsterdam, 1977

Madsen, B. W. and Wadsworth, M. E., "A Mixed Kinetics Dump Leaching Model for Ores Containing a Variety of Copper Sulfide Minerals", USBMRI 8547, U.S. Bureau of Mines.

McDougall, G. J., Hancock, R. D., Nicol, M. J., Wellington, O. L., and Copperthwaite, R. G., "The Mechanism of the Adsorption of Gold Cyanide on Activated Carbon", J. S. Afr. Inst. of Min. and Metall., 1980, pp. 344-356.

McKee, C. R. and Bumb, A. C., "The Importance of Unsaturated Flow Parameters in Designing a Monitoring System for a Hazardous Waste Site", National Conference on Hazardous Wastes and Environmental Emergencies", March, 1984.

Milligan, D. A., "Agglomeration - Design for Heap Leaching of Gold and Silver Ores", Precious Metals Recovery Seminar, June, 1984.

Muir, D. M., "Recovery of Gold from Cyanide Solutions Using Activated Carbon: A Review", Carbon-In-Pulp Seminar, The Aus. I.M.M. Perth and Kalgoorlie Branches and Murdock Univ., July 1982, pp. 7-22.

Mular, A. L. and Jergensen, G. V.,II, "Design and Installation of Comminution Circuits", AIME, 1982.

Nicol, M. J., Fleming, C. A. and Cromberge, G., "The Absorption of Gold Cyanide onto Activated Carbon. I.", J. S. Afr. Inst. Min. Metall. Vol. 84, No. 2, Feb. 1984, pp. 50-54.

Nicol, M. J., Fleming, C. A. and Cromberge, G., "The Absorption of Gold Cyanide onto Activated Carbon. II.", J. S. Afr. Inst. Min. Metall., Vol. 84, No. 3, March 1984, pp. 70-78.

O'Brien, R. J., "Agglomeration Pre-Treatment in the Heap Leaching of Gold and Silver", CIP Seminar, July, 1982, Aust. 1.M.M., pp 297-311.

Roman, R. J., "Heap Design with the Aid of Computer Simulation", Paper 26B, AICHE National Meeting, August, 1974.

Roman, R. J. and Olsen, C., "Theoretical Scale-up of Heap Leaching", Solution Mining Symposium, February, 1974, F. F. Aplan, AIME, pp. 211-229.

Sepulveda, J. E. and Herbst, J. A., "A Population Balance Approach to the Modeling of Multistave Continuous Leaching Systems", Fundamental Aspects of Hydrometallurgical Processes, AICHE Symposium Series. AIChE, 1978.

Splaine, M. et al, "The Effect of Head Grade on Recovery Efficiency in a Gold-Reduction Plant", Journal South African Institute of Mining and Metallurgy, January, 1982, pp 6-11.

Taylor, D. R., "Copper and Iron Impurities in the Cyanide Leaching of Gold", Hydrometallurgy of Copper: Its Byproducts and Rarer Metals, Haas and Weir - Editors, AIME, 1983.

van Zyl, R. M., "Computer Controlled Merrill-Crowe Processing", Engineering and Mining Journal, April, 1983, pp 58-59.

von Hahn, H. E. A. and Ingraham, T. R., "Kinetics of Silver Cementation on Zinc", Canadian Metallurgical Quarterly, 1967, pp 15-26.

Young, G. J. C., Douglas, W. D., and Hampshire, M. J., "Carbon in Pulp Process for Recovering Gold from Acid Plant Calcines Brand", Mining Engineering, March 1984, pp. 257-264.

Chapter 23

COST EFFECTIVE TAILINGS POND AND HEAP LEACH PAD LINERS: A CASE HISTORY

Mark W. Cadwallader, M.S.

Gundle Lining Systems, Inc.
Director of Research and Technical Development
Houston, Texas

Abstract. Heap leach mining requires liners in order to contain leaching solutions for both economic and environmental incentives. The use of HDPE geomembrane liners for heap leaching is discussed in this paper. Comparison is made to clay liners, with inclusion of economic considerations. Several heap leach projects incorporating geomembranes are discussed.

INTRODUCTION

Once considered uneconomical, many ore deposits are now economically feasible to process by a method called "heap leaching". Geomembrane liners are forming an important component of the heap leaching systems. They prevent the escape of leaching chemicals to the environment and maximize profit margins by retaining the solutions of precious metals for metal recovery downstream.

HEAP LEACH MINING REQUIRES LINERS

Several different processes for heap leaching are available. These differences are primarily in metal recovery techniques. As far as the heap leaching itself is concerned, however, the operation is quite standard and straightforward. Sulfuric acid solutions for copper mining, and cyanide solutions for gold and silver mining are percolated through a leveled "heap" of ore using a sprinkling system. The chemical solution dissolves the metal and collects on a "leach pad" underneath the heap. The pad is sloped to force the "pregnant" solution to flow into channels and then to the pregnant solution storage basin. The pregnant solution is then pumped through a metal recovery system and transferred to the "barren" solution basin. Figure 1 displays a flow chart showing these basic elements of a heap leaching operation.

Leaching solution seepage potential exists at all points along the leach pad, drainage system, and solution basins. Liners are therefore required in order to prevent the loss of valuable gold, silver, or copper bearing solution, as well as to protect the environment from contamination by the leaching chemicals.

A number of liner systems are available. Figure 2 displays some lining schemes in use today. Single liners are generally either made of clay or a geomembrane (polymeric membrane). Liners are often used in combination to provide for drainage (leak detection) and redundancy.

A composite liner, formed by a geomembrane in contact with a clay liner, offers the greatest degree of impermeability. This is because although the permeability of a good geomembrane is extremely low, the very low permeability that is there is due to diffusion. Unlike convection, diffusion is driven by a concentration gradient instead of a pressure gradient. The absorptivity and porosity of a clay liner directly underneath a geomembrane means that the chemical concentration gradient is reduced across the geomembrane since diffused chemical species accumulate in the clay pores at the clay/geomembrane interface. With chemical concentration roughly equivalent on either side of the geomembrane, the driving force for diffusion is eliminated. A geomembrane/clay liner works well because the geomembrane is a barrier to pressure driven mass transfer, while the underlying clay liner forms a barrier to concentration driven mass transfer.

The combination liner comprised of a geomembrane, drainage layer, geomembrane, and clay, is the liner system required for hazardous waste containment under Subtitle B of the Hazardous and Solid Waste Amendments (HSWA) to the Resource Conservation and Recovery Act (RCRA). The drainage layer can be either conventional sand or gravel, or drainage netting. Drainage netting is a net-shaped material made by overlapping high density polyethylene strands so as to conduct fluids in the plane of the net. Drainage net generally has a hundred times more flow capacity than do sand or gravel drainage layers.

CLAY LINERS

Clay liners have in the past been the most common way to contain chemical and waste solutions on a field scale. Recent studies, however, have shown that solutions of hydrocarbons can seriously affect the impermeability of clays (Brown, K.W. and Anderson, D.C., 1982 Kingsbury, G.L. and

FIG. 1 - FLOW CHART OF HEAP LEACHING OPERATION

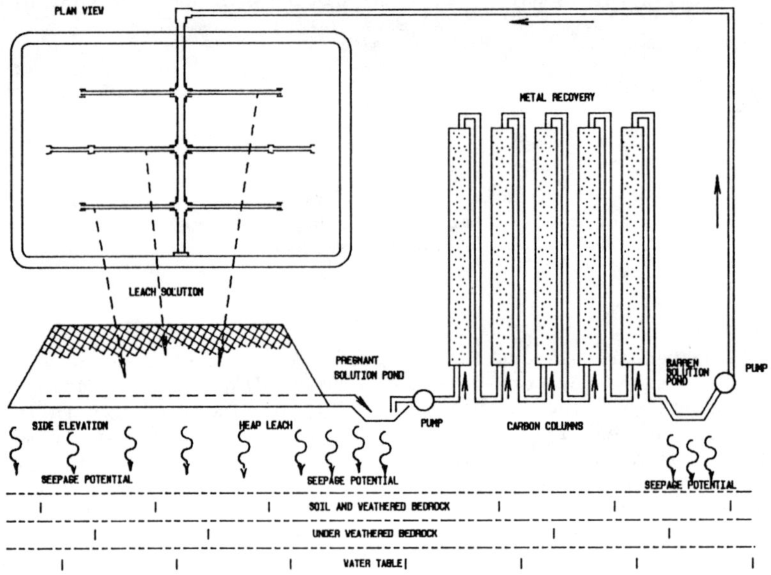

FIGURE 2 - LINER SYSTEMS AVAILABLE

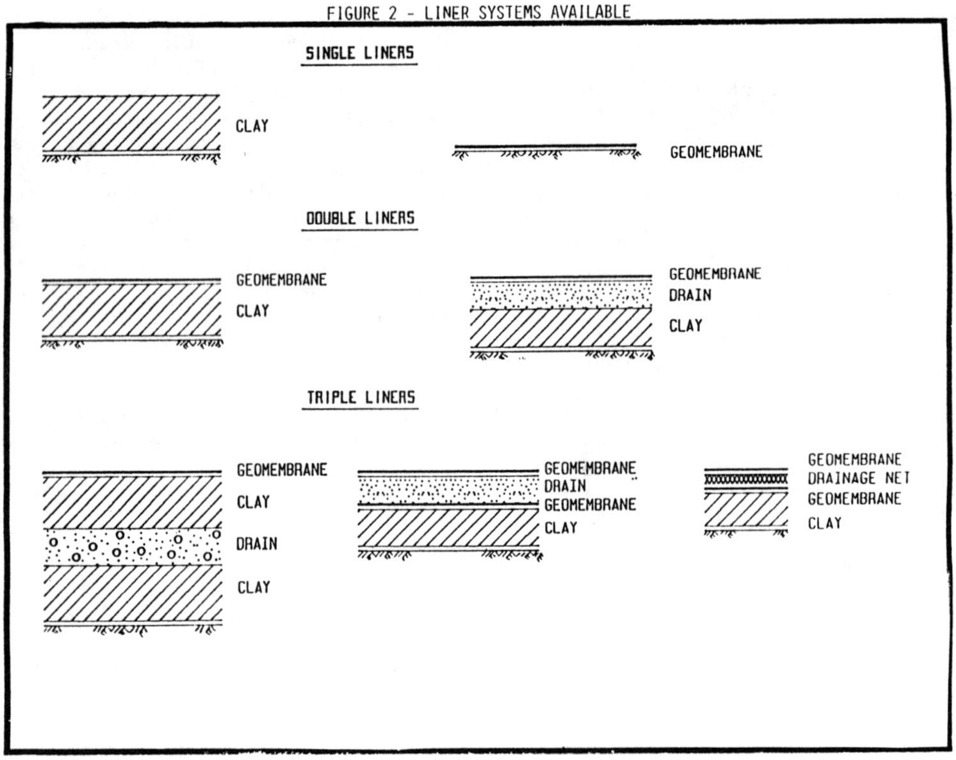

Truesdale, R.S., 1985). Conventional clay permeability tests using water with 0.01 N $CaSO_4$ qualify most clay soils for lining hazardous waste disposal facilities on the basis of their having permeabilities lower than 1 x 10-7 cm/sec. However, when the same clay soils were subjected to permeability tests with solutions of organic fluids, they underwent large permeability increases exceeding permeabilities of 1 x 10-7 cm/sec.

Acidic and caustic leachate have also been found to disrupt clay barriers with permeability increases (Anderson, D.C. and Jones, S.G., 1983). The mechanism in this case is dissolution of the soil binding agents in the clay followed by movement of the clay particles out of the barrier.

The problem is that clay minerals have a great deal of ion-exchange capacity and are subject to ion exchange reactions. For that reason, acids tend to adsorb on clay and to dissolve aluminum, iron, and silica particles in the lattice structure of the clays. Alkaline solutions have also been implicated in adsorption and dissolution of clay minerals by removing silica from the clay structure.

In addition, the actual field permeabilities of compacted clay liners can greatly exceed laboratory measured clay permeabilities (Day, S.R. and Daniel, D.E., 1985). Laboratory permeability tests do not take into account the presence of cracks, fissures, and inter-clod voids which appear by virtue of the larger scale in the field. Laboratory measured permeabilities might measure 10 -7 cm/sec while field permeabilities are actually 10 -5 cm/sec.

Clay permeabilities are also very sensitive to water content and degree of compaction. Clay liners are prone to differential settlement leading to localized cracking. Clay liners can dry out (desiccate), leading to the development of a large number of microcracks (van Zyl, D. and Smith, M.E., 1986).

Geomembrane liners are therefore highly attractive because of their negligible permeability and good chemical resistance. With HDPE liners, the technology can now boast of strength, toughness, durability, chemical resistance, and effective seaming technique, in addition to very low permeabilities. These properties are briefly defined as follows:

Strength - Tensile properties and elongation; stress/strain characteristics of the material.

Toughness - Resistance to tear and puncture.

Durability - U.V. and heat resistance of the material.

Chemical Resistance - No significant reaction with, or loss of properties from chemically agressive environment.

Effective Seaming Technique - True welding which results in a homogeneous fusion of liner sheets.

GEOMEMBRANES IN HEAP LEACHING

Geomembranes at heap leach sites minimize the loss of leaching fluids. This is highly desirable from both the environmental and economic points of view.

HDPE geomembrane liners are particularly resistant and impermeable toward dilute aqueous alkaline or acidic solutions. Polar, ionic aqueous solutions of cyanide or sulfuric acid have no extra affinity for HDPE, a nonpolar, nonionic substrate.

Table 1 compares seepage rates from natural clay liners versus HDPE geomembranes of varying thicknesses. In order to compare the pressure driven permeabilities of clay to the concentration driven permeabilities of HDPE, diffusion data for HDPE must first be converted to the coefficient of permeability suitable for use in Darcy's Law. This calculation is worked out in Appendix A.

Darcy's Law can be applied for soils where convection rates are low (flow regimes are laminar, not turbulent). The validity of using Darcy's Law for calculating infiltration through a low permeability soil liner is discussed in detail by Mitchell, 1976. The required form of Darcy's Law for the infiltration rate, or flow per unit area (Q/A) is (Mitchell, J.K., 1976):

$$\frac{Q}{A} = k \frac{\Delta h + t - h_i}{t} \qquad (1)$$

Where:
- Q = volumetric flow rate
- A = unit area
- k = coefficient of permeability
- Δh = liquid depth above liner
- t = liner thickness
- h_i = pressure head of water at bottom of liner

Since geomembrane thicknesses are relatively small compared to the head of water for almost all applications, the thickness term can be eliminated in the numerator of Darcy's Law applied to an HDPE liner. Also, the head at the bottom of the liner can be conservatively assumed as zero. Thus, Darcy's Law applied to a geomembrane can be simplified to:

$$\frac{Q}{A} = k \frac{\Delta h}{t} \qquad (2)$$

For clay liners the thickness term is large, and Darcy's Law should be written:

$$\frac{Q}{A} = k \frac{\Delta h + t}{t} \qquad (3)$$

Although the liquid head will not infiltrate the clay immediately due to absorption and adsorption into and onto the clay, the fact of absorption/adsorption prior to infiltration represents loss of leaching solution. And leaching solution loss means an undesirable economic as well as environmental impact.

HDPE geomembranes result in economic savings because no pregnant solution is lost through absorption/adsorption, and because infiltration rates are very low. In addition, capital expenditures for HDPE liners can be much less than for clay, depending upon local prices and haul distances. Material purchase and installation of a 1.5 mm (60 mil) HDPE liner were found to cost less than 60 cm (2 ft) of clay when the haul distance for the clay was greater than 10 miles (Roznowski, D., 1987).

As a result of the economic advantages, environmental requirements, and material superiority compared with other geomembranes, a large number of heap leach operators have chosen to install HDPE geomembranes at their mining facilities. Several of these projects are discussed below.

HEAP LEACHING COPPER

Magma Copper Company's San Manuel Mine offers an interesting example of the heap leaching process. The San Manuel Mine, located 40 miles northeast of Tucson, Arizona, uses the technique to extract copper from copper oxide rock. The 83 acre San Manuel leaching site includes a series of heap leaching systems consisting of a leach pad, leachate drainage system, and holding pond.

TABLE 1

SEEPAGE RATES THROUGH HDPE LINERS AND CLAY LINERS

HDPE Liner

	40 mil	60 mil	80 mil	100 mil
Permeability (k)	2.7×10^{-13} cm/sec	2.7×10^{-13} cm/sec	2.7×10^{-13} cm/sec	2.7×10^{-13} cm/sec
Thickness (t)	0.1 cm	0.15 cm	0.20 cm	0.25 cm
Liquid Head (Δh)	5 cm (2 in)	5 cm (2 in)	5 cm (2 in)	5 cm (2 in)
Infiltration [1] $\frac{Q}{A} = k \frac{\Delta h}{t}$	1.35×10^{-11} cm/sec	9.0×10^{-12} cm/sec	6.75×10^{-12} cm/sec	5.4×10^{-12} cm/sec
[2] Gal/Acre/Day	0.0125	0.0083	0.0062	0.0050

Clay Liner

	1	2	3	4
Permeability (k)	1×10^{-7} cm/sec	1×10^{-7} cm/sec	1×10^{-5} cm/sec	1×10^{-5} cm/sec
Thickness (t)	90 cm (3 ft)	60 cm (2 ft)	90 cm (3 ft)	60 cm (2 ft)
Liquid Head (Δh)	5 cm (2 in)	5 cm (2 in)	5 cm (2 in)	5 cm (2 in)
Infiltration [3] $\frac{Q}{A} = k \frac{\Delta h + t}{t}$	1.06×10^{-7} cm/sec	1.08×10^{-7} cm/sec	1.06×10^{-5} cm/sec	1.08×10^{-5} cm/sec
[2] Gal/Acre/Day	98	100	9,790	10,000

[1] Darcy's Law, assuming head at bottom of liner is zero; thickness of HDPE small, compared to liquid head.

[2] $\text{cm/sec} = \frac{\text{cm}^3}{\text{cm}^2 \cdot \text{sec}}$; $\frac{\text{cm}^3}{\text{cm}^2 \cdot \text{sec}} \times 9.2374 \times 10^{-8} = \text{gal/acre/day}$

[3] Darcy's Law, assuming head at bottom of liner is zero.

The heap leach process at San Manuel begins with sheets of copper oxide rock placed on the leach pad. The rock is sprayed with a weak sulfuric acid solution, separating the metal from the rock by oxidation. The copper, dissolved in the sulfuric acid solution, is then carried out of the pad into the holding pond via the drainage system. In the holding pond, the copper solution is treated with a solvent which ionizes the copper. Applying electrical energy then causes the copper to plate onto electroces dipped into the holding pond.

According to Magma officials, heap leaching of copper oxide presents a number of advantages over conventional processing methods. The favorable economics of heap leaching stem from the fact that such conventional and expensive mining procedures as crushing, milling, smelting, and refining can be eliminated. The result is a processing cost of approximately one-half that of conventional mining methods.

For the 3.6 million square feet of lined area at San Manuel, three different HDPE geomembrane thicknesses were deployed: 60, 80, and 100 mil sheet. The 100 mil liner was placed at the edges of the leach pad in the region of the anchor trench. This gave added tensile strength where tension forces were most likely to be experienced. The base of the leach pad utilized 60 mil liner, while the drainage system and holding pond were lined with the 80 mil. Unionized local laborers and operators assisted Gundle Lining Construction Corporation, which was subcontracted out to Bechtel Corporation for the construction of the site. Total manhours for the installation of the liners were only 18,000. The liner installation was completed in the spring of 1986.

HEAP LEACHING GOLD

Nine miles outside of the town of Chillagoe in North Queensland, Australia, is a gold extraction project known as Red Dome. The project is being developed by Elders Resources Limited, extracting gold by the heap leaching carbon absorption and desorption process.

Gold ore from the crushing and screening plant is pretreated in the pug mill. A cyanide solution and quick lime are mixed with water, and the ore is washed in this leaching solution. Cyanide functions as the gold leaching agent while lime helps bind the solution to the ore and provides necessary alkalinity. The pretreated ore is then stacked on the heap leach pads.

A system of pipes and spray nozzles is laid out on top of the ore heaps for irrigation by the heap solution. The cyanide solution filters through the ore, dissolving the gold along the way.

The ore pads at Red Dome are lined with HDPE geomembrane from Gundle Lining Systems, Inc., and are contoured, channeling the leach solution to the pregnant solution pond. The pregnant solution is pumped through a series of carbon column collectors where soluble gold is absorbed onto particles of activated carbon. The solution then gravitates to the barren solution pond.

The barren solution is then pumped to the spray feed tank where further quantities of cyanide, caustic and descaling agent are added to revitalize the leaching solution.

The cycle of heap leach/pregnant solution/carbon absorption/barren solution continues until the leaching of gold from each pad is complete.

One million square feet of 1 mm (40 mil) Gundline HD was installed for the Red Dome project. The HDPE membrane was delivered to site in 22.5 ft wide seamless rolls, the wide material minimizing the amount of field seaming required. As with all Gundle installations, on-site seaming was accomplished by extrusion welding, forming a homogeneous bond. The first two phases at Red Dome were completed in September, 1986, and are presently in operation.

HEAP LEACH TAILINGS POND

Another heap leach mining operation, this one for FMC Corporation at Gabbs, Nevada, utilized 60 and 100 mil HDPE geomembrane to line the containment area for spent gold-mining ore tainted with cyanide. The earthwork for the containment in this case simply consisted of building a "dam" between two hills to hold back the gold mining tailings. For this project, 4.6 million square feet (106 acres) of Gundline HD were deployed with completion taking place in the fall of 1985. Gusts of winds up to 80 mph added difficulty to the installation and mandated the use of heavy concrete bags by Gundle Lining Construction Corporation to hold the sheet in place.

Total extrusion welding seamed footage came to 41 miles for the project. Destructive testing in shear and peel modes was performed by Gundle on the seams. All of the destructive testing for quality control at the Gabbs, Nevada facility was cross-checked by a third party testing laboratory.

Destructive testing of the seams is very important because it provides the only direct evaluation of seam strength and bonding efficiency. Shear testing applies a tensile stress through the weld. Peel testing, on the other hand, peels the top sheet back against the overlapped edge of the bottom sheet in order to observe how the weld is coming apart. The important thing in the peel test is that polymer material tears, indicating a fully integrated connection between top and bottom sheets. The peel test is a very rigorous test and a very good indication of whether or not homogeneous bonding has occurred.

SUMMARY

Heap leaching allows a large area with ore potential to be mined systematically and efficiently. The procedure is much more cost-effective than conventional mining systems. Low-grade ores, mine waste dumps, and deposits that are too small to justify economically can now be processed quite satisfactorily. HDPE geomembranes enhance the favorable economics and simultaneously protect the environment by minimizing leach solution seepage into the groundwater.

APPENDIX A

CALCULATION OF EQUIVALENT HDPE PERMEABILITY FROM MOISTURE VAPOR TRANSMISSION DATA (ASTM D96)

At 23°C, p_s = 21.068 mmHg

where p_s = vapor pressure of liquid water

21.068 mmHg x 133.3 Pa/mmHg = 2808 Pa

For 60% relative humidity, p = 0.6 x 2808 Pa = 1685 Pa

where p = vapor pressure

Water Vapor Transmission, WVT = $0.04\ g/m^2 \cdot 24\ hr$ for 0.1 cm (40 mil) HDPE[1]

$$\text{Permeance, } \bar{w} = \frac{WVT}{P}$$

$$= \frac{0.04\ g/m^2 \cdot 24\ hr \times 10^{-3}\ kg/g}{1685\ N/m^2 \times 8.64 \times 10^4\ sec/24\ hr}$$

$$= 2.75 \times 10^{-13}\ sec/m$$

$$\text{Permittivity, } \psi = \bar{w} \times g$$

$$= 2.75 \times 10^{-13}\ sec/m \times 9.81\ m/sec^2$$

$$= 2.7 \times 10^{-12}/sec$$

where g = gravity constant

$$\text{Permeability, } k = \psi \times t$$

$$= 2.7 \times 10^{-12}/sec \times 0.1\ cm$$

where t = thickness

$$k = 2.7 \times 10^{-13}\ cm/sec$$

[1]Data by Matrecon Labs, Inc., Oakland, California

REFERENCES

Anderson, D.C. and Jones, S.G., 1983, Clay Barrier -- Leachate Interaction. National Conference on Management of Uncontrolled Hazardous Waste Sites, Washington, D.C., Nov.

Brown, K.W. and Anderson, D.C., 1982, Effects of Organic Solvents on the Permeability of Clay Soils, Report for Municipal Environmental Research Laboratory of EPA, Cincinnati, Ohio.

Day, S.R. and Daniel, D.E., 1985, Hydraulic Conductivity of Two Prototype Clay Liners, 11th Annual U.S. EPA Research Symposium, Cincinnati, Ohio, April.

Kingsbury, G.L. and Truesdale, R.S., 1985, Clay-Chemical Compatibility and Permeability Testing, 11th Annual U.S. EPA Research Symposium, Cincinnati, Ohio, April.

Mitchell, J.K., 1976, <u>Fundamentals of Soil Behavior</u>, John Wiley and Sons, New York, New York.

Roznowski, D., 1987, "Clay Liner Construction and Comparison to Synthetic", 4th Annual Solid Waste Seminar, Minnesota Pollution Control Agency, Bloomington, Minnesota, Feb.

van Zyl, D. and Smith, M.E., 1986, "Heap Leach Pad System Design", <u>Heap and Dump Leaching</u>, Vol. 3, No. 3, Sept.

Chapter 24

GOLD TAILINGS PROCESSING BY HEAP LEACHING

Gabriel E. Zárate (*)
Juan C. Guzmán (**)

(*) CIMM, Casilla 170, Santiago 10, Chile
(**) Univ. de Santiago, Depto. Ing. Metalúrgica, Santiago Chile

Abstract

In Chile there are significant amounts of gold in flotation tailings that could be recovered, being heap leaching the most promising alternative. In order to determine the feasibility of treating these tailings by this method, a detailed study is carried out using 4 representative samples.

After defining the best conditions an economic evaluation is done for a plant treating 200 tpd, sensibilizing with respect to parameters such us gold price, tailing grade and capital costs. The results show that it is technically feasible treat these gold tailings by heap leaching with gold extractions over 70 %. It is also economically feasible for gold grades higher than 0.5 g/t and gold prices over 350 US/oz.

INTRODUCTION

In Chile there are many tailings materials from former flotation plants that contain significant gold values. Most of these tailings resources are too low grade or too small to warrant the capital expenditure to construct a conventional agitated cyanide leaching circuit. However, heap leaching by cyanidation has a great potential, because it is a low capital and low operating cost process to recover gold from low grade ores and wastes (1-2). Since 1970 this process has been successfully applied to many ores in USA, Canada, and S. America. Additionally it has been applied to clayey ores since the USBM developed the agglomeration technology to improve its heap leaching performance (3-4).

On the other hand, most tailings materials are quite fine with no coarse material. The fine particles and clays, if present, prevent uniform solution flow and may blind the heap if agglomeration is not appropiately done.

The USBM has determined the best procedure for agglomerating finely ground tailings, evaluating variables such as binder addition, water addition, curing time and type of agglomerating equipment(5). In the same work it is mentioned the application in two commercial operations.

A larger commercial operation was also reported by Anaconda at its Darwin plant (6).

Considering that for chilean gold tailings, heap leaching is the most promising alternative, a detailed study was carried out using four representative samples in order to determine the feasibility of treating these tailings by heap leaching.

After defining the best conditions an economic evaluation is done for a plant treating 200 tpd, sensibilizing with respect to parameters such as gold price, tailing grade and capital cost.

EXPERIMENTAL

Tailings The as received tailings were prepared following methods standardized at CIMM. These are samples from a IV Region tailings materials, whose gold and silver content and particle size analysis are shown in Table 1 and Table 2.

Table 1 : Gold and Silver Analysis of Tailings Materials

Sample (Number)	Au (g/t)	Ag (g/t)
1	0,5	21
2	1,2	2
3	0,5	1
4	0,5	21

Table 2 : Particle Size Analysis of Tailings Materials

Sample (Number)	Particle Size (% - 200 Tyler Mesh)
1	36
2	38
3	27
4	40

Agglomeration

A rotating drum was used to agglomerate 4 kg to 50 kg batches of feed for leaching columns. The dried tailing charges were placed on the drum mixed with lime and cement. A controlled amount of liquid was slowly added from a graduated cylinder while the drum was running. The agglomerated feed was removed from the drum, placed in the leaching column and cured for a specified time at ambient conditions.

Column Leaching

Investigations to determine the best agglomeration parameters for fine tailings were conducted using 4 kg charges of material. After defining the best conditions final tests were run using 50 kg charge of material. The agglomerated material was leached in glass columns in the case of 4 kg charges and in a Plexiglass column for 50 kg charges.

The material was leached with an alkaline solution (pH 10,5) containing 0,5 or 1,0 gpl NaCN pumped on top of the charge, which was covered with a layer of coco matting.

The pregnant solution was daily analyzed throughout the leaching cycle. Calculated head analyses and gold recoveries were determinated by the amount of gold in pregnant solution and in the leached residue. All solids were analyzed by fire assay and solution samples by atomic absorption spectrophotometry.

Maximum percolation rate measurements were made after leaching was completed using a method similar to that mentioned in the work of the U.S. Bureau of Mines (5). Additionally to test the agglomerate strength an accelerated laboratory test, similar to that proposed in the work with Darwing tailings (6), was implemented. The agglomerate, in this test, was cured for 10 hours at 90°C, cooled, and dipped in water ten times.

The fines generated indicated the agglomerate strength.

Finally the degree of compactness of the bed, after finishing the leaching, was measured.

RESULTS AND DISCUSSION

Agglomeration

The conditions of agglomeration were defined using the accelerated laboratory test altogether with the results of measurement of percolation rate and degree of compactness.

Effect of Cement and Lime

A combination of lime and cement, in equal proportions, from 2 to 14 kg/ton was evaluated. The results are summarized in Figure 1 for the different tailings.

It is shown that the strength of the agglomerate increases as the amount of binder increases reaching its maximum for dosis between 11 and 15 kg/ton depending on the tailing. Generally in these tests the agglomerate kept its strength, with the exception of tailing # 2, for 8 kg/ton or more of lime and cement, generating an amount of fines of approximately 40 %.

The reasons for the different behaviour of tailing # 2 are now under investigation.

The results were compared with that obtained from the measurement of percolation rate and degree of compactness after leaching. Figure 2 and 3 show for tailing # 4 that increasing the amount of lime and cement up to 8 kg/ton markedly improved percolation rates and degree of compactness. Similar results were obtained for tailings # 1 and # 3, while for tailing # 2 the minimum amount of lime and cement was 11,5 kg/ton. These amounts of binder were added during subsequent experiments.

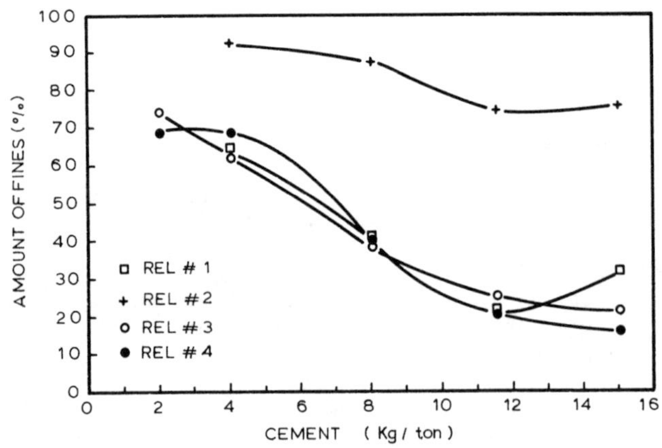

FIG. 1 : Effect of Cement and Lime Addition. Cement/Lime Ratio = 1. Accelerated Test

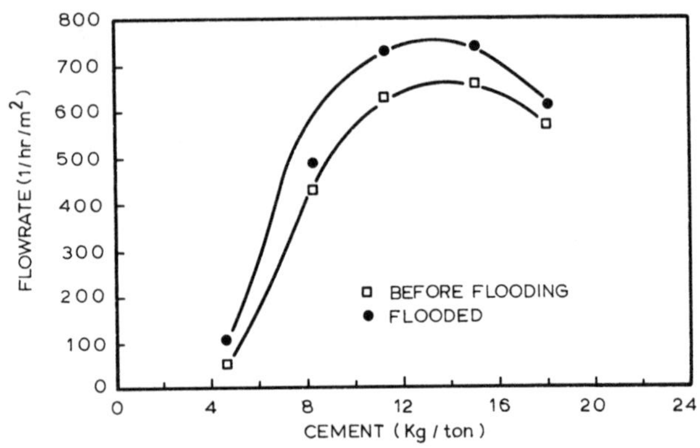

FIG. 2 : Effect of Cement and Lime on Flowrate. Tailing # 4 Moisture = 17,3 %, Cement/Lime Ratio = 1, Curing Time = 72 hrs.

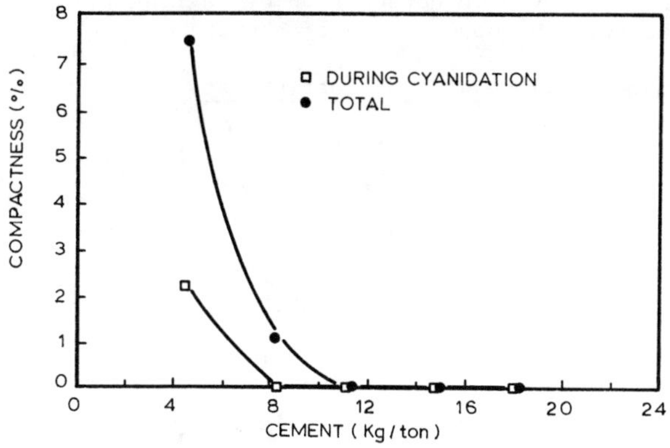

FIG. 3 : Effect of Lime and Cement on Compactness. Tailing # 4 Moisture = 17,3 %, Cement/Lime Ratio = 1 Curing Time = 72 hrs.

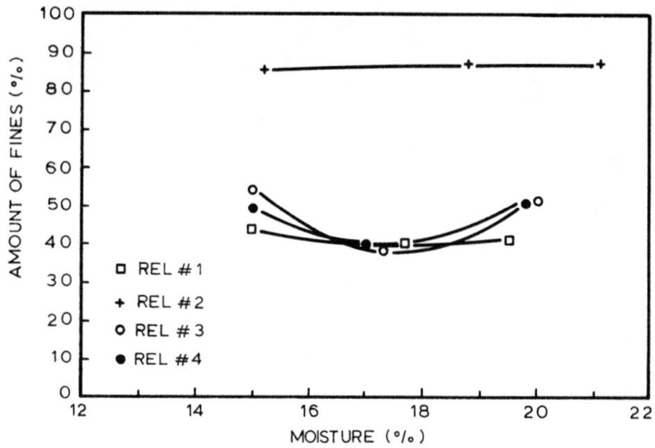

FIG. 4 : Effect of Moisture. Cement/Lime Ratio = 1. Cement = 8 kg/ton Accelerated Test

In these measurements it was also shown that cement alone was not an adequate binder for agglomerating tailings unless 18 or more kg/ton was used.

Effect of Moisture

Experiments were performed on charges mixed with 8 kg lime and 8 kg cement per ton of feed and cured for 72 hr. The amount of water ranged from 14 to 20 %. The results, summarized in Figure 4, show that the generation of fines is minimum for moisture of about 17 %, with the exception of tailing # 2.

On the other hand, the results of percolation rate and degree of compactness measurements show for tailing # 4 a maximum rate for 17 % moisture and minimum degree of compactness for even lower moistures (see Figures 5 and 6). Similar results were obtained for tailing # 1 and # 3, while for tailing # 2 the maximum rate was obtained at about 19 % moisture.

Effect of Curing Time

Only flowrate and degree of compactness measurements were done in this case. Four kg charges were mixed with 8 kg lime and 8 kg cement per ton of feed, wetted with 17 or 19 % water and agglomerated. Curing time ranged from 0 to 72 hrs.

The results, presented in Figures 7 and 8 for tailing # 4, show that curing time was important up to 24 hrs, with no further improvement after that time. Similar results were obtained for the other tailings.

Leaching

Preliminary bottle tests showed that all the samples were amenable to cyanidation. A summary of results is presented in Table 3.

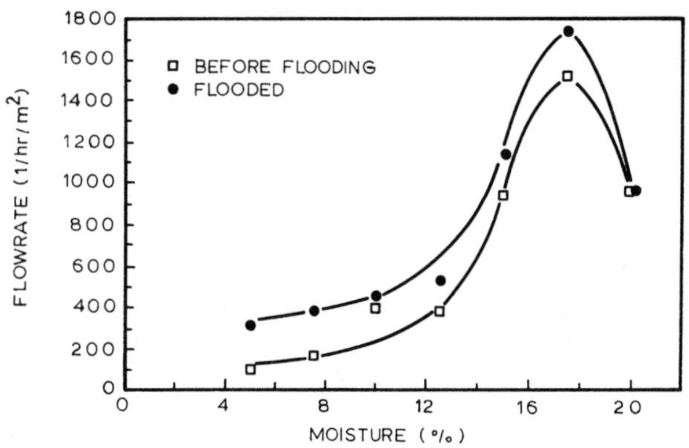

FIG. 5 : Effect of Moisture in Flowrate. Tailing # 4, Cement = 8 kg/ton Cement/Lime Ratio = 1, Curing Time = 72 hrs.

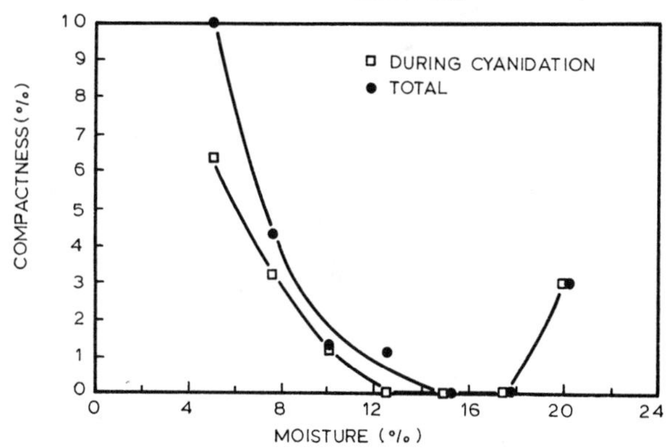

FIG. 6 : Effect of Moisture an Compactness. Tailing # 4, Cement = 8 kg/ton, Cement/Lime Ratio = 1, Curing Time = 72 hrs.

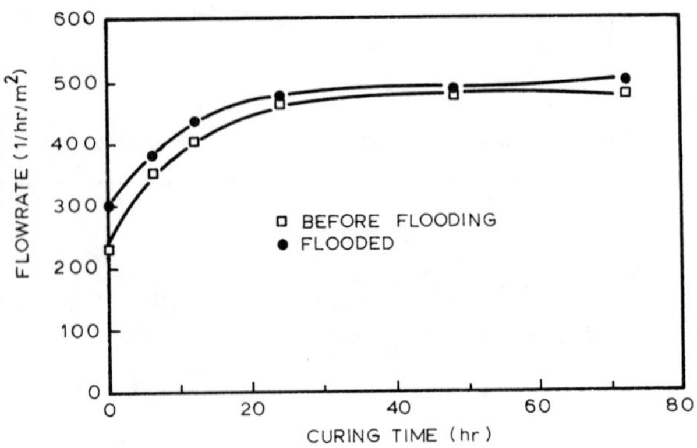

FIG. 7 : Effect of Curing Time in Flowrate. Tailing # 4, Moisture = 17 %, Cement/Lime Ratio = 1, Cement = 8 kg/ton

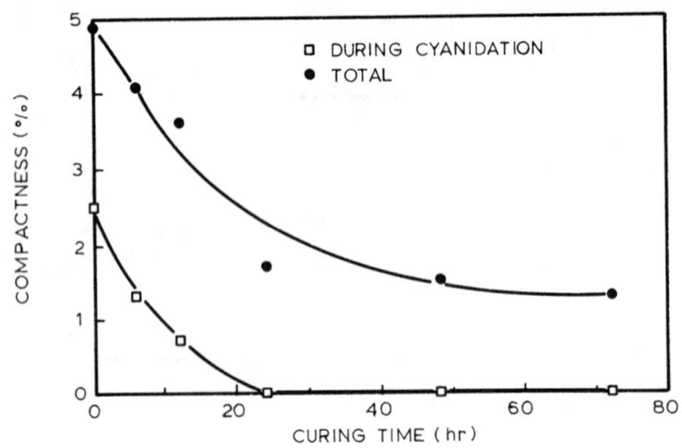

FIG. 8 : Effect of Curing Time in Compactness. Tailing # 4 Moisture = 17 %, Cement/Lime Ratio = 1, Cement = 8 kg/ton,

The high cyanide consumption of tailing # 2 is due to high copper content (about 1 %).

Table 3 : Results of Bottle Tests

Tailings	Extraction		Consumption
(#)	Au	Ag	NaCN
	(%)	(%)	(kg/ton)
1	91.3	65.7	0.4
2	92,2	60.0	4.3
3	89.9	21.0	0.2
4	84.0	59.4	0.4

In the case of small column leaching tests a summary of the best conditions determined experimentally and the gold extractions is presented in Table 4. The latter ones are within 10 % of those obtained by agitated-bottle leaching tests.

Table 4 : Summary of Experimental Results for Column Leaching

	Tailings			
	# 1	# 2	# 3	# 4
Agglomeration				
Lime : Cement (kg/ton)	8.0	11.5	8.0	8.0
Moisture (%)	17.7	18.8	17.3	17.3
Curing Time (hrs)	72.0	72.0	72.0	72.0
Leaching				
Leaching period (days)	5.0	5.0	5.0	5.0
Gold Extraction (%)	82.7	76.0	82.7	87.4

Finally in order to check the previous results two final tests were run with tailings # 1 and # 2 using 50 kg-charges. Agglomeration and leaching conditions were defined according to previous results. Conditions and results are shown in Table 5.

Results from Table 5 confirmed those obtained at small scale, in terms of agglomerated strength and stability and gold recoveries. And all the results demonstrated that these tailings can be treated by an agglomeration heap leaching process.

Table 5 : Final Tests with Tailings # 1 and # 2 50 kg charges

	Tailings # 1	Tailings # 2
Agglomeration		
Lime : Cement (kg/ton)	8.0	11.5
Moisture (%)	17.7	18.8
Curing Time (hrs)	72.0	72.0
Leaching		
Leaching Time (days)	7.0	7.0
Gold Extraction (%)	80.7	75.7

Table 5 : Cont.

	Tailings # 1	Tailings # 2
Silver Extraction (%)	68.2	N.D.
NaCN Consumption (kg/t)	0.35	1.0
Percolation rate (lt/hr x m^2)	1200	876
Compactness (%)	0	0

N.D. : No data

Economical Evaluation

A preliminary economical evaluation, for the following base case, was performed :

Capacity (tpd) : 200
Grade (g/t) : 0.5
Gold price (US$/oz) : 400

Agglomeration

Lime : cement (kg/ton) : 8
Moisture (%) : 17
Curing time (hrs) : 72

Leaching

Cycle (days) : 14
NaCN consumption (kg/ton) : 0.5

Recovery (by zinc precipitation)

Gold (%) : 75

Afterwards a sensitivity analysis considering different gold prices, gold recoveries, and investment was done.

The results showed that for the base case the investment is recovered after 3 years, the IRR is 52 % and the NPV is US$ 201.000. Concerning the sensitivity analysis a summary of results is shown in Table 6. From these ones it can be seen that depending on the case, the project will be economically feasible for gold grades higher than 0.5 g/t and gold prices over US$ 350/oz. These two are most sensible variables and the profitability of the project increased markedly with an increasement in both variables, specially gold grade.

Table 6 : Summary of Results of Sensitivity Analysis

	IRR (%)	NPV (US$)
Capital Investment		
Case a) Base	52	201.000
b) 70 % Base Case	158	302.000
Gold Price		
350*	5	-39.000
350**	24	65.000
400	52	201.000
450	143	477.000
Gold Grade		
0.5	52	201.000
0.75	455	1.140.000
1.0	878	2.076.000

* Investment case a)
** Investment case b)

CONCLUSIONS

Agglomeration pretreatment makes possible heap leaching of chilean flotation tailings. In this study it has been shown that it is technically feasible to recover more than 70 % of gold from these materials when agglomeration is carried out under appropiate conditions.

This process is also economically feasible for a plant treating 200 tpd with 75 % recovery when gold grades are higher than 0.5 g/t and gold prices over 350 US$/oz.

ACKNOWLEDGEMENTS

The authors wish to express their appreciation to the Executive Director of CIMM for their permission to publish this paper.

REFERENCES

1. Guzmán, J.C., 1985, "Tratamiento de Relaves por Cianuración en Pilas". Metallurgical Engineer Thesis, Universidad de Santiago, 166 pp.
2. Heinen, H.J.; Peterson D.G. and Lindstrom R.E. 1978, "Processing Gold Ores Using Heap Leach-Carbon Adsorption Methods". Information Circular IC 8770. U.S. Bureau of Mines
3. Heinen, H.J., Mc Clelland, G.E., 1979. "Enhancing Percolation Rates in Heap Leaching of Gold-Silver Ores" Report of Investigation RI8388, U.S. Bureau of Mines.
4. Mc Clelland, G.E., Eisele, J.A, 1982, "Improvements in Heap Leaching to Recover Silver and Gold from Low-Grade Resources". Report of Investigation RI8612, U.S. Bureau of Mines.
5. Mc Clelland, G.E. et al, 1985, "Agglomeration and Heap Leaching of Finely Ground Precious-Metal-Bearing Tailings". Information Circular IC 9034, U.S. Bureau of Mines.
6. Milligan, D.A. and Engelhardt, P.R., 1983, "Agglomerated Heap Leaching Anaconda's Darwin Silver Recovery Project", Proceedings SME-AIME, Au & Ag Heap and Dump Leaching Practice, pp. 29-39.

Chapter 25

DEVELOPMENT OF CYANIDE LEACHING PROCESSS FOR
SMALL HIGH GRADE GOLD AND SILVER DEPOSITS

by

Mr.David Mosch
Mosch Mining Company
Idaho springs Colorado

Mr.Michael Harris
New Mexico Bureau of Mines & Mineral Resources
Socorro New Mexico

ABSTRACT

The numerous small high grade gold and silver deposits of the Front Range Mineral Belt have been uneconomical because of the high costs of mill construction, transportation, and smeltering. Processing of these ores must be simple, cheap, with satisfactory recoveries, and be environmentally acceptable. Vat leach with direct electrolytic recovery satisfies these parameters. A battery of testing was conducted utilizing bottle and bucket tests. A five ton vat leach with direct electrolytic recovery gave excellent results while a five ton heap leach proved impractical due to channeling.

INTRODUCTION

The Front Range Mineral Belt of Colorado has numerous small high grade gold and silver deposits that have produced more than 855 million grams (27.5 million ounces) of gold with more than 3235 million grams (104 million ounces) of silver. Most of these mines ceased production during World War II. Post war efforts to re-open the mining districts have been largely unsuccessful due to the lack of ore processing capability.

The existing processing option is conventional toll milling and smelting at the nearest facility. At present milling costs are \$110 per tonne (\$100 per ton) with smelting costs at more than \$385.81 per tonne (\$350 per ton) for small lots. The nearest smelter to the Front Range is more than 1,207 kilometers (750 miles distance.) Using conventional milling and smelting, processing costs for these small high grade ore bodies is \$450 per ounce of gold, and this does not include mining costs.

Under these conditions it is impossible to profitably mine these deposits. This research program was conducted to investigate alternative processing techniques.

THE PROBLEM

A statistical evaluation, based on engineering judgement and past production records, resulted in the generation of a representative ore body for the Clear Creek County mining districts. The representative ore body has an average grade of 22gm/tonne (0.64 ounces per ton) gold and 256 gm/tonne (7.45 ounces per ton) silver with a production rate of 4.2 tons per day over 6 years, for a total of 8500 tons. These ores contain a sulfide assemblage of quartz, pyrite, galena, chalcopyrite, tennantite-tetrahedrite, and sphalerite. The gold ocurs generally as microcrystalline free gold occupying microcracks and ocassionally with telluride minerals. Silver in these ores occurs in the lattice of the tennantite-tetrahedrite or galena, and hence silver recovery is very difficult.

EXPERIMENTAL RESULTS

Bottle Test

Initial bottle testing of representative ore types gave excellent gold recoveries as shown in Table 1. These results show high reagent consumption due to high copper sulfide content and the generally acidic nature of these ores.

Table 1
BOTTLE TEST RESULTS

ORE TYPE	RECOVERY	CONSUMPTION LB/TON CN	CaO
Base Metal Sulfide - Pb, Cu, Zn	94.1%	25	1.00
Base Metal Sulfide - Cu, pyrite	82.5%	11	1.04
Sulfide - pyrite, quartz	96.2%	5	1.71
Oxidized Gold Ore - high grade	79.0%	2	0.64
Oxidized Gold Ore - low grade	93.7%	2	0.52

Bucket Test

Following the bottle testing, a series of bucket tests were conducted. Initial tests verified that the ores were amenable to perculation type leaching with gold recoveries in the 70% range. Further bucket tests were performed to investigate the relation of gold recovery, reagent consumption, and leaching time to particle size.

Results show that: a) gold recoveries increase with smaller particle size, b) reagent consumption increases with smaller particle size, and c) leach time increases with smaller particle size. Another interesting result was that very small volumes of solution could be used in the leaching process to produce a high grade pregnant solution (more than 31 gm/tonne or one ounce per ton of solution).

Electrowinning Test

Since a pregnant solution with a high gold content can be generated directly from leaching ore, direct electrowinning was investigated as a possible method to extract gold from the solutions. The high copper content favors the electrowinning process because the copper is removed from the solution with the gold thereby causing regeneration of cyanide.

The electrowinning experiment involved a small Zadra cell design with a typical high gold-copper content solution. Electrical current of 4.5 volts was applied to the cell over a ten hour period. Gold recovery relative to time is shown in Figure 1. The gold content of the cathode reached 9.4%, with a copper content of 60.8%. The current efficiency of the process is around 12%

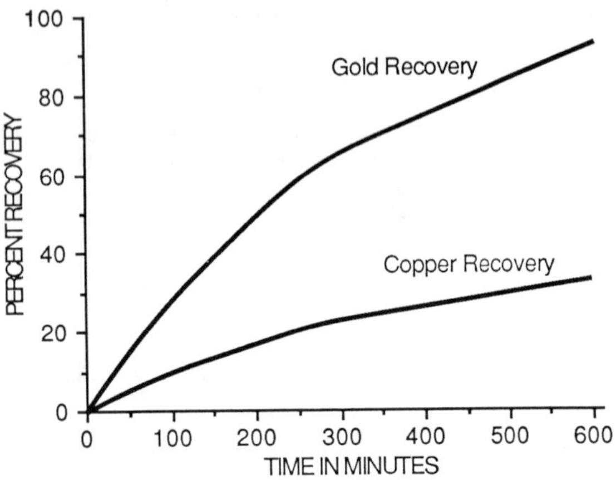

FIGURE .1 ELECTROWINNING TEST RESULTS

FIELD TESTING

Two five ton experiments, a heap leach and a vat leach, were conducted using representative ore samples.

Heap Leach

The heap leach was built on a small standard design pad. The ore was crushed to minus 10 mesh and placed on the pad with care to prevent channeling. The solution was applied to the heap by a standard ponding method, and cycled between the pregnant pond and the heap until loaded with gold. At this time the electrowinning cell was introduced into the circuit. The system was operated for 3.5 months and obtained a 74% (see Figure 2) gold recovery. Upon completion of the test the heap was systematically sampled. Some channeling did occur and it was found that high recoveries were obtained from the upper portions of the heap. The steel wool cathodes were loaded to a gold content of 12% near the beginning of the test, and less than 1% near the end. The copper did not significantly accumulate on the cathode until later in the test. Electrowinning efficiency was between 2 and 10 percent.

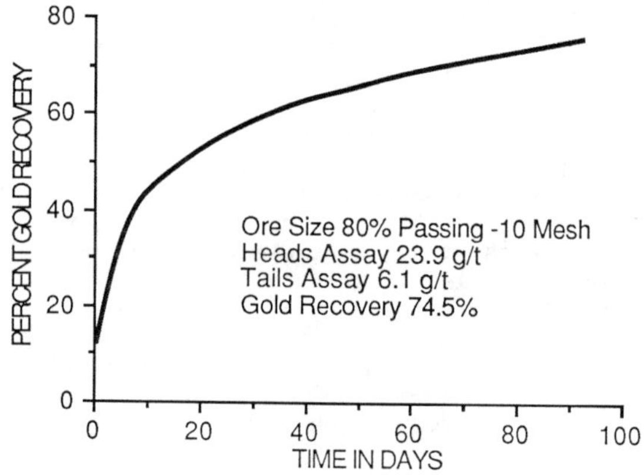

FIGURE 2. HEAP LEACH TEST RESULTS

Vat Leach

The vat leach was constructed with a filter bottom, and loaded with 5 tons of minus 10 mesh ore. Leaching began by rinsing the vat with caustic solution to neutralize any natural acids contained within the ore. The vat was then flooded with a solution containing 4.5 kg NaCN and 0.5 kg NaOH per tonne. The cyanide solution was then circulated downward through the ore and filter into a reservoir. From the reservoir the solution was pumped back to the top of the vat and sprayed over the ore, in an effort to introduce oxygen into the system. The concentration of gold in solution after 24 hours was found to be 30.3 gm/tonne. For the remainder of the test, the solution was cycled from the vat to a small Zadra cell system. Plated steel wool from the cell was removed daily and analyzed for its gold and copper content. As time passed the gold content decreased while the copper content of the plated material increased. Forty percent of the recoverable gold had been removed from the system in less than four days. The system was allowed to operate for a total of 13 days at which time the gold extraction rate was very low with a total overall gold recovery of 83% as shown in Figure 3.

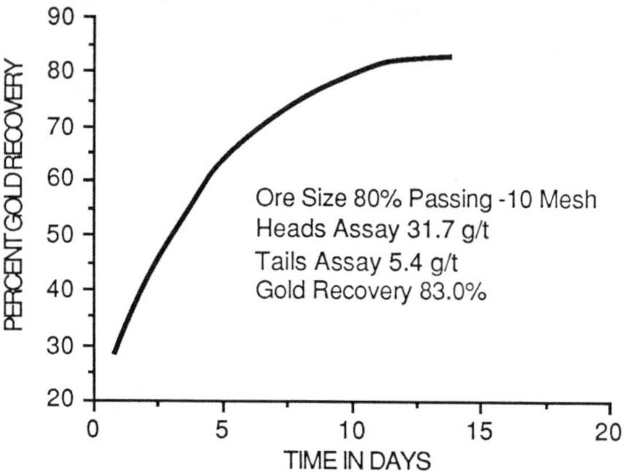

FIGURE 3. VAT LEACH TEST RESULTS

PROPOSED PROCESS

Based on the previous experimental results, the proposed method of processing small high grade ore deposits is a vat leach of crushed ore, and extraction of metals from the pregnant solution by electrowinning. Ore is crushed to a size suitable for maximum effective recovery at minimum leach time. The optimum size range must be determined for each deposit type. Then the ore is placed in a steel or concrete vat with a geotextile filter bottom, designed so a small front-end loader may load or unload the vat without damage to the filter. This can be accomplished by placing a steel grating over and under the filtering material. A small volume of concentrated cyanide solution is introduced into the vat and allowed to saturate the crushed ore. Initially, the solution is pumped from the bottom of the vat and sprinkled over the top to maintain the necessary oxygen content. Later, after the gold content has increased, the solution is pumped through a small electrowinning cell. When the maximum effective extraction of gold from the ore has occurred, the solution is drained and the ore is rinsed. Waste rock from the vat is removed with the loader and is disposed in accordance with local environmental regulations.

ECONOMICS

An economic study was performed to determine how a vat leaching process with direct electrowinning would compare with the conventional method of processing ore through a mill and smelter. The results of this comparison is shown in Figure 4.

The values in Figure 4 are based on a small mine with an 8000 tonne (8818 ton) reserve and an average grade of 22 gm/tonne (0.64 oz/ton). This hypothetical mine would have a production of 9.08 tonnes/day(10 tons/day). An examination of the data reveals that for a small mine, the vat leach with direct electrowinning is an economical method of ore processing, whereas the conventional system of marketing the ore is very costly. The low overall cost of the vat leach with direct electrowinning is mainly attributed to its low capital cost. The low capital cost of the electrowinning vat leach more than compensates for its higher operating costs.

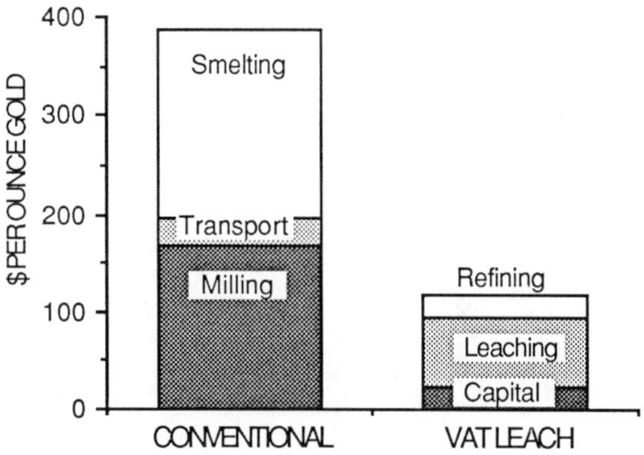

FIGURE 4. PROCESSING COST COMPARISON

CONCLUSION

A small simple vat leach with direct electrowinning to recover gold and other metals from solution is an economical option for certain very small high grade gold deposits. Application of this type of ore processing, along with good engineering and management controls, could make production from mines in the Front Range Mineral Belt economically viable.

Chapter 26

CLAY PROBLEMS ENCOUNTERED IN GOLD HEAP LEACHING

Jeffrey P. Kurtz, Phoebe L. Hauff, Paul D. Chamberlin

Geology Department, University of Colorado at Denver, CIRES/CSES, University of Colorado at Boulder, Chamberlin and Associates, Littleton, Colorado

ABSTRACT

Clay minerals can be a source of problems in heap leaching operations. The ore for many heap leach operations is commonly acid-sulfate alteration zone rocks which may be found in hot springs deposits as well as the supergene zone of other near-surface gold deposits. Clays can be important constituents of these ores. Swelling clays, such as smectite, can come loose from the agglomerated ore, migrate within the heap, concentrate in pore spaces and often plug up flow channels in the heap.

Cement is often added to the ore as a binder during the agglomeration process. Lime is added for pH control. Their presence, combined with the sodium cyanide commonly used to leach, can raise the pH of the leach solution to 10 or even 12. This very alkaline solution can cause dissolution of some mineral

Some clays, such as kaolinite, are not very stable in high pH solutions. They can be dissolved and reprecipitated as other clay species if sufficient silica and the appropriate cations are available in solution.

Silica gels and amorphous materials can also be created from the dissolved kaolinite and other minerals and act as retardants to the flow of leach solutions in the heap.

Therefore, before a heap leach operation is built, the operator should evaluate the clays in the ore and test the agglomeration procedure. It is important to know not only the mineralogy and the chemistry of the leach environment but how the minerals and solutions interact and what the products of this interaction are.

A hypothesis is presented in this paper that expanding smectite-bearing clays can be generated within the heap. How and why this happens will be explained.

OVERVIEW OF HEAP LEACHING PROCESSES

Ores that have a lot of fines are usually agglomerated before placing them in a heap. Agglomeration is the process whereby fines in the ore are caused to adhere to coarse ore fragments or to aggregate into coarser agglomerates. The objective of agglomeration is to uniformly distribute the fines and clays throughout the heap and to prevent them from migrating within the

Figure 1a is an idealistic portrayal of agglomerated material within the heap where the particles actually touch and support each other while leaving flow channels open for the leach solutions to percolate around each agglomerate.

The flow of leach solution is controlled so that fines are not washed off the surface of the coarse particles to which they adhere. Some of the solution may flow through the more porous agglomerates. Solubilized gold within the agglomerates works its way to the surface of the agglomerate, often by diffusion. Here it encounters the moving film of leach solution on the surface of the agglomerate and is carried out of the heap.

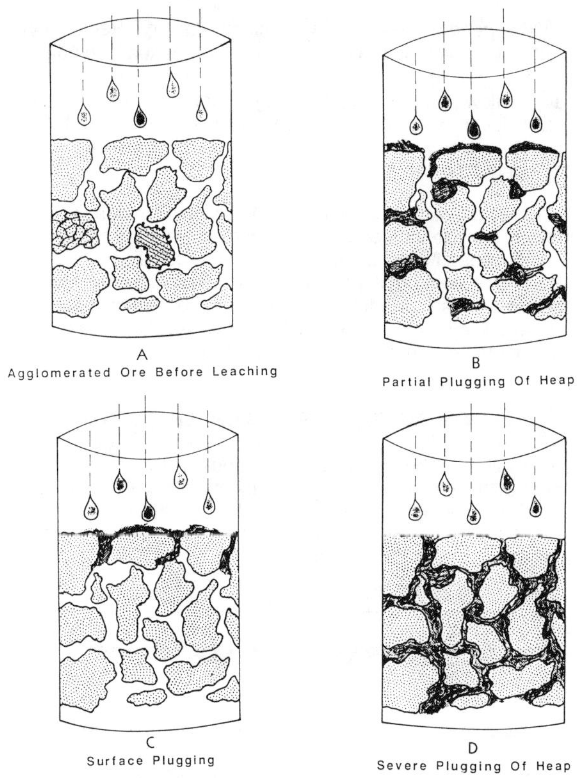

Figure 1.--Idealized Cross Sections of a Leach Heap.

Open flow channels allow solution to flow quickly around all the particles within the heap which means that the solubilized gold can be flushed quickly and completely from the heap.

Figure 1b is considered representative of what happens in most heaps. Some of the fines have broken loose from the agglomerates and have migrated downward. The quantity is small and the distance traveled not significant so general distribution of fines is still uniform. Some of the flow channels have been plugged by this migration but most are open and solution percolation rate is still relatively uniform within the heap. When agglomeration is done with water only, i.e. no binders or insufficient binder, the fines may migrate.

When the ore has a large portion of fines, the example shown in Figure 1c is created. The surface agglomerates have been degraded by the pounding of spray droplets and have plugged the surface of the heap. This causes ponding and potential runoff down the side slopes of the heap which erodes the heap and carries excessive suspended fines into the pregnant solution. In addition, the flow of the leach solution through the heap may be sufficiently impeded that extraction rate is slowed and the overall recovery may be reduced because not all agglomerates are contacted with enough leaching solution.

When the flow channels become plugged with fines and clays, the heap begins to look like Figure 1d. This is usually caused by inadequate agglomeration when an ore contains a large proportion of fines. When much of the material composing the agglomerates is fines, the agglomerate fails in compression from lack of strength and collapses in the heap filling the surrounding flow channels.

Case History

The ore from an operation contained 50-60% kaolinite with less than 10% illite. Metallurgical column testing indicated that agglomeration with about 3 pounds of cement per ton of ore would be enough to sufficiently bind the ore and give good percolation. These tests were not done on the actual ore but on material thought to be comparable.

Actual field experience, however, was not satisfactory. Percolation rate, predicted to be 0.004 to 0.005 gpm/ft^2, was actually 0.002 gpm/ft^2. This reduced the rate of extraction considerably and caused cash flow problems. Further agglomeration testing, this time on ore samples from the crushing plant, indicated that 20 pounds of cement per ton of ore was required for adequate agglomeration instead of only 3 pounds per ton.

Before this modification in agglomeration was employed, however, damage was done to the heap. It is probable that phase changes in the minerals caused more fines to be created with subsequent migration and plugging of the flow channels and collapsing of weak agglomerates. In this case the problem was solved with a much better understanding of the clays in the ore, better agglomeration test procedures, and much higher binder costs.

MINERALOGY OF A HEAP LEACH ENVIRONMENT

This paper will describe the mineralogy of the ore, the fresh agglomerate, the leached agglomerate, and interstitial material as defined in Table I as well as the effects of leach solutions on these samples. X-ray diffraction (XRD) and scanning electron microscopy (SEM) were used to analyze the samples.

TABLE I.-- SAMPLE CATEGORIES

ORE	Rock Material - Untreated
FRESH AGGLOMERATE	Ore + Cement + Lime
LEACHED AGGLOM	Agglomerate +NaCN Soln
INTERSTITIAL	Heap Pore Filling Material

The ore itself is untreated rock. The first step in the heap leaching process is crushing and agglomeration. Crushing reduces the ore to a manageable size without, hopefully, creating too many fines. A binder is added, commonly cement, to consolidate the ore into cohesive agglomerates. Lime is also added to raise the pH to greater than 10.0. A pH of greater than10.0 is required to prevent the generation of HCN (hydrogen cyanide gas). The agglomerates are stacked on the heap and a weak sodium cyanide solution, with an average concentration of 0.05% NaCN, is sprayed over them. The interstitial material is fines that have migrated and filled the pore spaces in the heap or weak agglomerates that have collapsed under compression load.

Table II shows the primary minerals, their formulae and approximate amounts present in the ore. Traces of hematite, possibly pyrite, jarosite (iron-potassium sulfate), rutile and probably some volcanic glass are also present.

TABLE II.-- MINERALOGY OF THE ORE

Mineral	%	Formula
KAOLINITE	~50%	$Al_2Si_2O_5(OH)_4$
QUARTZ	~30%	SiO_2
ALUNITE	~10%	$(K,Na)Al_3(OH)_6(SO_4)_2$
ILLITE	~10%	$K_{0.7}Al_2(Si,Al)_4O_{10}(OH)_2$

Table III summarizes the mineralogy of the four types of samples. The minerals appear to be the same for the ore, fresh agglomerate, and leached agglomerate samples. There is some minor carbonate probably in the form of calcite, in the fresh agglomerate and leached agglomerate samples. This is a function of the cement or lime additives in the agglomeration process.

TABLE III.-- GROSS MINERALOGY

	ORE	FRESH AGGLOM	LEACHED AGGLOM	INTERST. MATERIAL
KAOLINITE	Major	Major	Major	Major
QUARTZ	Major	Major	Major	(?)
ALUNITE	minor	minor	minor	(?)
ILLITE	minor	minor	minor	minor
Random I/S	trace	trace	trace	none
Regular I/S	none	none	minor	minor
CARBONATE	none	trace	trace	none
HEMATITE	trace	trace	trace	(?)
GIBBSITE	none	none	none	trace
AMORPHOUS SILICA	none (?)	none (?)	trace	minor

X-Ray Diffraction Identification Techniques

X-ray diffraction(XRD) was the principal identification technique used to characterize the samples in this study. XRD analysis of the fresh agglomerate, which is also representative of the ore and leached agglomerate samples, shows that the peaks for kaolinite, quartz, and alunite are sharp and well defined with slender profiles. This indicates good crystallinity. The illitic peaks are broad because these samples contain both discrete illite and another clay, randomly stacked illite/smectite(I/S).

The interstitial material, which can be considered pore filling fines, has different minerals. It is hypothesized that the interstitial material contains fines that have broken loose from the agglomerates and migrated to the pores where they were trapped. These fines could be unaltered ore minerals or they could be new species that were created on the surfaces of the agglomerates. XRD shows a much higher concentration of clay phases in the interstitial samples. The expanding nature of smectite bearing clays could disrupt some of the agglomerate surfaces, if not the entire agglomerate, thus setting fine particles in motion within the heap.

This hypothesis is supported by observations from the diffraction patterns. Quartz is no longer present as a crystalline phase. It is now in solution at the high pH within the heap. There is less kaolinite and it has lost some crystal structure integrity. Discrete illite is also present. New phases appear in the interstitial material. Some are not well enough characterized to identify. There is a poorly crystalline, almost amorphous gibbsite ($Al(OH)_3$) present. There are amorphous silica and alumina present. The most important new phase, however, is a regular interstratification of illite and smectite. This clay is thought to be a potential culprit in plugging up the flow channels in the heap.

Clay mineral identification techniques are quite involved. The methods used are discussed in detail elsewhere(Starkey et al, 1983, Hauff et al, 1984). One of the most definitive tests used to identify clay mineral species is solvation with ethylene glycol. The peak positions of swelling clays shift on the diffraction pattern when they have been solvated; those for discrete illite do not shift. Mixed layer clays react relative to their composition. Those containing a swelling clay will expand.

Figure 2 summarizes the reactions to glycolation of the illite, random illite/smectite, and regular illite/smectite from the four different samples in this study. The ore, fresh agglomerate and leached agglomerate all look similar. The interstitial material, however, is quite different. The 10 angstrom peak for discrete illite is present, but also present is a peak at 11.94 angstroms, and a smaller peak at about 24 angstroms,both belonging to the regular, mixed-layer illite/smectite.

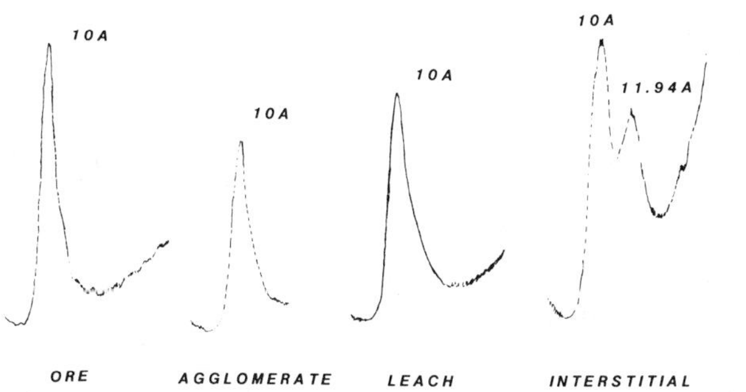

Figure 2.-- Comparison of Glycolated Samples.

A mixed-layer illite/smectite, is composed of layers of smectite and layers of illite. These layers can be arranged in a predictable, "regular" sequence in the structure or a random one. Each sequence produces a different x-ray pattern. Although composed of two minerals, an I/S clay is considered a discrete species as it has unique characteristics. An illite/smectite can contain different percentages of illite layers and smectite layers. These percentages can vary from the pure end members of illite and smectite through to a 50:50 arrangement which is known as rectorite. The mixed-layer clay in the interstitial material of this study contains 84% illite layers and 16% smectite layers. It is a regular interstratification , that is for every 5.25 layers of illite there is one layer of smectite present in a regular sequence. This becomes important to know as the more swelling layers in these clays, the more problems they can cause.

Illite/smectite is very fine grained, usually less than a micron in size, which means it can easily disaggregate from the agglomerate, migrate within the heap and plug up pores.

Scanning Electron Micrographs

One of the most effective ways to look at fine grained minerals, especially clays, is through the use of a scanning electron microscope (SEM). Chemical elements present can also be analyzed using an Energy Dispersive System (EDS).

Figure 3 is of kaolinite from the untreated ore. It is beautifully crystalline and very fine grained. The scale on the lower left side of the picture is 1 micron. This kaolinite is also authigenic, that is it was formed directly from solution. In this case it was probably precipitated from the hydrothermal ore solutions and is most likely the higher temperature pseudomorph, dickite.

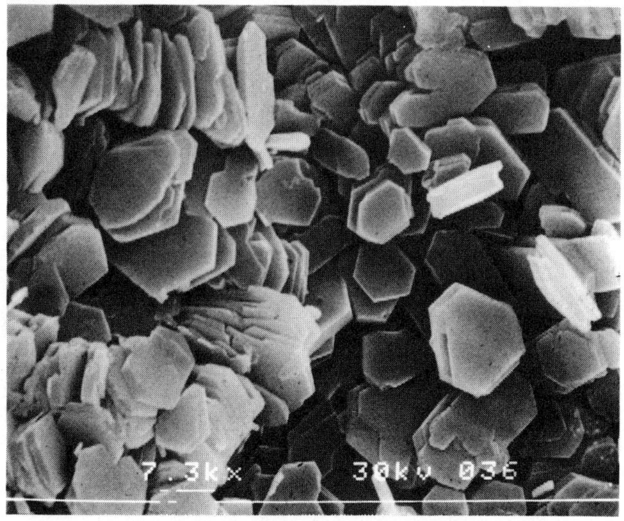

Figure 3.--SEM of kaolinite from untreated ore.

During crushing and agglomeration, this striking morphology of the kaolinite is degraded. The freshly agglomerated ore in Figure 4 shows the effects of agglomeration on the clay. It begins to look tattered. The crystal form is still recognizable but the crystals are broken with ragged edges and the stacking sequences of the books are disturbed. Leaching with sodium cyanide solution causes further deterioration of the kaolinite crystals.

Figure 4.--SEM of kaolinite from agglomerated ore.

TRANSFORMATION HYPOTHESIS

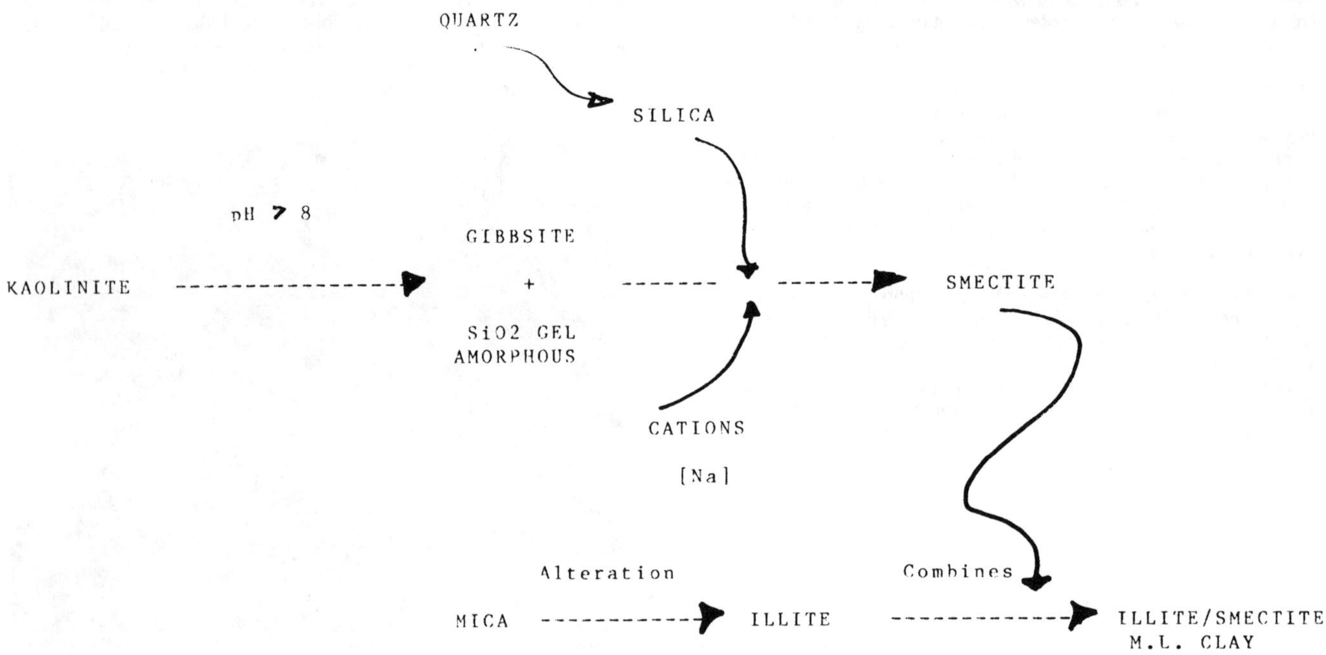

Figure 5.-- Chemical Transformations in Heap.

The observed changes to the clays are hypothesized to have occurred because of the chemical environment within the heap. This hypothesis is summarized in Figure 5.

The end product of the chemical transformations within the heap is considered to be an interstratification of illite and smectite. To achieve this interstratification, smectite layers must be formed. To form those layers, octahedral sheets requiring aluminum and hydroxides must be built along with tetrahedral sheets of silicon and oxygen and both of these components bonded together with interlayer cations.

Therefore, sources for aluminum, silicon, hydroxides, and interlayer cations must be found plus a host for these smectite layers once they are formed.

Kaolinite and quartz from the ore are not stable in a highly alkaline solution such as the NaCN leach solution(Drever, 1983). Kaolinite starts to deteriorate at a pH of 8 and becomes very unstable when the pH is greater than 10.0 (Drever, 1983).

Kaolinite will disassociate to an unstable gibbsite ($Al(OH)_3$) phase and silica. Velde (1977) suggests that silica gel is a logical product of this disassociation. Scanning electron micrographs of the interstitial material show an amorphous phase occurring in the interstitial areas of the heap. This phase has the appearance of silica gel. Gels are considered to be excellent incubators of clays.

At the high pH's found in the heap(~12), quartz goes easily into solution providing more available silica.

Alumina from the destruction of kaolinite is quite soluble but not very mobile. As silica activity increases, alumina will not stay in solution for very long but tends to form aluminosilicate minerals.

To form the smectite layers, the required silica comes from the kaolinite and dissolved quartz. The interlayer cations, which are hypothesized to be sodium and calcium, come from the sodium cyanide leach solution, the lime, and the cement. Free sodium and calcium ions bond with water molecules and enter the interlayer sites of the smectite layers creating the swelling characteristics of this clay.

It is also known from the x-ray diffraction patterns that random mixed layering of illite and smectite already exists. There are more smectite layers present in the illite/smectite regularly interstratified clay(after leaching) than in the random clay (before leaching) associated with the discrete illite.

The exact mechanism of the creation of the mixed-layer clays is not known. It theoretically does require less energy to nucleate the I/S clay on existent phases than for a discrete smectite to form.

The implication is that the longer the exposure to the leaching solutions, the more smectite layers will be formed.

Scanning Electron Micrographs of Transformations

SEM photographs further document the hypothesis. Figure 4 showed evidence of the chemical and mechanical degradation of the kaolinite from the agglomeration process.

Figure 6, kaolinite from the interstitial sample, shows that stacked crystals are blurred with single crystal outlines merged into each other. Many of these crystals are nestled in what are probably amorphous silica and alumina wrappings. These crystals have been attacked by the leach solution.

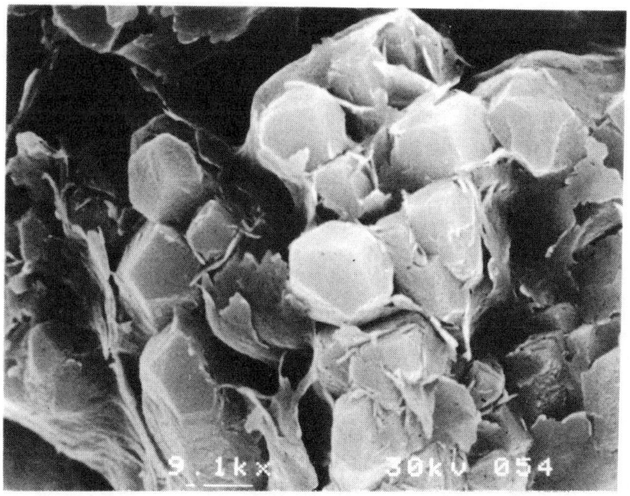

Figure 6.-- Kaolinite from the Interstitial Sample.

Figure 7 shows further development of amorphous or near amorphous material with wispy bridging structures beginning to appear.

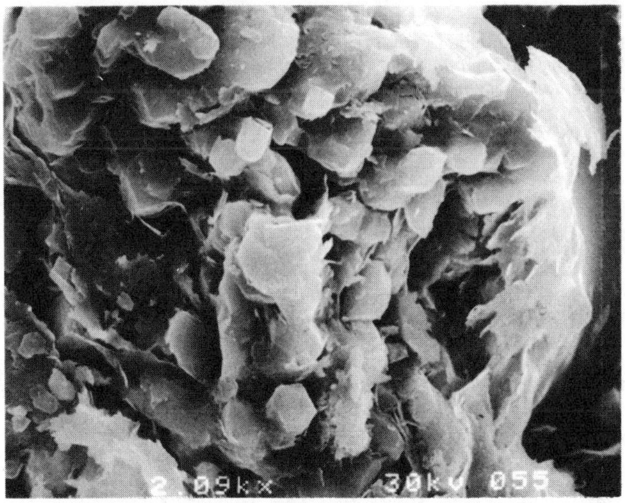

Figure 7.-- Development of Amorphous Material.

Figure 8 is a close up where one of the bridges can be seen in more detail as it forms a link between the kaolinite crystal and the surrounding amorphous material. This is direct evidence that kaolinite has been dissolved and another phase immediately precipitated.

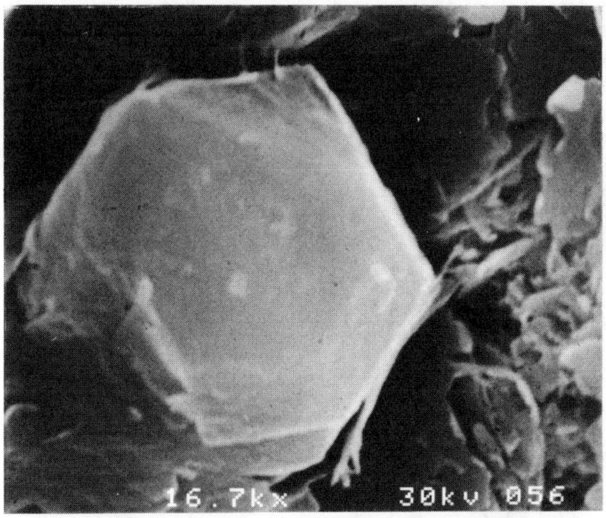

Figure 8.-- Solution Bridge from/to Kaolinite Crystal.

Figure 9 is a side view of a small book of crystals enfolded by their amorphous blanket, probably gibbsite with silica. Dissolved alumina is not highly mobile and would reprecipitate almost immediately as another phase because of the high silica activity in the solution.

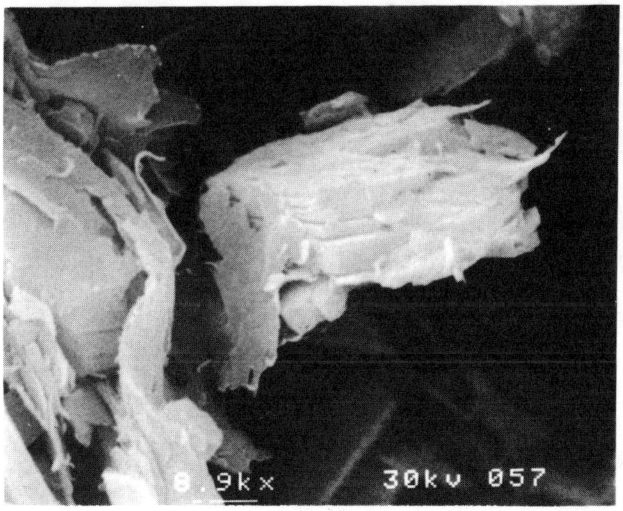

Figure 9.-- Kaolinite in an Amorphous Blanket.

These pictures show there is a progression from the sharp, euhedral crystals of the untreated ore to the mechanical abrading , to the dissolution of the kaolinite, and finally to the reprecipitation of a different, poorly crystalline phase.

The evolution of the mixed-layer clay is equally apparent. Figure 10 shows the edge of a mica flake. There is evidence of abrasion and deterioration of the edges as seen by the broken, fuzzy outline of the edge. Here the bonds are weakest and breached easier. As potassium is removed from the mica/illite structure, exfoliation occurs. There is now a charge imbalance and an ideal site for nucleation of another compatible species is created. It requires less energy to add smectite layers to an already degraded illite than to create a discrete smectite out of the solution.

The development of a mixed-layer clay at the very edge of the transformed mica flake is apparent in Figure 11,shown by a cabbage head morphology indicative of mixed-layering with a cornflake smectite/mixed-layer habit in the right foreground and the flame habit of discrete illite in the left background.

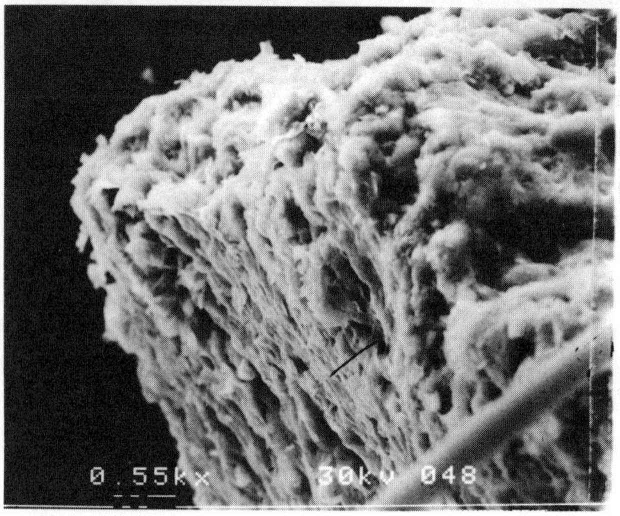

Figure 10.--The Edge of a Mica Flake.

Figure 11.-- Mixed-layer Illite/Smectite Clay.

These pictures graphically show a destruction and evolution cycle in which kaolinite is dissolved, its components reprecipitated in metastable phases which, in turn, are reformed into the interstratified, expanding illite/smectite clay.

EXPERIMENTAL

The authors are attempting to experimentally duplicate the clay species evolved in the heap. Five gram splits of the fresh agglomerates were soaked in 0.05%, 0.5% and 2% NaCN solutions over time. There is an almost immediate development of a random interstratification of illite/smectite layers at all strengths of NaCN. With time there is a progression towards a regular interstratification which is hypothesized to develop more fully with a longer exposure to leach solutions. Table IV is a compilation of values showing the shift in the 5 angstrom peak. This shift is used by Reynolds and Hower (1970) to show the percentage of smectite to illite layers existent in a mixed-layer clay. As the 5 angstrom peak shifts towards lower two theta values (i.e from 17.8 towards 15.75 degrees two-theta), the percentage of smectite layers within the I/S clay increases.

The experimental samples show a shift which indicates that smectization is indeed occurring. The original ore samples show about 5% smectite layers. This has increased to as much as 28% after 6 weeks in the leach solutions.

TABLE IV.-- % SMECTITE LAYERING PRESENT

TIME	NaCN .05%	0.5%	2%
1 week	8%	9.5%	9%
2 weeks	7%	12.5%	8.5%
3 weeks	9%	15%	8.5%
4 weeks	10.5%	17%	17.%
6 weeks	14%	18%	24%
8 weeks	18%	18%	28%

There is a definite increase in development of mixed layering with continued exposure to the leach solutions. This means smectite layer production, The laboratory beaker does not, obviously, simulate the exact conditions within the heap and yet , with exposure to the cyanide solutions, there is an almost immediate development of smectite layers which increases with time. These results are supportive of the general hypothesis advanced in this paper, however, they are preliminary and must be qualified as such until further work is completed.

DISCUSSION AND CONCLUSIONS

Not all ores have as much clay as the material that was examined in this study nor is the pH of the leach solutions usually at 11.5-12.0. However, the presence of a small amount of clay may potentially cause a slowdown of the extraction rate and lower recovery because of a loss of heap permeability.

It is likely that ore changes through the heap leaching cycle and may cause percolation problems. The problems arise when kaolinite is subjected to high pH leaching solutions causing it to disassociate and contribute components which reform into a swelling component of a mixed layer clay.

This expanding clay may tear apart the agglomerates, creating fines which can migrate and fill in the open flow channels within the heap.

Specific remedies to the problem can only be speculative until further investigations are completed. One remedy, which is not speculative, is to lower the pH. The solubilities of the SiO_2 and Al^{3+}, which are key products in the creation of an expanding clay, are relatively low and constant from pH 7 to 10. At pH 11, however, the solubility is 10 times that at pH 10 and at pH 12, it is 100 times as great(Drever, 1983)! Therefore, lowering the pH to about 10 (as 10.3 is thought to be optimum for gold solubilization), should at least retard the dissolution of the kaolinite and formation of smectite.

Kaolinite is stable in acidic environments. It is, therefore, theoretically possible that an acid leaching reagent might be used such as thiourea when the amount of kaolinite in the ore is very large.

A method which has been demonstrated to be effective and which appeared to have solved the problem in the case history quoted earlier in this paper is to add more binder. More binder, with very careful and detailed testing of the agglomeration process, can hold the pellets together despite the clays. Again, caution is advised, as the more binder(containing lime) added, the greater the probability of increasing the pH with subsequent degradation of vulnerable phases.

The final conclusion which cannot be stated firmly enough is to know the components of your ore, particularly the clays.

REFERENCES

J.I., Drever, 1983, *The Geochemistry of Natural Waters*, Prentice Hall, 388p.

Hauff, P.L., Reynolds, R.C., and Giese, R.F., Jr., 1984, *Clay Mineral Identification Techniques*, The Clayschool Conifer, CO., 546 p.

Starkey, H.C., Blackmon, P.D., and Hauff, P.L., 1984, "Procedure for the Routine Mineralogical Analysis of Clay-bearing Samples", U.S. Geological Survey Bulletin 1583, 32 p.

Reynolds, R.C., and Hower, J., 1970, "The Nature of Interlayering in Mixed-layer Illite-montmorillonite", *Clays and Clay Minerals,* Vol. 18, pp. 26-36.

Velde, B., 1977, *Clays and Clay Minerals in Natural and Synthetic Systems*, Elsevier, Amsterdam.

5

Financing and Economics That Work for Small Mines

R. B. Bhappu
Mountain States Research & Development
Vail, AZ

Chapter 27

LARGE OPERATOR'S APPROACH TO SMALL OPERATOR'S HEAP LEACH PROPERTIES

G.A. Lang
Echo Bay Mines, Ltd.
Denver, Colorado

T.M. Plouf
Minproc (U.S.A.) Inc.
Denver, Colorado

INTRODUCTION

The finding or acquisition of ore bodies containing 200,000 opt gold and greater is becoming more difficult. There are more investors and small heap leach operators who have or can raise the capital to make the investments in the large ore bodies. Echo Bay Mines' approach has been to regionally develop small ore bodies, two or three in the same geographic location, which when considered together contain over 200,000 opt gold.

CONCEPT

The "world class" ore body syndrome, which many large gold producing companies maintain, can restrict company growth if the finding or acquisition of these large ore bodies is not forthcoming. Another approach has been taken by Echo Bay Mines. The concept is to regionally develop two or three small ore bodies suitable for processing with portable equipment. The systematic scheduling of mining, processing with a single flexible processing plant using the same management/operator team can reduce your overall investment. The initial selection of mining and crushing equipment suitable for small open pit operations is essential. A very viable option in the mining sequence is to use contract mining/crushing/screening/agglomerating and stacking. This option also lowers your initial capital investment and increases the flexibility in mining as you move from one mining operation to another.

LOCATION AND SEQUENCING

One single small ore body of itself is not attractive enough for the large producer. The recovery of the initial capital investment, although very rapid, does not make a viable project. The only alternative is to find two or more small ore bodies close enough geographically to each other so that the same management/operator team can be used to run the operations as each is put into operation. Echo Bay Mines has regionally located several ore bodies ranging in recoverable ounces up to 100,000 oz. These small separate ore bodies, located within a 30-mile radius of each other, for planning purposes have been considered as one single ore body.

The first small ore body has the initial burden to recover the capital invested in the office, laboratory, power plant and processing plant. Other capital investments such as water supply, access and haul road, fencing, septic system, pads and ponds are reoccurring at each mine site. The first small ore body must recover the initial capital investment in the movable structures as well as the reoccurring capital investment in the fixed structures very rapidly. A total capital recovery time of 6 to 9 months should be used as an economic decision point for both choosing which small ore body should be processed first, and what the plant design capability should be.

The sequencing of the small mines is very critical. All the fixed structures mentioned above must be installed at the second mine site before the first mine has been mined out. The ideal sequencing will allow for only one month between shutdown of the first small mine and production from the second.

SPECIFIC OPERATION

Echo Bay Mines has completed the construction of the Illipah Mine located in White Pine County, Nevada, with this small mine sequence approach. The Illipah Mine is a 40,000 ounce recoverable gold ore body. This ore body met all the necessary requirements.

The Illipah project could support the capital investment in the movable structures which will be used at the other two mines and the capital investment in the fixed structures needed to put the operation into production. The total capital investment of 2.5 million could be recovered in less than 9 months of operation. The capital investment for the Illipah project is shown in Table I.

TABLE I

ILLIPAH CAPITAL INVESTMENT

	($000's)	
Movable Structures:		
Office	52	
Laboratory	105	
Power Plants & Dist.	140	
Processing Plant	300	
Misc. & Eng.	192	
Communications	16	
		805
Fixed Structures:		
Pre Production Costs	280	
Access & Haul Road	110	
Fencing	36	
Septic System	12	
Water Supply	195	
Pads & Ponds	640	
		1.273
Working Capital	460	460
TOTAL CAPITAL INVESTMENT		2.538

COST ANALYSIS

The cost analysis of the entire project consisted of estimating operating costs as well as the fixed structure capital investment for the second two small mines.

It was not difficult to estimate the fixed structure capital investment because of the geographic location of the small properties. Contract mining/crushing/screening/agglomerating/stacking was selected because it lowered the capital burden on the first of the ore bodies. Using contract mining and the same management/operator team for each property helps in estimating the operating cost for the successive properties. The only operating adjustments that were needed were grade, recovery, reagent consumption and strip ratio. The cost analysis for three small properties is shown in Table 2.

TABLE 2

COST ANALYSIS

(000's)

	1st Mine	2nd Mine	3rd Mine	Total
Estimated Capital:				
Movable	805	-	-	805
Fixed	1,273	1,463	1,910	4,646
Total	2,078	1,463	1,910	5,451
Recovered Gold Ounces:	40	70	90	200
Operating Months:	18	26	30	74
Estimated Cash Cost: ($/oz)	247	212	218	222

SUMMARY AND CONCLUSIONS

Echo Bay Mines has taken the large operator's approach to small operator's heap leach properties. They have successfully found or acquired small ore bodies which normally would not fall within their scope to be put into operation. By grouping regional and geographic properties into a single mining unit the economics of producing from several small ore bodies is enhanced. The first of three ore bodies is in operation. The second two are being drilled and metallurgically tested. The "world-class" ore body syndrome has been addressed and an alternate approach has been taken.

Chapter 28

FINANCIAL CONSIDERATIONS FOR NEGOTIATING THE GOLD SALES CONTRACT

William Mayrsohn

Asamera Minerals (U.S.) Inc.

INTRODUCTION

The net smelter return (NSR) calculations made on various smelter-refiner alternatives, is usually done by taking the gross metal value less metal deductions, smelter-refiner charges and transportation-insurance costs.

An important further step in the calculations is the financial charges imputed by refining time, transportation delays, to the time gold is returned (if tolled) or paid for if bought by the refiner.

Metal that is priced in advance of payment also requires a forward premium that the mine may or may not be getting. This is a hidden cost that is often omitted from the calculations.

Segmentation of each of the above costs are vital to see where improvements can be made through negotiations, better choice of metal pricing or transportation alternatives.

BASIC NSR CALCULATIONS SMELTER-REFINER CHARGES

Payable Metals - Treatment Refining Charges:

Since these can vary from one smelter to another, the effect of a $25 per ounce price increase/decrease should be calculated and kept in mind.

This effect is compared to differences in refining/treatment charge and a total smelter/refiner cost on a cost per ounce of dore, gold, or a dry ton of concentrate.

Transportation - Insurance:

Research has to be done on transportation alternatives perhaps with some assistance from the smelters, railroads or trucking companies. For dore, the U.S. postal service is viable and inexpensive. For railroads, deal with the regional office ie., B&N El Paso, if production warrants regular bulk shipments. Spot rates tend to be higher.

A small mine producing concentrates usually goes to the nearest smelter due to extremely high transportation costs. Here is where a refiner can help.

Insurance-Quotes for rates on dore/concentrates are easily obtainable, but the mine has to shop for the best rates. Your insurance broker is the best source of this information.

Differences between Mine and Smelter/Refiners

Metal losses can be caused by the following:

- Concentrate dry weight differences caused by moisture determination or faulty scales (often at the mine).
- Sample bias by poor preparation practices on either side ie., grinding of the sample removing gold; or a poorly mixed concentrate. Dore bullion has no such difficulties.

- Assay bias can be caused by uncorrected assays. Try to get the smelter to correct for cupel absorption and slag loss, if not, do sufficent assay work to determine the dollar loss, as part of the negotiations, and comparison between refiners/smelters.

Pricing the Metal:

An often overlooked item is to compare various pricing/payment schemes to see which fits your company best.

For example, if you price before shipment, you want to obtain that forward premium gained to the day cash is paid. For gold that is about 9 cents per day, at today's market.

Thus a common offer (by a merchant) that has a very low T/C on dore, ie., 25 cents an ounce with the usual metal return can be very poor, if the metal is priced during the month of production, but paid for upon outturn or some time after delivery to a refiner.

It calculates like this: price each ounce of gold about 45 days on average, prior to outturn/payment, and the $3.75 premium ($0.09 x 45) leaves $2.75 as profit. Not bad for a little paperwork!

How do you handle this? By tolling your metal and price it according to your needs and risk perception.

Refiners will return metal (for a fee) as prompt as 3 to 5 days after melting and sampling at their facility.

Early Gold Return (prior to actual Refining)

Most mines need a prompt cash flow prior to actual metal return. This action requires a financing charge either built into treatment charges or listed separately. Merchants and brokers are usually more competitive and may return metal as it is shipped from the mine. Gold brokers/merchants are better suited than smelters for the early return of gold in concentrates.

	Offers I	Offers II	Offers III
Payable Metals:	99.8%	99.6%	99.5%
Gold @ $450 60% Au Dore	= 269.46	= 268.92	= 268.65
Silver @ $7.80	98%	98%	98%
@ 40% Ag Dore	= 3.06	3.06	3.06
	$272.52	$271.98	$271.71
Treatment/Refg. Charge/oz dore	(1.20)	(1.00)	(0.50)
Transport. Insurance	(2.00)	(3.50)	(3.50)
NSR Before Financing/oz dore	$269.32	$267.48	$268.21

Figure 1 CONVENTIONAL NET SMELTER RETURN (NSR). This is without time value of money, and pricing at spot (current market gold price) for offers I-III.

Offer I is best due to a combination of better gold payable, and transportation. If the price of gold were to increase by $50.00 an ounce, offer I improves by $0.27 per ounce per 0.1% gold payable over II and III.

The offers have further refinements; Offer I returns gold in 5 days after arrival/sampling, II returns gold in 30 days and III in 21 days. Let's look at Figure 2.

We take Figure 1 NSR and add the followi

	Offers I	Offers II	Offers III
NSR:	$269.32	$267.48	$268.21
Days to Metal Return:	5 Day	30 Days	21 Days
Cost Option of early Metal:	None	15¢/oz for 5 Day Return	0.1% for 5 Day Return
Net Finance cost Early Metal/oz	-0-	($ 0.09)	($ 0.27)
Adjusted NSR	$269.32	$267.24	$267.94

Figure 2 ADJUSTED NET SMELTER RETURN (NSR). Early gold charge included in offer I, offered as options in II and III.

Now we take the above and apply various pricing schemes (forward pricing) on a conservative basis, assuming gold return in 5 days by a refiner/merchant:

PRICING PLAN	**DAYS FORWARD**
A. Price after shipment	2 + 5 = 7
B. Price during production period	7 + 15 = 22
C. Price a month prior to production	0 + 22 = 52

Converting days forward into premium at 9 cents per day we now have:

Days	Forward Premium	Broker Fee	Net Premium Per Oz.
A. 7 x $0.09	$0.63	(0.50)	$0.13
B. 22 x $0.09	$1.98	(0.50)	$1.48
C. 52 x $0.09	$4.68	(0.50)	$4.18

Figure 3 FORWARD PRICING PLANS. If done on a regular basis you get an average price plus the above premium.

A usual plan is to price gold-silver every market day to avoid speculation.

The result is forward sales which are roughly the average annual price, plus this forward premium (less some $0.50 fee for the merchant/broker) and you have the above net premium that goes back to the mine.

A middle of the road, non-speculative plan, would be produce and price simultaneously and get gold/cash 5 days after arrival to earn $1.48 per ounce over the spot price. Add this to our adjusted NSR before forward selling, and the results look like Figure 4.

	Offers I	Offers II	Offers III
Adjusted NSR/oz with 5 day payment	$269.32	$267.24	$267.94
Net Improvement /oz, due to forward sale	+ 1.48	+ 1.48	+ 1.48
NSR Adjusted 5 day payment/ forward sale/oz	$274.80	$267.72	$269.42

Figure 4 NSR ADJUSTED FOR SELLING FORWARD AS YOU PRODUCE AND, GET EARLY GOLD/CASH 5 DAYS AFTER SAMPLING DORE AT THE WORKS.

By pricing 30 days earlier you can pick up an additional $2.70 an ounce, at todays prices, and COMEX forward premium.

A fourth offer comes in looking very attractive: 99.9% gold return and $0.25 and ounce treatment-refinery charge and pricing at spot, the month before you ship. You are to receive cash upon delivery, see Figure 5.

Payables		Value Per Ounce
Gold @ 99.9% .6 x .999 x $450	=	$269.73
Silver @ 98% .4 x .98 x 7.8	=	3.06
Total Payables		$272.79
T/C		(0.25)
Transport/insurance etc.		(2.00)
Finance Charges - Early Gold		-0-
Forward Premium Foregone 45 days x $0.09 less 50¢/oz	=	($2.55)
NSR		$266.99

Figure 5 FOURTH OFFER, PRE-PRICING BUT CASH PAYMENT OCCURS UPON DELIVER OF DORE.

Despite the apparent "bargain", you could have done better by doing it yourself, for example, by taking offer I, toll the gold, sell to a merchant/broker and take the premium for the mine's additional revenue.

Lets go back to Figure 4 where we got an adjusted NSR $274.40 which is $7.41 better than above figure.

It could be well worth it, providing there is enough ounces of gold to provide additional revenue offsetting the additional bother.

Effect of Debt on the NSR

We could go on and on about various strategies, but it should become obvious that all offers have to be put under close scrutiny before a rational business decision can be made.

Since most companies have debt, at least in the initial stages of the mine, cash flow is important. When you are offered great terms but later payment, the cost of money has to be added to the NSR.

Assuming a metal value (at $450 gold) valued at $269 per ounce of dore. A payment delayed 30 days at 10% costs the mine $2.24 per ounce of dore. The advantage of any early gold return/payment is that same cash return should be financed from 20 to 50 cents an ounce per month. Saving the above interest charges, is done at a modest comparative cost.

Now you see this business of getting your cash or gold early takes on a different meaning and deserves careful consideration when the mining company has debt.

Dealing - Negotiating Direct

You can negotiate direct with several refiners first and later merchant-brokers to get a variety of terms to evaluate, without actually committing until the final NSR's are calculated.

Occasionally, a merchant offer will equal the best you can do, with a whole lot less work on your part. If you like the refiner, then take the deal.

As started in my previous paper, when you deal direct you have control over the situation and know a great deal more about your options. However, the longer the anticipated mine life (more than five years) the more attractive doing it yourself becomes. Too many miners make a commitment to refine and sell their gold based on a single factor like treatment charges alone, and regret it later, when it is too late.

Generally refiners will let you cancel a contract but merchants will not.

Don't ever be afraid of putting in a 90-day cancellation clause in your contracts. If this is refused, take another offer. Remember though, forward sales already made are not cancelable.

The golden rule still applies, "He who has the gold, makes the rules". If you give it up, you have to abide by others rules.

Chapter 29

Simplified Estimating Guide for the Evaluation of Small Scale Gold Mining Operations

Sukhendu L. Barua and Roshan B. Bhappu

Mountain States Mineral Enterprises, Inc.

ABSTRACT

The economics of developing and operating small precious metal mines are unique and require much more scrutiny than larger mines since a limited tonnage must carry a heavy burden of capital investment as well as operating cost per unit ton to be mined and processed. Moreover, potential operators of small mines do not have the required expertise available to critically evaluate the technical and economic aspects of their projects.

In this paper, the authors have attempted to provide technical and economic guidelines for developing and operating small precious metal mines. Capital and operating cost estimates as well as return on investment are provided for various scenarios utilizing different mining and processing techniques.

INTRODUCTION

In recent years under favorable metal prices, gold mining has enjoyed a high degree of activity throughout the world during the last decade. Ghost towns of yester-years which thrived on the gold mining boom are once again experiencing the golden days and new orebodies are being discovered practically every day throughout the world. Moreover, because of higher metal prices many of the sub-marginal deposits have become economical overnight and with increasing metal prices and development of new extraction technology, the economical grade is constantly being lowered. Furthermore, the favorable economic climate has encouraged the development of small mines which previously could not be operated at a profit.

By definition, there is no agreement on what constitutes a small mine. Nature of the metal or mineral occurrence, institutional actors, and the degree of industrialization all affect the definition. In general, a small mine constitutes any single-unit mining operation having an annual production of 50,000 metric tons of ore or less. Other universal definitions call for little or no investment in infrastructure and fixed plant, involvement of concentrators or mills of a movable type, production up to 250 short tons per day, heavy reliance on manual labor, and employment of less than 100 people.

Although the public view of mining in recent years has been focused on the large-scale mining operations, the small miners continue to quietly make a vital contribution to world exploration for, and supply of, essential mineral commodities. Even in the highly-developed United States current census of mining industry indicates that two-thirds of all mining operators employ fewer than 50 persons.

Currently, under the fluctuating gold metal prices, ore grades and economics, it is very difficult to evaluate the economic viability of gold mining operations, especially the small ones. The small mine owner or exploration manager faces a real dilemma of whether to or not to proceed with the project. This problem is further complicated by the applicability of a large variety of processing technique and unit operations, each of which presents its characteristic technical and economic complexities. Accordingly, the economics of developing and operating small precious metal mines are unique and require much more scrutiny than larger mines since a limited tonnage must carry a heavy burden of capital investment as well as operating cost per unit ton to be mined and processed. Moreover, majority of the small mine operators do not have the required expertise available to critically evaluate the technical and economic aspects of their projects.

In this paper, the authors have attempted to provide technical and economic guidelines for developing and operating small gold mines. Capital and operating cost estimates as well as return on investment are provided for various scenarios utilizing different mining and processing techniques. For this paper, the authors have extended the production capacity of small gold mining operations up to 500 short tons per day.

EXTRACTION PROCESSES

The commonly used extraction processes used for treatment of gold/silver ores include:

1. Gravity (jigs, tables, spirals, Reichert cones)
2. Flotation, with or without cyanidation of concentrates or tails
3. Heap Leaching - Merrill Crowe or Carbon Adsorption-Electrowinning
4. Conventional Cyanidation (Leach-CCD-Merrill-Crowe)
5. Cyanidation Carbon-in-Pulp (CIP)
6. Pressure Leaching of refractory ores followed by cyanidation
7. Bacteria Leaching of ores or concentrates
8. Leaching with alternate lixiviants

The gravity methods are used generally for recovery of gold from placer deposits as well as an aid in flotation and cyanidation treatment of non-placer type deposits. Such methods include jigs, Reichert cones, tables, moving belt separator, corduroy or rubber-lined riffles and a wide variety of traps. These are ideal for recovery of relatively coarse particles which do not respond readily to flotation and cyanidation processing.

In the treatment of some ores and tailings, flotation has been found to be quite effective for recovery of precious metal values contained in sulfides and as native gold and silver. More importantly, flotation is an ideal treatment procedure for those ores in which all or part of the gold values are contained in carbonaceous material. The flotation concentrates are sent to the smelter or are further processed by cyanidation, perhaps after regrinding or roasting.

Conventional cyanidation will continue to be the standard process for treating precious metal ores of sufficient grade to warrant such treatment. The ground ore, either by conventional multi-stage crushing and ball-mill grinding or by semi-autogenous grinding (SAG), is leached either in agitator tanks or in Pachuca-type tanks with lime and cyanide. If the ores contain carbonaceous material, a pre-treatment with an oxidizer such as air, oxygen or chlorine is necessary to decompose the organic component prior to cyanidation. The leached pulp is then subjected to solid-liquid separation using the countercurrent-decantation (CCD) system or the filtration circuit, either drum or belt. In recent years, the filtration system has been shown to be environmentally and economically more attractive. The precious metal values are finally won from the pregnant leach liquor by Merrill-Crowe (zinc dust precipitation) process.

For lower grade ores or for ores containing excessive slimes, the carbon-in-pulp (CIP) or the carbon-in-leach (CIL) technique has proved to be a very viable alternative to conventional cyanidation. Such a system may be built and operated for an estimated 65 to 75% of the cost of a conventional CCD or filtration unit. Moreover, the carbon adsorption system has a chance to recover nearly all of the soluble values normally escaping the CCD or filtration plant. The technique consists of contacting the leached pulp, usually in a separate step, with coarse (6 by 16 mesh) activated carbon to adsorb gold and silver values. The carbon is next separated by screening at each stage (usually five stages are employed) and advanced countercurrently as it is loaded. Finally, the loaded carbon is stripped by hot caustic-cyanide solution (at atmospheric pressure or at higher pressures with or without the use of alcohol) and the concentrated gold/silver values electrowon or precipitated with zinc dust.

For much lower-grade ores, heap leaching, if technically applicable, is the lowest cost method. In this process, the crushed ore (3/8" or coarser) is piled on an impervious surface and sprinkled or percolated with dilute alkaline cyanide solution. The pregnant solution is then passed through activated carbon columns to adsorb the values or is treated by the conventional Merrill-Crowe system. The selection of the recovery system is usually based on the quality of pregnant solution, the precious metal content and the comparative economics. High silver values and low solution flows probably favor the zinc dust precipitation; while higher flows of predominantly gold-bearing solutions are advantageously processed by activated carbon. For high clay-bearing ores which would present problems during percolation in heap leaching, the agglomeration pretreatment of the crushed ore has been found to be very helpful.

Refractory ores and concentrates show poor response to conventional cyanidation techniques. The refractoriness, in this case, is primarily due to the complex nature of the association of precious metals in valuable as well as gangue minerals. Finely disseminated gold or gold contained in pyrite, pyrrhotite, arsenopyrite, tellurides or other base metal sulfides do not respond to conventional cyanidation treatment. Such refractory ores are usually treated by flotation followed by roasting or pressure leaching of concentrates and their cyanidation of the decomposed product. Carbonaceous ores may be treated by flotation or by decomposition of the carbonaceous ores by preconditioning with chlorine prior to cyanidation. In recent years, efforts have been made to apply bacterial oxidation to the treatment of arsenopyrite ores.

Several lixiviants other than cyanide have been suggested and/or applied for dissolution of precious metals. Aqueous chlorine, thiourea, and thiosulfate (hypo) have all been under investigation as environmentally attractive alternatives to cyanide. Most of these leaching schemes are not commercially developed yet, but with further testing and piloting, they may prove to be effective reagents in the treatment of gold/silver ores, especially those which do not respond well to conventional cyanidation.

ECONOMIC EVALUATION METHODOLOGY

Economics of a mining venture, like any other enterprise, depend on capital costs, operating costs, and revenues. It has been found through long experience that both capital and operating costs for a specific mining property can be expressed as functions of tons per day of ore production. Annual revenues are, of course, a function of tonnage, grade, and percent recovery. Empirical equations can be derived for these functions by using the curve fitting technique.

MSME-MSRD has conducted feasibility studies of various small mines for both mining and processing. The preliminary cost estimates obtained in these studies and data gathered from field were summarized and plotted to generate empirical equation for capital and operating costs. Figure 1 shows these equations for various options of mining and processing.

Since the aim of this paper is to develop a technique for economic evaluation of operating small precious metal mines, the capital and operating costs can be calculated from the respective equations in Figure 1. Assuming a specific operating life, a typical expenditure schedule during design and construction, etc., a cash flow analysis can be generated to yield a DCF rate of return for any given combination of tonnage and grade. A more explicit analysis, however, is obtained by setting the required rate of return and calculating the grade required for various tonnages. This technique of economic evaluation is best illustrated by an example.

<u>Mining Method</u>

1. Open-pit Mining
   - i) Capital Cost : $0.0789 \times (TPD)^{0.6365}$
   - ii) Operating Cost : $25.53 \times (TPD)^{-0.0919}$
2. Underground Mining (cut & fill method, shaft depth 500')
   - i) Capital Cost : $0.475 \times (TPD)^{0.459}$
   - ii) Operating Cost : $148.47 \times (TPD)^{-0.191}$

<u>Processing Technique</u>

1. Crushed Heap Leach-CA-EW
   - i) Capital Cost : $0.001377 \times (TPD)^{1.2057}$
   - ii) Operating Cost : $17.0 \times (TPD)^{-0.1814}$
2. Flotation-Cyanidation of Concentrates
   - i) Capital Cost : $0.0213 \times (TPD)^{0.926}$
   - ii) Operating Cost : $157.96 \times (TPD)^{-0.423}$
3. Cyanidation, CIP
   - i) Capital Cost : $0.01147 \times (TPD)^{0.98}$
   - ii) Operating Cost : $110.25 \times (TPD)^{0.34}$

NOTE: Capital Cost is shown in million \$, and Operating Cost is shown in \$/ton.

FIGURE 1. CAPITAL AND OEPRATING COST ESTIMATES

Suppose a mine operator wants to mine 250 short tons/day (227 metric tons/day) gold ore from an open-pit mine, crush the ore, and use heap leach-carbon adsorption-electrowinning technique to recover the gold. Assume that the mine life is 5 years, gold value is \$400/oz. (\$14.11/gram), and gold recovery is 70% for the proposed heap-leach-CA-EW-process. Following the equations in Figure 1, the capital and operating costs are calculated below:

<u>Mining Costs</u>

i) Capital $= 0.0789 \times (250)^{0.6365} = 2.651$ (in million \$)

ii) Operating $= 25.53 \times (250)^{-0.0919} = 15.37$ (in \$/ton)

<u>Processing Costs</u>

i) Capital $= 0.001377 \times (250)^{1.2057} = 1.072$ (in million \$)

ii) Operating $= 17.0 \times (250)^{-0.1814} = 6.244$ (in \$/ton)

<u>Total Costs</u>

i) Capital $= 2.651 + 1.072 = 3.723$ (in million \$)

ii) Operating $= 15.37 + 6.244 = 21.614$ (in \$/ton)

If we assume 360 working days in a year, then the annual operating cost $= 21.614 \times 360 \times 250$
= \$1,945,260. per year

For the economic analysis, we will consider the following equal-payment-series cash flow analysis where a deposit of amount P is made now at an annual interest rate of i. The depositor wishes to withdraw his principal plus earned interest in a series of equal year-end amounts (A) over the next n years. When the last withdrawal is made there should be no funds left on deposit. The cash flow diagram for this situation is shown below.

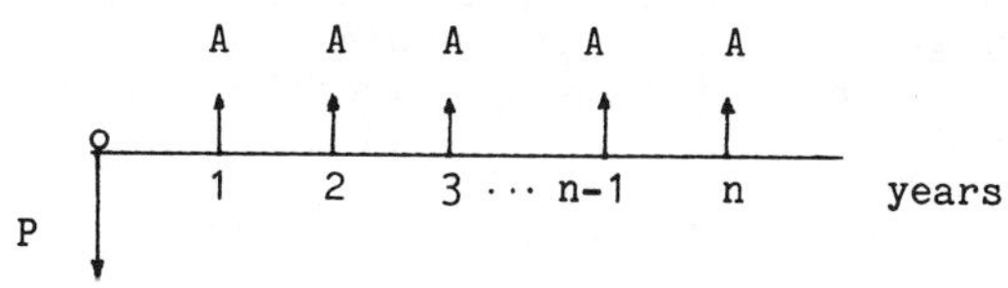

The relationship among A, P, and i is

$$A = P \frac{i(1+i)^n}{(1+i)^n - 1}$$

In the present exercise, if the mine operator wants to realize 30% rate of return on his investment, then the year-end-amount A is calculated as:

$$A = 3.723 \times \frac{.3\,(1+.3)^5}{(1+.3)^5 - 1} = 1.5286 \text{ million dollars}$$

Therefore,

Annual revenue = year-end-amount A + annual operating cost

= 1.5286 + 1.945 = 3.4736 million dollars

The annual revenue is generated from the sale of gold. Given the percent recovery and the current selling price of gold, the annual revenue equation can be written as:

Annual revenue = 250 X 360 X grade X 0.7 X 400

Thus, 3.4736 = 25.2 X grade

$$\text{or} \quad \text{Grade} = \frac{3.4736}{25.2} = 0.138 \text{ oz/T}$$

It should be noted that in the above calculations the salvage value of the plant is not taken into consideration. This should provide additional funding for the subsequent mining project.

The above technique is used in calculation of grade for various scenarios utilizing different ROI, production level, mining, method and processing techniques. The results are summarized in Tables 1-6.

Table 1 = Open-Pit Mining - Heap Leach- CA - EW
Table 2 = Open-Pit Mining - Flotation - Cyanidation
Table 3 = Open-Pit Mining - CIP Cyanidation
Table 4 = Underground Mining - Heap Leach - CA - EW
Table 5 = Underground Mining-Flotation-Cyanidation
Table 6 = Underground Mining - CIP Cyanidation

ECONOMIC ALTERNATIVES

Since in small mining ventures, a restricted tonnage must confront a high burden of capital and operating costs per unit ton mined and processed, any modifications which lower these costs are eagerly sought by small operators. Some of these alternatives include:

- Contract mining of the entire orebody at one time to avoid capital investment and realize lower operating cost associated with higher tonnage. Such a scheme appears very attractive for dump or heap leaching operation.

- Use of portable or movable processing facilities and unitized or modular construction techniques. The deployment of such systems allows reuse of the same plant at more than one location (mine) resulting in considerable savings in capital investments and making the small operations economically attractive.

The cost of a 75 mtpd Caravan Mill mounted on trailers and incorporating a complete CIP-Cyanidation circuit is estimated at about US$ 2.5 million. A comparable non-portable facility would cost about US$ 0.8 million. The primary advantage of the portable facility is that it can be moved to a new location in days and the new operation started more or less immediately with generation of instant cash flow. This is exactly what a small gold mining operator wants. With a 3 to 5 year mine life involved in smaller operations, a mobile plant can be used in 3 to 5 separate mining operations before replacing the plant.

From the foregoing presentation it is apparent that small mining operations can be run at a profit using innovative technology and cost saving alternatives. It is hoped that the capital and operating estimating formulas as well as the economic evaluation tables presented herein will provide economic guidelines for developing and operating small precious metal mines.

TABLE 1

GOLD EXPLORATION
PROJECT EVALUATION: OPEN-PIT/HEAP LEACH

MINING TYPE = OPEN-PIT MINING
PROCESS = HEAP LEACH: CRUSHED ORE-CA-EW
ORE RECOVERY FACTOR = 70.00
MINE LIFE = 5
WORKING DAYS/YEAR = 360.00
ORE VALUE IN DOLLAR = 400.00

DAILY TONNAGE AND GRADE

ROR	100.	150.	200.	250.	300.	350.	400.	450.	500.	PAYOUT YR
10.	.134	.126	.120	.116	.113	.111	.109	.107	.106	3.79
15.	.140	.131	.125	.121	.118	.116	.113	.112	.110	3.25
20.	.147	.137	.131	.127	.123	.121	.118	.117	.115	2.99
25.	.154	.143	.137	.132	.129	.126	.123	.121	.120	2.69
30.	.161	.150	.143	.138	.134	.131	.129	.127	.125	2.44
35.	.168	.156	.149	.144	.140	.137	.134	.132	.130	2.22
40.	.175	.163	.155	.150	.146	.142	.140	.137	.135	2.04
45.	.183	.170	.162	.156	.152	.148	.145	.143	.141	1.88
50.	.191	.177	.168	.162	.158	.154	.151	.149	.146	1.74
D:	.2M	.3M	.4M	.4M	.5M	.6M	.7M	.8M	.9M	

RESERVE IN MILLION TONS

TABLE 2

GOLD EXPLORATION
PROJECT EVALUATION: OPEN-PIT/FLOTATION—CYANIDATION

MINING TYPE = OPEN-PIT MINING
PROCESS = FLOTATION-CYANIDATION
ORE RECOVERY FACTOR = 80.00
MINE LIFE = 5
WORKING DAYS/YEAR = 360.00
ORE VALUE IN DOLLAR = 400.00

DAILY TONNAGE AND GRADE

ROR	100.	150.	200.	250.	300.	350.	400.	450.	500.	PAYOUT YR
10.	.191	.173	.161	.152	.146	.141	.137	.133	.130	3.79
15.	.200	.181	.169	.160	.153	.148	.144	.140	.137	3.35
20.	.210	.189	.177	.168	.161	.155	.151	.147	.144	2.99
25.	.219	.198	.185	.176	.169	.163	.158	.154	.151	2.69
30.	.229	.208	.194	.184	.177	.171	.166	.162	.158	2.44
35.	.240	.217	.203	.193	.185	.179	.174	.169	.166	2.22
40.	.250	.227	.212	.201	.193	.187	.182	.177	.173	2.04
45.	.261	.237	.221	.210	.202	.195	.190	.185	.181	1.88
50.	.272	.247	.231	.220	.211	.204	.198	.193	.189	1.74
D:	.2M	.3M	.4M	.4M	.5M	.6M	.7M	.8M	.9M	

RESERVE IN MILLION TONS

TABLE 3

GOLD EXPLORATION
PROJECT EVALUATION: OPEN-PIT/CYANIDATION-CIP

MINING TYPE = OPEN-PIT MINING
PROCESS = CYANIDATION: CARBON-IN-PULP-CIP
ORE RECOVERY FACTOR = 90.00
MINE LIFE = 5
WORKING DAYS/YEAR = 360.00
ORE VALUE IN DOLLAR = 400.00

DAILY TONNAGE AND GRADE

ROR	100.	150.	200.	250.	300.	350.	400.	450.	500.	PAYOUT YR
10.	.162	.148	.139	.132	.127	.123	.120	.117	.114	3.79
15.	.169	.154	.144	.138	.132	.128	.125	.122	.119	3.35
20.	.176	.160	.150	.143	.138	.134	.130	.127	.124	2.99
25.	.183	.167	.157	.149	.144	.139	.136	.133	.130	2.69
30.	.190	.174	.163	.156	.150	.145	.141	.138	.135	2.44
35.	.198	.181	.170	.162	.156	.151	.147	.144	.141	2.22
40.	.206	.188	.177	.169	.162	.157	.153	.150	.147	2.04
45.	.214	.196	.184	.175	.169	.164	.159	.156	.153	1.88
50.	.223	.203	.191	.182	.176	.170	.166	.162	.159	1.74
D:	.2M	.3M	.4M	.4M	.5M	.6M	.7M	.8M	.9M	

RESERVE IN MILLION TONS

TABLE 4

GOLD EXPLORATION
PROJECT EVALUATION: UNDERGROUND/HEAP LEACH

MINING TYPE = UNDERGROUND MINING
PROCESS = HEAP LEACH: CRUSHED ORE-CA-EW
ORE RECOVERY FACTOR = 70.00
MINE LIFE = 5
WORKING DAYS/YEAR – 360.00
ORE VALUE IN DOLLAR = 400.00

DAILY TONNAGE AND GRADE

ROR	100.	150.	200.	250.	300.	350.	400.	450.	500.	PAYOUT YR
10.	.359	.321	.297	.281	.268	.258	.250	.243	.237	3.79
15.	.373	.333	.308	.291	.277	.267	.258	.251	.245	3.35
20.	.389	.346	.319	.301	.287	.276	.267	.259	.252	2.99
25.	.405	.359	.331	.311	.296	.285	.275	.267	.260	2.69
30.	.421	.372	.343	.322	.306	.294	.284	.276	.269	2.44
35.	.438	.386	.355	.333	.317	.304	.293	.285	.277	2.22
40.	.455	.401	.368	.345	.327	.314	.303	.294	.286	2.04
45.	.473	.416	.381	.356	.338	.324	.313	.303	.295	1.88
50.	.491	.431	.394	.368	.349	.335	.322	.312	.304	1.74
D:	.2M	.3M	.4M	.4M	.5M	.6M	.7M	.8M	.9M	

RESERVE IN MILLION TONS

TABLE 5

GOLD EXPLORATION
PROJECT EVALUATION: UNDERGROUND/FLOTATION-CYANIDATION

MINING TYPE = UNDERGROUND MINING
PROCESS = FLOTATION-CYANIDATION
ORE RECOVERY FACTOR = 80.00
MINE LIFE = 5
WORKING DAYS/YEAR = 360.00
ORE VALUE IN DOLLAR = 400.00

DAILY TONNAGE AND GRADE

ROR	100.	150.	200.	250.	300.	350.	400.	450.	500.	PAYOUT YR
10.	.388	.343	.316	.297	.282	.270	.261	.253	.246	3.79
15.	.404	.357	.328	.308	.293	.280	.270	.262	.255	3.35
20.	.421	.372	.341	.320	.304	.291	.280	.272	.264	2.99
25.	.439	.387	.355	.332	.315	.302	.291	.282	.274	2.69
30.	.457	.402	.369	.345	.327	.313	.302	.292	.284	2.44
35.	.476	.418	.383	.358	.340	.325	.313	.303	.294	2.22
40.	.495	.435	.398	.372	.352	.337	.324	.314	.305	2.04
45.	.515	.452	.413	.386	.365	.349	.336	.325	.316	1.88
50.	.535	.469	.428	.400	.379	.362	.348	.337	.327	1.74
D:	.2M	.3M	.4M	.4M	.5M	.6M	.7M	.8M	.9M	

RESERVE IN MILLION TONS

TABLE 6

GOLD EXPLORATION
PROJECT EVALUATION: UNDERGROUND/CYANIDATION-CIP

MINING TYPE = UNDERGROUND MINING
PROCESS = CYANIDATION: CARBON-IN-PULP (CIP)
ORE RECOVERY FACTOR = 90.00
MINE LIFE = 5
WORKING DAYS/YEAR = 360.00
ORE VALUE IN DOLLAR = 400.00

DAILY TONNAGE AND GRADE

ROR	100.	150.	200.	250.	300.	350.	400.	450.	500.	PAYOUT YR
10.	.336	.300	.276	.260	.248	.238	.230	.223	.217	3.79
15.	.350	.311	.286	.269	.256	.246	.237	.230	.224	3.35
20.	.364	.322	.297	.279	.265	.254	.245	.238	.231	2.99
25.	.378	.335	.308	.289	.274	.263	.254	.246	.239	2.69
30.	.393	.347	.319	.299	.284	.272	.262	.254	.247	2.44
35.	.408	.360	.330	.309	.294	.281	.271	.263	.255	2.22
40.	.424	.373	.342	.320	.304	.291	.280	.271	.264	2.04
45.	.440	.387	.354	.331	.314	.301	.290	.280	.272	1.88
50.	.456	.401	.366	.343	.325	.311	.299	.289	.281	1.74
D:	.2M	.3M	.4M	.4M	.5M	.6M	.7M	.8M	.9M	

RESERVE IN MILLION TONS

Chapter 30

OPTIMIZATION OF SMALL SCALE MINING IN LESS DEVELOPED NATIONS: GOLD AND PRECIOUS METALS

by

Michael Rieber

The University of Arizona
Professor of Mineral Economics
Tucson, Arizona

Abstract. Capable of establishing a net present value of zero over a short period of time small mines, though not simply scalers of large operations, minimize market risks and reduce investment capital. Particularly now they are relevant to LDC development. Small mine concepts are reviewed with emphasis on sizing, speed, and LDC problems. The special problem of gold and the precious metals is examined with respect to mine evaluation. Small mine analysis is introduced with emphasis on the financial model including gold price forecast considerations.

THE LDC PROBLEM

Historically, mining has been one of the principal routes to economic development. But, the current recession in mineral commodities has led to unprofitable mining in both developed and less developed countries (LDCs). Recently the LDC problem has been exacerbated by the reduction in international lending to levels designed primarily to permit debt rescheduling and by the relative strength of the major currencies which increased debt service cost in borrowing country terms. Small mines provide one way out of this problem.

A small mine is an operation which, whether greenfield or utilizing an existing mine infrastructure, is capable of establishing a net present value of zero, over a short period of time. The small mine requires low capital outlay; rapid access to and high productivity exploitation of a carefully defined orebody; minimum development in, or removal of, country rock and a high level of planning design and relevant mechanization. By implication small mines will be mainly, but not exclusively, underground operations.

These mines are neither scaled down versions of larger mines nor necessarily labor intensive operations. In development terms, they offer a low risk, low cost entry into mining. The product sales provide both foreign exchange and the ability to expand or start new mining projects from the cash flow. Such projects can be better integrated with the LDC economy than can large mining enterprises (usually enclaves); they may avoid reliance on large international corporations, but still provide the necessary mining training and technology transfer needed to properly handle mining complexities. The experience gained is an asset in the bargaining or negotiation with multinationals in other mining developments.

Small Mines Concepts

A rise in real interest rates, especially in the LDCs, can be seen as reversing the trend to large scale mines. A high level of these rates makes smaller ore deposits amenable to smaller scales of mining which may be optimally depleted using smaller scales of equipment or alternative mine designs. This suggests that optimizing capital factor proportions due to rising capital costs may offset the technical economies of scale and depletion.

For any undeveloped deposit to be mined optimally by small scale methods, the rising interest rate must create financial diseconomies of scale of sufficient magnitude to dominate technical economies of scale, and raise the average costs at a rate faster than that of interest, reversing the discounting effects on future rents which would normally require increasing rates of extraction. The first component is the more important. Where capital investment is high relative to other factors of production, smaller scale operations requiring less investment will tend toward lower costs and higher profits. This suggests that new capital using techniques are required for small scale mines which require less investment per unit of output and which do not result in large increases in unit operating costs.

Correct mechanization gives optimum productivity and maximum cost effectiveness. But units must be optimally sized and equipment should be transferable to another site or alternative use. Introduction of mechanization modifies the mining method. For example, in underground mines, the major effect of using LHD with or without truck haulage is that ramps and spirals connecting levels at 25 - 30 m or less, replace raises in the conventional layout. The limiting inclination for LHD equipment is 8-10°.

To combine development with exploration and to reduce drifting costs, there is now a tendency to drive development drifts in the ore body where this is feasible. But if the ore body is valuable, development may have to be in the country rock to ensure maximum recovery. All of the traditional underground mining methods may be modified for in-the-ore body development; there are even examples of in-the-ore body development in sublevel and block caving operations.

The use of rock mechanics methods in underground design, usually a combination of analytical numerical methods to isolate potential weaknesses or zones of stress concentration combined with observation, monitoring and empirical modification, may be particularly useful in small mines where rapid development is essential. Use of numerical methods in design can be related - through probabilistic methods - to a decision support system, modelling the technical and financial environment of a mine in order to assess the effects of various alternatives on the overall economics of the mine system.

Gold and the Precious Metals

Irrespective of whether or not a mineral commodity sold on the international market is denominated in dollars or sold for a particular currency, to the producer, profits and project feasibility can only be assessed in terms of its own domestic currency and its purchasing power parity. This requires that current and anticipated exchange rates be considered. If the producing country's currency becomes weaker, the value of the exported mineral increases in local terms, its internally related costs fall relatively, but its foreign based costs (imports and, usually, capital related) rise. Profit and loss statements on a purchasing power parity basis take relative productivity and inflation into account.

As an extracted commodity, gold has some unique characteristics. When extracted in an LDC, it confers advantages fortunately not often necessary elsewhere. The uniqueness, whether argued on the basis of intrinsic value, history, or religious example, stems from the direct use of gold as money, as an internationally accepted medium of exchange. The precious metals include gold, silver, and the platinum group. Today, they are expensive, but except for gold are no longer precious. Although silver, platinum, and palladium prices are linked to gold during short periods of extreme excess demand for gold, all but gold are now essentially industrial metals responding to the same supply/demand determinants as any other depletable mineral and should be so analyzed. There is now no direct linkage between changes in the price of gold and changes in the price of any other precious metal. In fact, for the platinum group, dealer-producer price spreads among the metals differed widely and 1985-86 inter-year price spread changes were not similar. Furthermore, percent changes in the dealer price over the two years were widely different with platinum, palladium and rhodium increasing while indium, ruthenium and osmium decreased (U.S. Bureau of Mines, 1987), consistent with changes in industrial demand and supply.

The contrapuntal dance of gold and the dollar shows that both are alternative reserve currencies. Reserve currency status cannot be sought; it is thrust by all others on a usually unwilling nation, the value of whose currency now swings with trade events beyond its control. The result is export/import swings and imbalances with painful domestic transitions (e.g., U.S. mining). But as a reserve, gold holdings reduce exchange risks. The ability of governments and ordinary investors to hold gold as an unrestricted, freely traded, commodity reduces the risk of a fall in the dollar or in the value of an LDC's own currency with respect to the dollar or other hard currency. Finally, as an alternative monetary asset, gold, like the currencies of other nations with lower inflation rates, may be held as a hedge against domestic inflation. It is also true that if new gold holdings are treated as currency, both domestic inflation and an import imbalance will ensue.

For all of the above, the impact of new gold will be most strongly evident in LDCs and least in the U.S., the reserve currency nation. But one other aspect is still more important to the gold producing LDC. For an extracted mineral, including the other precious metals, to be valuable it must be sold. The proceeds of the sale are recorded in the current account of the balance of payments, netted against imports, and the positive balance, if any, is held as an addition to foreign exchange reserves in the capital account. At one remove a stockpile of any mineral could be used as loan collateral. Gold can be treated in the same way. Singularly, however, new gold holdings of governments need not be sold or exported to create value. As the counterpart to a reserve currency its very existence increases reserves and is the basis for new loans, if desired. Gold held in private hoards will not directly serve these functions (though increased gold prices and devalued currencies will cause dishoarding), but for most LDCs mining is at least semi-public. Proportionally, this aspect of gold is paramount for the LDCs.

SMALL MINE ANALYSIS

Four functional models should be developed: an ore reserve model to define the ore body, a rock mechanics model to determine the optimum mining method, an operational model to determine equipment requirements and a financial model to monitor profitability. The last is of interest here.

The financial model determines at what equivalent uniform grade of the mineral the mining venture becomes profitable. First, the initial investment in land, facilities, and equipment is determined. Then the costs of production, overhead, depreciation and reclamation are calculated. From these data a discounted cash flow analysis determines an equivalent uniform grade of mineral that makes the venture profitable. The financial model determines an appropriate discount rate which reflects the risk (or apparent lack of risk) in small, short-lived projects. Comparing the minimum ore grade, as determined by the model, to the estimated grades of small deposits, overlooked deposits which are feasible to mine can be identified.

To determine the minimum uniform equivalent ore grade we may start with net present value (NPV). Given: (1) metal price and the prices of by-product minerals, (2) the appropriate discount rate, (3) rate(s) of production over the entire project, (4) inflation rate, and (5) unit costs of equipment, labor, etc., it is possible to determine: (a) initial development and construction costs, (b) equipment replacement costs over the project life, (c) post-production closure and reclamation costs, (d) operating costs (i.e., equipment, labor, overhead), (e) depreciation schedule, (f) depletion schedule, (g) a pro forma income statement, and (h) an NPV cash flow analysis. It should be noted that all revenue figures and estimates are functions of the ore grades, mineral prices and ore quantity. The impact of both costs and prices on the financial program is country and currency specific.

Mine feasibility studies are not independent of the mine design. The latter is crucially related to the optimum production rate. As rules of thumb are insufficient for small mines, some of which may have an optimum life of 2 - 3 years, and may lead to an unwarranted "uneconomic" appraisal, the model should include a generic determination of the optimum production rate and a sensitivity analysis.

Among the parameters to be considered in the determination of optimum mine life (which divided by estimated reserves yields the optimum production rate) are: possible production rates, capital and operating costs (a potential trade-off as a function of size), economic cutoff grades, discount rates (risk and the time value of money), and commodity market prices. The object is to maximize net present value, either before or after taxes. The capital cost estimates, for various small mine sizes and production rates, include items in common with large mines, but in LDCs also include provision for workers where required. Operating cost items are similarly known.

Mineral prices are a significant factor in both mine feasibility and optimal mine life determinations. For small mines the price forecast period cannot be long, but it may be long enough for some significant changes. This suggests that price forecasts should be based on structural market conditions rather than projection from a long-term econometrically based analysis. Of more importance is the effect of market conventions on small mines' product pricing. This involves the kind of contracts, type of market structure, comparative prices received, and degree to which buyers can be differentiated from those purchasing from large producers.

Forecast prices of extracted commodities are usually estimated for the refined form. This is difficult for any mineral; the monetary aspect of gold makes the task more formidable. A forthcoming article (Newcomb and Tsuji, 1987) shows that for forecasting purposes both the microeconomic aspects of fabricated gold demand and producers' incremental costs as well as the movements of monetary gold stocks due to macroeconomic policies are inadequate to explain past shifts in gold prices. It is necessary to consider gold as an asset the price of which is determined in concert and competition with other monetary asset prices. It is a liquid, zero interest, asset which substitutes for cash or risk free bonds in an overall investment portfolio and therefore is subject to speculative and precautionary demands. This is not the usual way to estimate future commodity prices, but its main thesis has been supported elsewhere (Anon., 1987a).

Obviously, gold prices can spike, but this is not a basis for mine development, large or small. Long-run, changes in the price of gold are likely to track the real rate of interest in the reserve currency nation. Output will increase due to shifts to gold by major base metal producers (Anon., 1987) and increases in capacity and production by new and established producers. Prices will be capped by advances in both exploration and recovery technology. In the mid-term, the life of a small mine, gold prices are subject to investors' behavior which, in turn is affected by interest and exchange rates, inflation, and substitute investments. Short-run, even spikes are limited by decreases in fabricated gold demand, recycling to bullion, and movements from public and private inventories (which may total 90 times current annual new gold production). For small mine development there seems no warrant for the expectation of large, rapid, permanent, or real price increases (in reserve currency terms).

Project Analysis

Project analysis starts with generic simplifications based on geology and rock mechanics to allow for specific complexities of design as needed. In addition to first phase formulations (and second phase computer programming with testing), the first phase also develops the data requirements for inputting into computer programs in order to provide for specific mine layout, costing and evaluation. There may be three computer modules: mine design, costing/profitability, and command or management. The last is capable of a feasibility statement and/or alteration of the other two. It also allows for a sensitivity analysis so that the effect of differing mining and/or economic conditions or assumptions can be assessed prior to development.

Input into the model's functions (and to its major submodels) may be broadly characterized as

(1) Geological and geomechanical considerations to be supplied by a user.

(2) Capital equipment, mine design and their linkages which are supplied by the program. With respect to the equipment lists, these must be kept current. Additions can be supplied by potential suppliers. Programmatically, this section is based on (1) above.

(3) Capital costs - as part of the equipment list - should be included in the program, but must be continually updated. Reference data or suppliers' bids may be used. These costs must be translated from the originators' currencies to

local currency via exchange rates or purchasing power parity. Included in the costing is the monetization of all bidders' sales conditions (e.g., payment conditions, and share of commodity values produced). Though the leading edge of modern mining technology may be assumed, only existing technology should be assumed. Therefore, bids and sales lists may be utilized. As an alternative, however, costs may be hypothesized as an aid to choosing among suppliers or for the alteration, within limits, of equipment selection. Capital costs include the necessary infrastructure. This is likely to be crucial in remote areas. As much of the equipment may be expected to outlast the productive life of a small mine, salvage value or remaining value for transfer to other mines or uses must be estimated as an offset to initial capital costs.

(4) Basic engineering is developed from (1) and (2) as modified by (3) above to obtain minimum costs on a capital (not operating) basis. This provides designs and equipment lists, with costs, as an integrated whole.

(5) Project feasibility is developed in a command module. The input is (a) the mine and facilities plan, costs, and alternatives, (b) financing arrangements related to local currencies and/or purchasing power parities to determine repayments in local terms, (c) the costs of capital (government or private), (d) special conditions including environmental factors, joint facilities with other mines or projects, the possibility of sequential mines or expansions, and other uses for equipment and structures, (e) the pay-out period which yields production rates, operating costs and the return to the mine plan and investment.

To determine mining feasibility within the mine plan and costs, an objective must be stated for input. This may be net present value, wealth growth rate, or other. Alternatively, given the objective, capital costs and operating costs, the module may be used to alter the mine plan and equipment (hence capital costs) as well as alter the production rates.

Mine Design Model

The basis of any successful mining operation is a mine design which represents a realistic working plan on which financial decisions can be made. The fundamental requirements for mine design are an accurate knowledge of the geometry and grade of the ore body and of the mechanical characteristics of the ore body and the surrounding rocks.

Four basic geomechanical conditions can lead to difficulties in mine development:

(a) the presence of weathered and/or swelling rocks, which may be susceptible to changes in environment and particularly moisture content during mining;

(b) the presence of geological structures such as faults, multiple joint structures, shear zones associated with faults, dikes, veins or bedding;

(c) the presence of high rock stresses, usually in rocks at depth, but sometimes in shallow, large excavations;

(d) the presence of groundwater at significant pressures or quantities.

Any of these conditions may require extensive support or relocation of the main mine openings, or backfilling as part of the mining process. In a small mine, where rapid return on limited capital is a major prerequisite, it is essential that major support and stability problems be determined in advance. This will require collection of simple rock mechanics data such as drill core strength and fracture frequency, as well as ore grade data during exploration.

A functional model must therefore incorporate geological, geometrical, geophysical and rock mechanics data. Efficient processing and storing of this data and recall for the design requires ideally an interactive graphics program capable of producing sections through the ore body incorporating grade and likely rock behavior. The basic requirements are:

(i) Ore body profiles based on diamond drilling to give a range of options available at various average grades. Parallel vertical sections at between 10 and 20 m intervals will be required depending on the complexity of the ore body and the vertical extent of stopes;

(ii) Stope profiles superimposed on the ore body profiles and using finite element analysis to determine probability stress distributions and to estimate short and long term deformations during excavation and likely support and/or fill requirements with increasing span. A wide range of design options based on the selected ore body profile may be tested at this stage to determine the most efficient stope design and the minimum pillar, fill and support requirements.

(iii) Selection of the optimum stope profiles allows consideration of access drifts and shafts to give optimum access in the shortest possible time, and to plan development stages. Finite element or empirically modified analytical methods may be used to test the design of drifts and shafts and to relate support options to equipment selection.

(iv) A data base of available equipment for mining and development, updated for costs, should be incorporated into stage (iii) to allow a further range of development options. Manpower costs are also incorporated to allow a range of cost/production rate relations to be produced in order to aid selection of the optimum mining method.

(v) Where complex ore bodies or rock deformation conditions occur a degree of flexibility may be desirable allowing several different options to be selected for later decision during development stages.

The basis of the methodology is to produce an interactive program to allow comparison of ore body geometry and grade, mine design, equipment and manpower costs and development costs as a series of options, allowing comparison of overall cost and profit data before final decisions are made. During early mining operations, feedback from observation and simple instrumentation allows for constant calibration of the model and refinement of later design and options.

Financial Analysis

A methodology for evaluating the profitability of an ore body is outlined. It is assumed that the geological work has been completed, a "trial" mine plan with its associated costs developed, and product prices have been estimated. It is also assumed that all prices and costs are denominated in dollars although appropriate translation into local currency terms crucially affects the domestic profitability outcome. Thus, exchange rates and purchasing power parity must also be estimated over the mine life. The effects of uncertainty can be incorporated by risk analysis. For concentrate values, including by-product credits and deductions, a reasonable starting point is the processor's (e.g., smelter's) toll schedule, costed if mineral processing is done locally. If the concentrate is exported, competing toll charges become a part of the evaluation problem as this helps determine desired export destinations and the quantities exported. Cost estimation requires a mine production rate. Here it is assumed that the rate is known. A widely accepted method of mine cost estimation has been developed by the U.S. Bureau of Mines (Minerals Availability System). But, such costing may be inapplicable to the specialized mining techniques used by small mines and may be inapplicable for use in LDCs given both infrastructure and currency considerations.

Capital costs are usually estimated by analogy from known capital costs scaled from existing mines. It is usually assumed that these costs vary with capacity.

$$\text{Capital cost} = k(\text{capacity})^{\text{scale factor}}$$

and

$$\frac{\text{Cost A}}{\text{Cost B}} = \left(\frac{\text{Capacity A}}{\text{Capacity B}}\right)^{\text{scale factor}}$$

Again, if the small mines are designed differently than their large counterparts, such relations will not hold. Thus, a system for determining both capital and associated operating costs from the ground up rather than relying on scaled estimates must be developed. Following traditional principles capital costs are divided into fixed and working capital. The former is divided into depreciable and depletable assets.

Cash flows (annual) can be estimated on the basis of estimated product prices and mining plus beneficiation costs. A simplified mine income statement is presented:

Gross Revenues
less transport costs and royalties equals

Gross Mining Income
less operating costs and depreciation equals

Net Income Before Depletion
less depletion equals

Net Operating profit
plus other income but less expenses equals

Net Income Before Taxes
less taxes equals

Net Profit (annual)

Cash Flow Equals Net Profit
plus depreciation and depletion

The gross revenues depend on the amount mined and beneficiated and on the estimated product prices. Transport costs include mine to beneficiation plant and should include shipment to further processing or an export point. Also included are other non-mining and milling costs. Royalty payments are subject to negotiation or possible elimination if the facility is government owned. Depreciation, straightline or some form of accelerated depreciation, is subject to negotiation. For a small mine with small reserves, even an accelerated rate may yield a time period longer than mine life. Thus, salvage value or transfer worth becomes important. Depletion is usually determined by taking the larger of the two following calculations: (1) Cost of depletable assets (capital cost estimation) divided by the total tonnage in reserve and multiplied by the production rate. Note that the last varies as the optimum rate is estimated and that this impacts the assets required. (2) Gross mining income times a required percentage depletion rate (as determined for taxation). The product must be less than a given fraction of net income before depletion (as determined for taxation purposes). Other income less expenses includes interest payments, dividends, and interest received as well as other items which may be subject to negotiation. Taxes are subject to negotiation. As indicated here they do not include those import or sales taxes which accrue to capital or operating costs.

The evaluation of the cash flow assumes that it is estimated over the life of the mine (the latter subject to variation via the production rate) and that an appropriate or desired discount rate has been determined. Of the evaluation methods available, three may be noted here.

Net present value is the sum of the discounted net annual cash inflows less the sum of the discounted cash outflows. NPV should be greater than zero. The benefit/cost ratio is the ratio of the sum of the discounted net annual cash inflows to that of the cash outflows. The ratio should be greater than one. The internal rate of return (also known as the discounted cash flow-rate of return) estimates the discount rate required so that the sum of the discounted cash inflows less the sum of the discounted cash outflows equals zero. The investor's discount rate (private or social) should be less than that calculated for the project.

As all such analyses require the use of an acceptable discount rate, methods for its calculation (where it is not provided) must be included in the overall analysis. For private firms two estimation methods are useful for illustration: the Quirin and the CAPM. The second, with modification, can be adapted for a (private) profit maximizing government entity.

A Quirin normalized long-run marginal cost curve analysis requires the calculation of the specific costs of capital.

For bonds,

$$k_d = (1 - T)\left[\frac{rP + 1/n(P - Q)}{(P + Q)/2}\right]$$

where

T = tax rate
r = coupon rate
P = face value
Q = current market value.

For common stock,

$$k_c = D/P + g$$

where

D = current dividend/share
P = current price/share
g = growth in earnings/share over time.

For preferred stock,

$$k_p = D/P$$

The spot long run marginal cost (k_t) may be calculated as:

$$k_t = \frac{[(\text{\# of bonds})(\text{current price of bonds})(k_d) + (\text{\# of common stock})(P_c)(k_c) + (\text{\# of preferred stock})(P_p)(k_p)]}{[(\text{\# of bonds})(Q) + (\text{\# of common stock})(P_c) + (\text{\# of preferred stock})(P_p)]}$$

To remove cyclical variations k_t is normalized using a weight (w).

$$k_t(\text{normalized}) = w(k_t) + (1 - w)(k_{t-1})$$

The normalized k_t is used as the discount rate.

The CAPM method is calculated by the equation:

$$R_j = RF + \beta(R_m - RF)$$

where

R_j is the discount rate
RF is the risk free rate
R_m is a proxy for the market rate
β is the firm's volatility with respect to the market rate.

RF may be estimated using long-term government bonds. R_m may be equated with an average rate of return for a representative group of firms. β can be calculated in two ways:

$$\beta = \frac{\text{cov}(R_m, R_j)}{\text{var}(R_m)}, \text{ or}$$

$$R_j = \alpha_0 + \alpha_1(R_m - RF).$$

Using a least squares regression, the resulting α_1 is the firm specific β.

Optimization for the investment decision involves at least mine life, cutoff grade, average grade, and ore tonnage, all of which are interrelated and all affect costs per pound of product mined. A feasibility decision requires the simultaneous solution of the interrelationships yielding the associated variations in cost.

Conceptually, if

P = price/pound
C = cost/pound
y = annual production (ore)
$\bar{q}$ = average ore grade (as a proportion)
r = required return (as a decimal fraction)
n = mine life (years)

then profit (π) may be defined as:

$$\pi = 2000(P - C) \cdot y \cdot \bar{q}\left[\frac{1 - e^{-rn}}{r}\right].$$

The terms can be defined functionally.

$C = \phi(\bar{q}, t)$
$t = f(\bar{q})$ (ore tonnage)
$y = \psi(t, \bar{q})$
$n = \theta(t, y)$
$q = \lambda(q')$ (q' = cutoff grade)

Therefore, profit can be rewritten as

$$\pi = 2000[P - \phi(\bar{q}, t)] \cdot \psi(t, \bar{q}) \cdot \bar{q}\left[\frac{1 - e^{-r\theta(t,y)}}{r}\right].$$

This may be expanded and simplified using the functions for t and $\bar{q}$. Then setting $d\pi/dq' = 0$ provides q', the cutoff grade which, in turn, provides t, y, $\bar{q}$, n and C. Clearly, the work involved is to set up the means for specifying the generalized functional relations.

From the above it may be noted that tonnage-ore grade relationships are desirable. Considerable work has been done in this area over the past thirty years in developing equations which show how changes in the average grade of an ore body affect its tonnage. In the general form the relation is expressed by

$$\ln t = \alpha_0 + \alpha_1(\bar{q})$$

An example is the well known Lasky relation for porphyry copper:

$$\ln t = 21.22 - 1.645\,\bar{q}$$

where

$\bar{q}$ is in percentage terms and

$\bar{q} = q' + 1/\alpha$, identifying q' as the cutoff grade.

CONCLUSION

Small mines in LDCs differ from their large counterparts in the industrialized nations. Both because they are not simply scalers of large mines and because they are small (with a short time horizon), the approach to the physical aspects of mining differs. This includes exploration, the mine plan and the degree and type of mechanization. An LDC location requires the provision of infrastructure. It also requires that costing and profitability be considered both in the lenders' and the borrowers' terms. This leads to a necessary consideration of both relative inflation rates and exchange rates. For example, to maximize profit a small mine contemplated for Venezuela, on borrowed capital from Japan, selling the output (denominated in dollars) on the European market, might well wish to buy equipment from Brazil (assuming that the Japanese loan is not tied).

Excluding gold, the mining of precious metals adds little to the feasibility arguments. It should be recognized, however, that if metals are precious due to scarcity, the opening of even a small mine may affect price. Gold is a more unique commodity.

Mine feasibility studies require a commodity price forecast. For an LDC, as the counterpart to foreign exchange reserves gold, even if it is merely stockpiled domestically, enhances borrowing ability and reduces country risk. These externalities are in addition to the commodity value. However, the monetary character of gold poses problems in the mine valuation or feasibility exercise. This can be seen in the discount rate calculation. For the lender or for a concessionaire mining in an LDC, current market value (Q), coupon rate (r), and current price per share (P) in the Quirin calculation and the discount rate (R_j), risk free rate (RF) and the market rate (R_m) in the CAPM calculation, are all likely to exhibit colinearity with the price of gold. Furthermore, the production of gold, while it may lead to domestic inflation, will lead to an alteration of exchange rates. Mine feasibility analysis is simply not the same as for other metallic minerals.

REFERENCES

Anon., 1987, "Are Gold Bugs Digging Too Deep," Economist (London), March 28.

Anon., 1987a, "Gold Finally Becomes a Respectable Holding for Investors in U.S.," Wall Street Journal, pp. 1, 31, May 5.

Newcomb, R. T. and Tsuji, K., 1987, "The Price of Gold," Department of Mining and Geological Engineering, University of Arizona, Tucson.

U.S. Bureau of Mines, 1987, "Platinum Group Metals," Mineral Industry Surveys, Table 8, p. 12, March.

6

Understanding the Environmental and Permit Maze

G. M. L. Robinson
EnecoTech Inc.
Denver, CO

Chapter 31

THE REGULATORY REQUIREMENTS AND A PERMITTING STRATEGY FOR SMALL HEAP LEACH MINING OPERATIONS

Jeffrey D. Thatcher, Senior Project Manager
Debra W. Struhsacker, Project Coordinator

STEFFEN ROBERTSON AND KIRSTEN (COLORADO) INC.
Lakewood, Colorado

Abstract. With the burgeoning development of precious metals heap leach mining projects throughout the country, regulatory agencies are beginning to take a closer look at the short-and long-term environmental consequences of heap leach mining. This increase in regulatory review is occurring on federal, state and local levels, and is affecting both large- and small-scale mining developments.

Small-scale mining developments are receiving especially critical regulatory scrutiny because in the eyes of some regulatory agencies, small developers lack the credibility automatically accorded to major mining companies. This is due in part to a bad image of small miners based upon past environmental problems and permit violations caused by a few, small operators.

The proponent of a small mining development thus faces special problems during the project permitting process, and must convince the regulatory authorities of his company's ability and commitment to plan, develop, and decommission a mining operation in compliance with environmental regulations. Not only must the small developer have a thorough understanding of the regulatory requirements affecting his project, he must also devote considerable effort to careful internal planning in order to produce a well organized approach to project permitting.

The importance of proper planning cannot be overstated. Project planning is key to convincing the regulatory agencies that the small miner is capable of developing a viable project. Moreover, the economics of a small mining project demand permit acquisition at modest cost and in a timely fashion. A well planned permitting effort will thus ensure that permits are received within budget and on schedule.

This paper will review the federal, state, and local permitting requirements for heap leach projects, and the variability of these requirements from state to state. The environmental assessment process for projects developed on public lands, and the acquisition of zero-discharge water quality permits, air quality permits, and local special use permits as they apply to small mining developments will be discussed. The paper will conclude with a review of project decommissioning and cyanide neutralization requirements and their implications for small-scale heap leach projects.

INTRODUCTION

As smaller heap leach precious metal projects are becoming increasingly common in the United States, federal, state and local regulatory agencies are beginning to take a closer look at the short- and long-term environmental consequences of this recovery technology. From a small operator's point of view, there is seemingly a confusing regulatory maze to be navigated and a plethora of permits and approvals to be obtained prior to commissioning a heap leach operation. The degree of difficulty associated with obtaining permits for a small heap leach project depends largely upon the state in which the project is located. Permit acquisition is generally easiest in those states most familiar with heap leaching technology. For example, heap leaching is well understood in Nevada, and heap leach operations are relatively easy to permit, although permitting conditions in Nevada have recently become more stringent.

However, many states are just beginning to become familiar with heap leach operations. Many of these states such as Idaho, Colorado, South Dakota, Montana, Washington and South Carolina are in the process of drafting regulations and establishing guidelines for heap leach projects. In most cases, these regulations specify design and operating criteria as well as reclamation and decommissioning procedures.

The recent increase in the price of gold has prompted a 1980's gold rush, with the vast majority of new projects being developed as small to medium sized heap leaches. With the considerable increase in new projects, the regulatory agencies are faced with a complex set of new problems. Not only are increasing numbers of new projects being developed, they are being developed by new companies. Very few of the major mining companies survived the recent mining depression and, as a result, new companies are being formed for the express purpose of developing only one or two projects. Gold has become a magnet for attracting investors, and companies are quickly formed and financed to develop the smaller properties.

For the purposes of this paper, small mines are categorized in the less than three million tons (total reserves), with the next category in the three to ten million tons.

Small mining operations are also, for the most part, new companies with little or no operating experience. This lack of a track record often creates a credibility problem with the regulatory agencies. As a result, the small operator is viewed very skeptically and is, therefore, scrutinized more closely.

This paper provides an overview of the regulatory and permitting aspects facing smaller heap leach operations. Permitting requirements for projects developed on federal land are discussed in detail. The state and local regulatory atmosphere and permitting requirements for various states are also presented. The paper concludes with a generic permitting strategy.

Key Environmental Concerns and Political Awareness

The primary concern of most local, state and federal regulatory agencies is the use, storage and transport of sodium cyanide (NaCn), and potential cyanide contamination of surface and groundwater resources. The toxicity of cyanide to people, fish and wildlife is also an important issue. Unfortunately, in the public eye, cyanide heap leaching has some emotional impacts similar to those associated with radioactivity and uranium mining. This perception and the general lack of understanding of cyanide can lead to unnecessary concern about cyanide contamination, further complicating the regulatory environment.

Thus the main objective in permitting a heap leach operation is to convince the regulators (and the public) that the facility is designed to contain and control cyanide. If an applicant can convince the regulatory agencies that the facility can contain cyanide in most emergencies, the agencies will probably issue the permits with a minimum of fuss.

The small operator must size up the permitting process and people involved very quickly and react accordingly. If the operator senses that his credibility is in question, he must react decisively to counter that perspective. Often, simply spending time initially with the key regulatory people will convince them that the operator knows the business. More often than not, the regulatory agency personnel are far less experienced than the operator.

Guiding and Coordinating the Permitting Process

The procedures for obtaining permits and licenses for a heap leach project vary from state to state and are strongly influenced by the regulatory agency or agencies involved. The land status of the proposed project also influences permit acquisition. Projects developed on public land (i.e., land administered by either the United States Bureau of Land Management, BLM or the United States Forest Service) have different permitting scenarios than projects on state or private lands.

In some cases, one regulatory agency is designated as the lead agency responsible for guiding and coordinating the permitting process for a specific project. The commonly designated lead agency varies from state to state. For example, in California, the County Planning Commission and Board of Supervisors is the lead agency for projects on public lands. In some other states, there is no lead agency *per se*, but one agency will be the primary group responsible for issuing most of the required permits. In this situation, it is the applicant's responsibility to determine what permits are necessary and to coordinate the permitting process.

For projects located on both BLM- and Forest Service-administered land, only one of these federal agencies will act as the lead agency. Generally a Memo of Understanding between the two organizations will be drafted to designate which entity has the responsibility of acting as the lead agency. Similarly, in states such as California in which the county plays a major role in the permitting and environmental assessment process, a Memorandum of Understanding between the county and the federal agency will also be prepared.

THE ENVIRONMENTAL ASSESSMENT PROCESS FOR PROJECTS DEVELOPED ON PUBLIC LANDS

Governing Laws and Statutes

The development of mining projects on public lands is regulated by a number of laws including:

1. The Mining Law of 1872;
2. The National Environmental Policy Act of 1969 (NEPA);
3. The Federal Land Policy and Management Act of 1976 (FLPMA);
4. 43 CFR 3809, Surface Management of Public Lands Under U.S. Mining Laws (BLM);
5. 36 CFR 228, Surface Management of Public Lands Under U.S. Mining Laws, (USFS);
6. The Federal Water Pollution Control Act of 1972 as amended by the Clean Water Act of 1977; and
7. The Clean Air Act.

In addition to these regulations, there are other federal, state and local permits and approvals necessary for the construction and operation of most heap leach mining projects on public lands.

Submitting a Plan of Operation

Any mining project causing a surface disturbance of more than five acres per year on public lands is required to submit a Plan of Operations to either the BLM or the Forest Service, whichever agency is responsible for administering the lands in question. The Plan of Operations must describe the proposed mining operation; supply a map showing project facilities, access and areas of surface disturbance; provide the name and serial numbers of the claims involved; indicate a schedule for the proposed activities; and prepare a reclamation plan.

The Environmental Assessment Process

Submitting a Plan of Operations initiates the environmental assessment process. In order to review a Plan of Operations, the BLM or Forest Service must prepare an Environmental Assessment (EA) analyzing the potential environmental impacts associated with the proposed project. For many heap leach projects, preparation of the EA is relatively straight forward.

Contents and Preparation of an Environmental Assessment

An EA generally consists of the following elements:

1. A detailed discussion of the proposed activity;
2. A detailed description of the environment that would be affected by implementation of the proposed activities;
3. A presentation of other feasible alternatives to the proposed activities;
4. An analysis of alternative actions and their predictable short- and long-term environmental effects including physical, biological, economic, and social factors and their interactions; and
5. A discussion of mitigation measures that would minimize the environmental impacts of the proposed activities.

Preparing an EA requires a number of baseline environmental studies in order to evaluate the potential impact of the proposed project on the site. An EA typically addresses the geologic, hydrologic, soils, vegetation, wildlife, aquatic, and cultural (historical and archaeological) resources of the proposed project area. Other parameters such as air quality, noise, visual resources (aesthetics), land use and socioeconomics are usually considered as well. Table 1 lists the types of data commonly collected for preparation of an EA.

For many heap leach projects, the BLM or the Forest Service will perform all or some of the site resource evaluations and will prepare the EA. For large-scale more complex heap leach projects, or when the federal agency has too few personnel to perform the site studies and prepare the EA in a timely fashion, it may be expedient or necessary to contract a third-party contractor for this purpose. This can be done through a third-party agreement and a Memorandum of Understanding stating that the contractor is responsible for performing some or all of the environmental studies and preparing the EA under the supervision of the lead federal agency.

Deciding whether to use a third-party contractor to prepare the EA is usually at the discretion of the applicant with the concurrence of the agency. The advantage of using a contractor is that in their position as a third party, the contractor can also act as a liaison and negotiator between the project proponent and the federal agency. The contractor can thus play a critical role in promoting communication and in facilitating discussions between the two sides. The possible disadvantage of using a contractor is that preparing the EA may cost a bit more than if it were prepared by the BLM or the Forest Service. However, the time frame for preparation and approval is generally considerably less using a third-party contractor.

It should be noted that the project proponent generally pays for the EA in either case. This is especially true for larger projects and more involved EA's. Sometimes the BLM or the USFS will absorb the costs of preparing a small-scale EA. However, for most projects if the EA is prepared by the BLM or the USFS, a budget for EA preparation will be agreed upon, and the mining company will be charged for this amount.

The NEPA mandates that the environmental assessment process includes a forum for public notice and comment. This is generally accomplished by publishing a notice in local newspapers. This notice should display a preliminary project map showing the areas that will be affected by the proposed project. The notice must also request comments for a three-week period from all interested parties. This notice will also be mailed to various state and local agencies (i.e., agencies responsible for managing water and wildlife resources, county and city officials, etc.), to individuals living in or near the project boundary, and to any other interested organization or party.

At the close of the public comment period, the project proponent and the EA preparers meet to discuss any concerns voiced by the public. The NEPA stipulates that any public concern or issue identified during the public comment period must be addressed in the EA. As mentioned previously, public concerns about cyanide contamination are frequently expressed during the comment period.

The EA process also requires that a number of reasonable project alternatives to the proposed activities be evaluated. The alternatives provide different ways to address public concerns as well as issues significant to the federal land management agency and the project proponent. For most heap leach operations, project economics severely limit the options feasible to the project proponent.

TABLE 1

DATA COMMONLY COLLECTED DURING ENVIRONMENTAL BASELINE STUDIES FOR AN EA

Physical Factors

1. Location
2. Geomorphic/physiographic
   a. Geologic hazards
   b. Unique land forms
3. Climate
4. Soils
   a. Productivity
   b. Capability
   c. Hazard
      (1) Erosion characteristics
      (2) Mass failure
5. Minerals and energy resources
   a. Locatable minerals
   b. Leasable minerals
   c. Energy sources
6. Visual resources
7. Cultural resources
   a. Archaeological
   b. Historical
   c. Architectural
8. Wilderness resources
9. Wild and scenic rivers
10. Water resources
    a. Water quality
    b. Streamflow regimes
    c. Flood plains
    d. Wetlands
    e. Ground water recharge areas
11. Air quality
12. Noise
13. Fire
    a. Potential wildfire hazard
    b. Role of fire in the ecosystem

Biological Factors

1. Vegetation
   a. Forest, including diversity of tree species
   b. Rangeland, including conditions and trends
   c. Other major vegetation types
   d. Threatened or endangered plants
   e. Research natural area (RNA) potentials
   f. Unique ecosystems (other than RNA's
   g. Diversity of plant communities
   h. Noxious weeds
2. Wildlife
   a. Habitat
   b. Populations
   c. Threatened or endangered species
   d. Diversity of animal communities
   e. Animal damage control
3. Fish
   a. Habitat
   b. Poplulations
   c. Threatened or endangered species, including State-listed species
4. Recreation resources (usually a combination of physical and biological factors)
5. Insects and diseases
6. Exotic organisms; for example Russian thistle, Siberian ibex

14. Land use including prime farm, timber, and rangeland
15. Infrastructure improvements
    a. Roads
    b. Trails
    c. Utility corridors and distribution
    d. Water collection, storage, and distribution
    e. Communications system
    f. Solid waste collection
    g. Sanitary waste collection

1. Population dynamics
   a. Size (growth, stability, decline)
   b. Composition (age, sex, minority)
   c. Distribution and density
   d. Mobility
   e. Military
   f. Religious
   g. Recreation/leisure
2. Special concerns
   a. Minority (civil rights)
   b. Historic/archaeological/cultural

TABLE 1
DATA COMMONLY COLLECTED DURING ENVIRONMENTAL
BASELINE STUDIES FOR AN EA

Physical Factors	Biological Factors
15. Infrastructure improvements	3. Ways of life-defined by:
	a. Subcultural variation
	b. Leisure and cultural opportunities
	c. Personal security
	d. Stability and change
	e. Basic values
	f. Symbolic meaning
	g. Cohesion and conflict
	h. Community identity
	i. Health and safety
	4. Land tenure and land use
	5. Legal considerations

Approval of the Plan of Operations

Again it should be emphasized that the scope of an EA is greatly influenced by the size and location of the project. EA preparation for a small- to middle-sized project in a remote area is generally fairly simple, and a "Finding of No Significant Impact" will result from the environmental process. The BLM or Forest Service will specify mitigation measures to minimize environmental impacts to be included in the approved Plan of Operations.

If, however, in the eyes of the federal regulatory agency there is "substantial public interest" in the project or the project may have significant environmental impact, an Environmental Impact Statement (EIS) may be required. Other factors which can cause delays in approving a Plan of Operations include the need to perform a cultural resource inventory per section 106 of the National Historic Preservation Act (NHPA) or compliance with Section 7 of the Endangered Species Act.

Barring these complications, the federal regulatory agency has up to 30 days to approve a Plan of Operations, and may upon notifying the applicant take an additional 60 days in which to evaluate the project and prepare the EA. If it is deemed that an EIS is necessary, the Plan of Operations cannot be approved until 30 days after the final EIS is prepared and filed with the Environmental Protection Agency.

In reality, the EA - Plan of Operations process for most heap leach projects takes about a minimum of four to six months to complete. This schedule assumes that there are no significant environmental effects and that baseline environmental data have already been collected. Collec- of baseline data must be scheduled with weather and snow conditions in mind. This can also impact the EA schedule. In addition, the mine plans must be sufficiently finalized to include firm locations for all proposed mine facilities. Conceptual plans which do not specify the project facility layout are not ade-quate to perform the EA.

STATE ENVIRONMENTAL ASSESSMENT REQUIREMENTS

In addition to the federal environmental assessment regulations mandated by NEPA, some states also have NEPA-type environmental assessment acts which apply to projects developed on state, and in some cases, public land. For example, the California Environmental Quality Act (CEQA) stipulates that the lead agency (usually the county) must prepare an Environmental Review (ER) of a proposed project. If the project is found to have no significant environmental consequences, a Negative Declaration is prepared. If, however, the project is judged to have significant environmental impacts, an Environmental Impact Report (EIR) is required. The EIR can be prepared either by the lead agency or by a contractor in coordination with the lead agency. It usually takes from 60 to 120 days to complete an ER. An EIR generally requires six months to one year to complete. Washington and Montana also have state environmental policy acts which mandate preparation of an ER.

The environmental assessment requirements in most states are patterned after federal requirements. One environmental assessment document can generally be used to satisfy both state and federal environmental assessment requirements.

In addition to the EA procedures required by some states and the federal government, there are many other federal, state, and local permits which must be obtained for a heap leach project. These permits are discussed in the following sections.

AIR QUALITY PERMITS

All heap leach mining operations must comply with Federal and State air quality standards, including the Prevention of Significant Deterioration (PSD) requirements under the Federal Clean Air Act. The PSD regulations establish New Source Performance Standards (NSPS) criteria for the following air quality parameters:

1. Particulate matter - both Total Suspended Particulates (TSP) and PM-10 (fine particulate standard);
2. Ozone;
3. Lead;
4. Hydrocarbons;
5. Sulfur oxides;
6. Nitrogen dioxide; and
7. Carbon monoxide.

The regulatory authority to enforce these standards generally rests with the state or the county.

Emissions from a heap leach project must be quantified and evaluated against these NSPS criteria in order to assess the air quality impact of the project. Air quality regulations also stipulate that a project must use the Best Available Control Technology (BACT) to minimize emissions. Projects with emissions exceeding one or more NSPS criteria may have to apply to the EPA for a PSD permit, in addition to filing for the state and county air quality permits. Some states also have a PSD-type permit required for projects exceeding certain NSPS parameters.

Air quality permits specify design and operational criteria and stipulate emission sampling monitoring requirements. These permits will require emissions data from all aspects of the heap leach operation including the crushing, grinding, agglomerating, roasting, milling and smelting facilities, the stripping, loading, hauling, stockpiling and power generating operations, and will also evaluate hydrogen cyanide gas concentrations near the pad, solution ponds and mix tank areas. Operational monitoring programs for particulate matter and hydrogen cyanide gas may be required as part of the permit conditions. Operating emissions criteria may also be specified by the air quality permits.

SURFACE AND GROUNDWATER QUALITY PERMITS

Zero Discharge Operations

Protecting surface and groundwater from cyanide contamination is probably the foremost concern to regulatory agencies and the general public. Every effort should thus be made to locate cyanide facilities away from surface waters (especially fishable streams and water supplies) and to isolate facilities from groundwaters. A heap leach project proponent must also be prepared to submit sufficient geotechnical, hydraulic and engineering design reports to document total cyanide containment in the event of an emergency.

The importance of adequate design criteria cannot be overstated. Cyanide containment is an engineering task with the goal of controlling cyanide spills and preventing or minimizing cyanide leaks. Wherever hydrologic and climatological conditions allow, a heap leach facility should be designed as a zero-discharge operation, meaning that there is no discharge of processing solutions or effluent to surface waters or groundwater.

All states, the BLM and the Forest Service require engineering design reports documenting design criteria sufficient to ensure a zero-discharge operation. In some states the regulatory agency with jurisdiction over surface and groundwater quality exerts a very strong influence upon the heap leach permitting process. For example, in California the regional office of the Water Quality Control Board (WQCB) plays an equal if not larger role than the county in the permitting scenario.

State water quality permits will generally specify design and operational criteria and will stipulate operational monitoring and decommissioning requirements for heap leach projects. These permits typically specify what types of liners and leak detection systems are to be used, outline an operational monitoring program, detail cyanide detoxification levels and reclamation procedures for project decommissioning, and establish bonding requirements. Most states are also now requiring the submittal of "as-built" construction documents stamped by a registered professional engineer.

These permit conditions can be quite stringent in some states and specify the use of double- or even triple-lined facilities and elaborate leak detection monitoring systems. Some states are more lenient and will evaluate a project based on the proximity of downgradient, high-quality (i.e., fishable) streams and valuable (i.e., potable) groundwater. The operating and design criteria specified in the permit will reflect these factors. In some states, the water quality and discharge permitting process involves a public notice and a public hearing. Most state water quality and discharge permits take from two to four months to obtain.

Project with Surface Discharges

For projects located in areas where precipitation exceeds evaporation, designing a zero-discharge facility may not be possible. Under these circumstances, cyanide-bearing processing solutions must be detoxified and discharged to the environment, generally into a nearby stream. The Federal Water Pollution Control Act of 1972 as amended by the Clean Water Act of 1977, as administered by the National Pollution Discharge Elimination System permit program (NPDES) mandate that an NPDES permit be obtained from the EPA for all discharges of wastewater to surface waters of the United States. In many states, the EPA has delegated the authority to review and issue NPDES permits to a state agency.

The NPDES permit application asks for a complete description of the effluent treatment plant and the effluent characteristics including the concentration of any pollutants in the discharge. The permit application also requires specific engineering studies, water quality studies and a limited environmental assessment. As part of the NPDES review process, a public notice is placed in a newspaper of general circulation in the area of the proposed discharge. Comments are accepted for a period of 30 days. A public hearing may also be requested by the applicant, the EPA, any public agency or any interested individual. An NPDES permit generally takes about six months to obtain. A permit application must be submitted at least six months before commencement of the discharge. NPDES permits may also be necessary for other project facilities. For example, discharge of non-process related waters from sediment control structures requires an NPDES permit.

COUNTY LAND USE OR ZONING PERMITS

In many counties throughout the country, mining is a permitted use of the land and requires a land use or zoning permit from the county. The application will be evaluated by the county in terms of the projects' compatibility with existing county land use planning and zoning.

The degree of effort required to obtain a county land use permit is quite variable depending upon location. In some states, the permit application is routine and not particularly involved. However, in other states like California, the application for a Conditional Use Permit must be accompanied by an ER. Most county land use permits are subject to public review and comment. The amount of time required to obtain a land use permit is variable and ranges from about two to four months depending upon the county. The CUP is also the most abstract of all the permits. Compatible use is subject to arbitrary opinions and often requires the testimony of various land use experts such as appraisers.

U.S. ARMY CORPS OF ENGINEERS WETLANDS PERMIT

Heap leach gold facilities located in non-arid environments such as South Carolina and Maine (both sites of recent exploration activity) may require a 404 Permit from the U.S. Army Corps of Engineers (COE). A 404 permit is required for any excavation or fill in or immediately adjacent to "wetlands" in the project area. The term wetland refers to "areas that are inundated or saturated by surface or groundwater at a frequency and duration sufficient to support vegetation typically adapted for life in saturated soil conditions." Wetlands generally include swamps, marshes, bogs and similar areas.

A 404 permit application must include detailed engineering designs and substantive documentation. This permit application can also trigger an EIS if it is deemed that there may be a significant environmental impact associated with the project. Assuming an EIS is not required, a 404 permit takes about two months to obtain.

STATE SURFACE MINING AND RECLAMATION PERMITS

Because most heap leach projects involve open-pit mining, many states require an operator to submit a reclamation plan simultaneous with applying for a permit to operate a mine. In general, the reclamation plan must address current land use, rehabilitated land uses, grading, drainage and erosion control, salvaging and reapplication of topsoil, revegetation, monitoring and control of pollutants and safety hazards. Many states also require the applicant to obtain a reclamation bond.

For heap leach projects, most states require that the ponds be backfilled, retopsoiled and revegetated, and that the heaps be recontoured, retopsoiled and revegetated. Some states also require that the heap be covered with an impermeable cap to prevent infiltration of water through the heap and possible groundwater contamination. Most states also have general reclamation requirements for the haul and access roads, open pits and waste rock dumps.

HAZARDOUS WASTE PERMITS AND CLASSIFICATIONS

Many states have hazardous waste statutes similar to the Federal Resource Conservation and Recovery Act (RCRA). These statutes designate certain types of cyanided mine wastes as toxic or hazardous waste subject to taxation and stringent design and operational criteria. Most heap leach projects cannot afford to pay a hazardous waste tax on their spent ore heaps, or to design hazardous waste disposal facilities. It is thus critical to convince the regulatory agencies that spent, cyanided ore on heaps or in spent ore waste dumps is not a hazardous waste and should not be taxed as such, nor should the disposal site dump areas for these products have to be designed using hazardous waste disposal site criteria.

The best way to avoid the hazardous waste designation is to specify a cyanide neutralization and detoxification program for the spent ore. In California, the precedent has been set by two Water Quality Control Boards to regard spent, cyanided wastes as Class C wastes under Subchapter 15 regulations if they meet certain specified cyanide detoxification parameters. A Class C waste facility does not require a double liner and there is no hazardous waste tax. Thus using the proper cyanide detoxification techniques, it is possible to build a single-lined pad and waste the heaps in-situ, or move the spent ore to a single-lined dump area. Cyanide neu-tralization requirements are discussed in the next section.

Some states have hazardous waste facility permits which might be necessary for a heap leach operation. These permits commonly address generation, transportation, storage, treatment and disposal of hazardous wastes, specify spills reporting and clean up procedures and establish financial responsibility for accidents and final closure. The application process may involve a public hearing.

It should be noted that residual cyanide is not the only issue considered when determining a hazardous waste classification. Spent ore heaps and waste rock dumps may be classified as hazardous waste based upon their heavy metal content. This may be a serious problem for operators with certain types of ores containing arsenic, antimony, mercury, etc.

The EPA has recently decided to classify mining wastes as non-hazardous wastes subject to Subtitle D regulations. Existing Subtitle D regulations establish environmental performance standards aimed primarily at municipal and industrial solid wastes. EPA will be developing a body of regulations specifically for mining wastes under Subtitle D. These regulations are likely to address high-acid generation potential, radioactivity, and asbestos and cyanide wastes. Hopefully the new regulations will be flexible enough to accommodate variations in ore mineralogy, processing techniques and site conditions. The EPA will also be developing additional methods of enforcement authority to implement the new Subtitle D regulations.

CYANIDE NEUTRALIZATION REQUIREMENTS

The cyanide neutralization requirements for cyanided heaps vary considerably from state to state. Many states are just beginning to consider this question while some states are currently drafting regulations. Table 2 lists the agencies responsible for establishing cyanide standards for 14 states. In the absence of regulations dealing specifically with cyanide detoxification levels for heap leach operations, most states adopt the EPA-drinking water standard of 0.2 mg/l free cyanide as a default value. It may be very difficult if not impossible for an operator to satisfy this requirement. This is especially true for agglomerated ores (Harry van Drielen, verbal communication).

Generally speaking, those states with some experience in heap leach operations have more reasonable standards and typically analyze each project on a site-by-site basis. The required cyanide detoxification level will be based in part upon the environmental sensitivity of the project. Projects with nearby, downgradient fishable streams and/or potable surface and groundwaters will have more stringent requirements than projects isolated from valuable surface and groundwater resources. Many of these states are more concerned with the design controls to ensure cyanide contaminent rather than specific cyanide neutralization levels *per se*.

Some states (e.g., California and Nevada) specify detailed heap rinsing procedures and analytical methods for cyanide detection as operating conditions in their water quality or waste discharge permits. Some states are not yet this experienced in dealing with heap leach projects and have less specific regulations which may be ambiguous. As mentioned above, the less experienced states may also impose drinking water or fish toxicity standards on heap leach projects. Table 3 summarizes cyanide neutralization requirements for selected states.

The best strategy for avoiding unrealistically stringent and possibly unattainable cyanide neutralization requirements is to convince the regulatory authorities that the project is designed to ensure cyanide containment and to prevent cyanide migration and surface and groundwater contamination during and after project operations. Using this approach, an operator may be able to negotiate for more lenient neutralization requirements while agreeing to diligently attempt neutralization to a more stringent, "target" cyanide concentration.

PERMITTING STRATEGY

The first step in obtaining permits for a heap leach project is to carefully and quietly check out the regulatory requirements. This should be done at a fairly early stage, and can normally be accomplished in several days by personnel with experience in heap leach projects and the regulatory framework of the respective state. Once the regulatory requirements have been identified and evaluated, an operator should prepare conceptual designs and a preliminary plan of operations to be presented informally to the appropriate regulatory agencies. These initial, informal meetings will give the agencies an opportunity to voice any specific concerns about the proposed operation. This meeting will also help set the pace and tone of the ensuing permitting activities. These meetings also help establish the credibility of the operator to the regulatory personnel. If possible, agency personnel should be encouraged to visit other heap leach operations and to talk to other, more experienced regulatory people.

It is important to remember that agency personnel have to answer to their superiors and thus to the public. Every effort should be made to incorporate (where reasonable) known or suspected concerns into the overall design. Cutting corners in a simple process like permitting a heap leach project is not a good idea. It is well worth the effort to spend an extra day or two explaining the process to a key agency person who can then explain it to the concerned public. In addition, various regulators (particularly those who do not understand heap leaching) often ask for what appear to be unnecessary types of information and/or considerations. Do not dismiss these requests as superfluous. It often pays to incorporate some inexpensive items into an application or design if it serves to make the permitting process easier.

TABLE 2

STATE AGENCIES RESPONSIBLE FOR ESTABLISHING CYANIDE STANDARDS

STATE	ORGANIZATION
Alaska	Department of Environmental Conservation (907/425-1714)
Arizona	State Mine Inspector's Office (602/255-5971)
California	Regional Water Quality Control Board: North Coast Region (707/545-2620) San Francisco Bay Region(415/464-1255) Central Coast Region (805/549-3147) Los Angeles Region (213/620-4460) Central Valley Region (916/445-0270) Lahontan Region (916/544-3481) Colorado River Basin Region (714/346-7491) Santa Ana Region (714/684-9330) San Diego Region (714/286-5114)
Colorado	Dept. of Natural Resources/Mined Land Reclamation Division (303/866-3567)
Idaho	Dept. of Health and Welfare Division of Environment (208/334-4784)
Montana	Dept. of State Lands (406/444-2074)
Nevada	Dept. of Conservation and Natural Resources/Division of Environmental Protection (702/885-4670)
New Mexico	Groundwater Section New Mexico Environmental Improvement Division
North Dakota	Health Dept - Water Supply and Pollution Control (701/224-2354)
Oregon	Dept. of Environmental Quality - Water Quality Division (503/229-5325)
South Carolina	South Carolina Land Resources Commission Div. of Mining and Environment (803/758-2823)
Utah	Div. of Environmental Health/Water Pollution Control (801/533-6146)
Washington	Div. of Geology and Earth Resources/ Dept. of Natural Resources (206/459-6372)
Wyoming	Water Quality Division (307/777-7756)

TABLE 3

SUMMARY OF STATE CYANIDE REGULATIONS

STATE	CYANIDE REGULATIONS	COMMENTS
Alaska	No specified standards	No experience with cyanide heap leach mining
Arizona	0.2 mg/l free CN^-	State health standards apply. Assume zero discharge. Discourage hypochlorite neutralization.
California		
Northern Lahontan Region	0.4 mg/l total CN^- 0.2 mg/l free CN^-	Standards apply to remaining liquid in ponds, heaps, etc. Also have solids extraction test of 4.0 mg total CN^-/kg spent ore and 1.6 mg/kg free CN^-/kg spent ore.
Southern Lahontan Region	0.2 mg/l free CN^-	Standard applies to remaining liquids in ponds, heaps, etc. Also have solids extraction standard of 10 mg total CN^-/kg tailings.
Colorado River Basin Region	no specified standard (precedent of 10 mg/l free CN^- for Mesquite)	Statistical approach: 90 percent of at least 10 samples must contain 10 mg/l free CN^-; no sample can contain 20 mg/l Testing and sampling procedures specified for 5:1 extraction, analytical method specified.
Colorado	No set standards (will be redrafting regulations)	Evaluate each site for quality of receiving surface and groundwater. Emphasize zero-discharge, design controls to ensure containment.
Idaho	No set standards (in process of drafting regulations)*	Regulations will not specify one, universal CN^- standard. Will evaluate each site-proximity to surface water, quality of receiving water, etc. Stibnite Mine permit specifies 0.2 mg/l free CN^-.
Montana	No set standards (nondegradation rule)	Determined by quality of receiving water. Nondegradation implies a drinking water standard (0.2 mg/l free CN^-) if high-quality receiving waters. Permits could be more lenient for lined facilities with caps and diversions to prevent drainage into streams.
New Mexico	0.2 mg/l total CN^-	Not an effluent standard _per se_. Will take into consideration attenuation and dilution (i.e., can have 0.2 mg/l total CN^- in effluent). Ortiz rinse test: 100 lb sample rinsed with 1 gallon fresh water; filtrate must be 100 ppm free CN^-. Might accept this for other lined facilities.
Nevada	No set standards 0.2 mg/l free CN^- is a "target" concentration (most operators can't meet)	Evaluate each site for proximity to valuable surface water and modify "target" concentration accordingly. Rinsing and testing procedures differ for agglomerated vs. nonagglomerated ore. Discourage hypochlorite and/or peroxide neutralization methods.

TABLE 3 (cont'd)
SUMMARY OF STATE CYANIDE REGULATIONS

STATE	CYANIDE REGULATIONS	COMMENTS
South Dakota	No set standards	Current standards range from .75 ppm total CN to .2 ppm free CN. Standards vary according to proximity to surface and ground waters.
North Dakota	No set standards	No heap leach mines. Would probably impose surface water quality standards.
Oregon	.01 mg/l total CN^-	Standard based on results of bench-scale test. No active heap leach projects in state. Advocate hypochlorite rinse.
South Carolina	No set standards (Haile precedent of 10 mg/l free CN^-	Will evaluate each site. After closure heaps at Haile will drain into stream w/pH of 4 and elevated CN^- content.
Utah	No set standards (Mercur precedent of 5 ppm free CN^-)	Would look at Mercur permit as a precedent but not a standard. Require double-lined pads w/leak detection system. Mercur uses permanent/expandable pads (i.e., heaps wasted in situ).
Washington	No set standards (in process of drafting regulations)	Will probably evaluate each site, no measurable CN^- allowed in effluent from reclaimed facilities.
Wyoming	.02 mg/l free CN^-	Based on quality of receiving water, effluent can't exceed ambient water quality. Would be more lenient for zero-discharge facilities (i.e., if no surface of groundwater impacts)

*In process

Chapter 32

EXPEDITING THE ENVIRONMENTAL PLANNING PROCESS FOR SMALL MINE PERMIT APPLICATIONS

by

Gary J. Fletcher
Project Manager
Fletcher Associates
Aurora, Colorado

Mining and processing activities which are proposed on private land, state leases, or unpatented mining claims are governed by numerous local, state, and federal laws which identify the procedures and conditions which must be met before resources can be developed. The respective government agencies responsible for the administration and enforcement of these laws have promulgated supporting regulations and guidelines, and have implemented certain permit and approval processes to ensure compliance.

Each exploration project has a unique environmental setting and location which further dictates the types of air, water, hazardous waste, and wildlife laws and regulations which must be complied with before development commences.

The environmental planning process should be the first step in the evaluation of existing company properties and the potential acquisition of other mineral properties. However, many companies don't consider the environmental requirements for a project until the decision is made to develop the property for mining and processing.

If a mining company wants to expedite the environmental planning process, it must first develop an in-house process for evaluating the project and determining the environmental approval procedures in the state where the project is located.

The first stage of environmental planning involves general information which can set the overall company philosophy for development of projects.

I. GENERAL INFORMATION

A. Prepare a conceptualized description of the proposed project, including the types of mining and possible alternative processing methods.

B. Establish a time schedule for production. There may be critical delivery dates or financing commitments which can not be postponed.

C. Conduct a site tour to generally familiarize the development team with the site and check for fatal flaws which could change or delay the proposed project.

D. Select one person to act as liaison with the regulatory agencies responsible for issuing environmental permits.

E. Determine the company policy for the level of environmental protection and studies that would be required to satisfy stockholders.

After the company has set the general parameters for environmental planning and has selected a person to coordinate the process, specific information must then be acquired. The following information can be used for planning on all types of environmental work. The step-by-step process is essential to determine if a project can be expedited, and particularly which specific steps can be shortened.

II. SPECIFIC INFORMATION

A. Analysis of the exact location of the project is the first step in determining which agencies might be involved. This step should identify surface and mineral owners and adjacent land owners.

B. A preliminary development plan should be prepared. This will be a coordinated effort among the exploration, mining, and processing staffs. This plan should be of sufficient detail for discussions with the regulatory agencies and for determining environmental studies.

C. Review and evaluate the local, state, and federal laws, regulations and guidelines, application requirements, and mandatory review time schedules pertinent to the proposed project.

D. Meet with the local, state, and federal agencies which will have approval authority to discuss the preliminary plan and obtain agency input to the overall planning process. Identify special permit conditions and past experience the agencies have had in approving other similar projects.

E. Identify other proposed or pending laws and/or regulations which, if enacted, might materially affect the proposed project.

F. Collect and evaluate existing available environmental baseline data which would satisfy permit requirements.

G. Prepare a draft Plan of Study to obtain additional environmental baseline data which will be required to prepare complete and adequate permit applications.

H. Meet with the local, state or federal agencies which will take the lead in the approval process, review the Plan of Study, and obtain agency concurrence.

I. Identify special interest groups which will be "pro" or "con" regarding the proposed development, and determine the best method to approach the group.

J. Develop and maintain a close working relationship with the agencies and key staff members.

K. Obtain proposals from consulting firms to obtain baseline data and prepare reports which will aid in the permit process.

L. Review field data with consultants to identify potential problems and adjust the scope of work to obtain adequate baseline data.

M. Prepare and submit permit applications.

The environmental planning process can be expedited to save time and expenses for certain types of mining development. The process can also be expedited to meet critical time schedules, but there will not necessarily be a saving in expenses.

The methods which have been successfully used are listed below, and each project must be reviewed to determine if any or all of the steps will be required.

A. The company must make the decision that all phases of the project will be expedited.

B. One person in the company should be appointed as the spokesman to interface with the regulatory agencies.

C. The individual responsible for interfacing with the agencies must have a complete understanding of the laws, regulations, and guidelines which dictate the process for approval.

D. A personal and sincere approach to understanding the regulatory process, work load of the agencies, and realistic time schedule is essential.

E. The development program must be designed to have a minimal impact on the environment or contain mitigating measures which will offset the impacts.

F. All critical information should be hand-carried to its destination and explained.

G. All requests or questions must be responded to promptly and clearly.

Each project can be expedited in the environmental planning process. Reducing the time in the preparation, review, submittal, and response steps of the permit schedule will mean increased earnings to the mining company.

Mr. Fletcher prepares Regulatory Handbooks for mining companies in the western states to expedite the environmental approval process for specific sites. Mr. Fletcher can be reached at (303) 693-2516.

Chapter 33

GEOCHEMICAL ASPECTS OF ENVIRONMENTAL PERMITTING

by

Mr. Jeffrey A. Holcomb

EnecoTech, Inc.
Senior Engineer
Denver, Colorado

INTRODUCTION

The passage of Comprehensive Environmental Response Compensation and Liability Act (CERCLA/Superfund) as well as the Resource Conservation and Recovery Act (RCRA) and state laws, have placed added regulatory pressures on the mining industry. These laws which focus on the regulation of hazardous substances and wastes for the protection of the public health and welfare and the environment, present the potential for tremendous financial impacts and liabilities on existing mine operators and owners and/or potential mine operators/owners.

This paper takes a brief look at these laws and presents an environmental audit procedure to identify environmental liabilities prior to mine land transactions. The environmental audit approach presented is an overview of the procedures that could be used to document, evaluate and inventory existing and/or potential environmental concerns. This audit procedure can be applied to developed or undeveloped properties and at active or inactive mining sites.

REGULATORY OVERVIEW

Mining environmental liabilities have been addressed in RCRA, Superfund, the Clean Air Act (CAA) and the Clean Water Act (CWA). Even though the U.S. Environmental Protection Agency (EPA) has determined that mining wastes will be treated as solid wastes under RCRA and not as hazardous wastes, these wastes could still bring about a great deal of financial liability under Superfund. The catch is if the mine wastes generated contain hazardous substances such as the heavy metals arsenic, beryllium, cadmium, chromium, copper, lead, mercury, nickel, selenium, silver, thallium, zinc and others as well as cyanide. Thus, it is not only important for a mine operator/owner or a potential mine operator/owner to know if the mine wastes generated are hazardous, but it is important to know if the wastes contain hazardous substances. Following is a brief overview of RCRA and Superfund and their potential implications.

RCRA

RCRA, which provides the basis for the regulation of solid and hazardous wastes, is applicable to active mining and milling sites. Within RCRA the definition of a hazardous waste is provided. This definition applies to owners and operators of mining and milling sites if the operator has stored, disposed or treated hazardous wastes on-site. If this is the case and if strict process control and waste management activities are not adhered to, environmental liabilities may be associated with the site.

Superfund

The 1980 Superfund law can be applicable to inactive mining sites or to inactive sections of an active facility. The provisions of Superfund call for the regulation of hazardous wastes and hazardous substances disposal and release not only by the existing owner, but also for any previous owner. This concept of joint and several liability means that one party, as among many, can be held totally responsible for the costs of hazardous waste cleanup (Haller, 1987). Furthermore, liabilities under Superfund have also been applied back to past acts; thus, several mining properties, active or inactive prior to the 1980 Superfund law, are at risk of being placed on the EPA's National Priority List (NPL) for cleanup.

Last years amendments to Superfund, known as the Superfund Reauthorization Act of 1986 (SARA), provided some significant developments for the mining industry. The major requirement of the amendments as it pertains to owners or operators of mining sites is that the EPA must revise its present Hazard Ranking System (HRS). This is important because in most cases the primary reason for several mining sites having been included on the NPL was due to the large volume of wastes generated. As is typical at most mining sites, large volumes of low toxicity wastes are generated and the current HRS model prejudices against this. It should be noted that this provision only applies to sites that are not already on the NPL or have not been proposed for inclusion on the list.

However, this does not mean that mining site owners/operators will be able to escape environmental liability and responsibility. The joint and several liability clause as well as retroactive liability has not been changed. The only way an owner/operator can escape liability under the 1986 amendments is to prove several things by a preponderance of evidence (Haller, 1987).

Land owner liability is a provision of the 1986 amendments that allows an owner/operator to escape liability if he/she can provide evidence that he/she did not know of the presence of hazardous substances at the property when it was acquired. In order to do this, the owner must show that he/she undertook an adequate investigation to determine the presence of hazardous substances at the site. What this provision does is place a premium on the buyers efforts to thoroughly investigate any mining site he/she is contemplating buying. Thus, SARA places the burden on the buyer to undertake an adequate environmental investigation of the property. The following sections describe a general environmental auditing procedure that can be used to identify these environmental liabilities.

AUDIT APPROACH

As a result of Superfund and SARA, any potential buyer of a mining property, whether it is a developed or undeveloped property or an inactive or active site, should perform an environmental audit to identify the potential environmental liabilities and risks that may be associated with the property or site. The audit is a thorough investigation of a facility's operations, procedures and records, environmental practices and should include the gathering of information pertaining to compliance with federal, state and local environmental regulatory requirements. The audit should also identify not only the presence of hazardous wastes, but also hazardous substances. The audit can be performed in-house, but it should be performed by a third party with no financial interest in the property.

As far as performing the audit at developed versus undeveloped properties and at activie versus inactive sites, each property and/or site is unique and not all aspects of the audit will be necessary to perform. The general objectives of the environmental audit are as follows:

- To determine the presence and levels of hazardous wastes on-site;
- To determine the presence and levels of hazardous substances on-site; and
- To determine the potential for off-site release of hazardous wastes or hazardous substances to the environment.

It should be noted that this type of audit is not a total environmental assessment or site characterization effort. The primary objective is to identify "hot spots" or areas of concern at the site under question and to make an evaluation based on the existing data (if available) and data generated during the audit. It is a quick in and out effort to minimize costs, but yet generate the maximum amount of information needed to make a fair assessment of the property or site.

The auditing procedure is conducted using a phased approach. Phase I consists of activities performed before the on-site audit; Phase II consists of conducting the audit field work; and Phase III is work completed after the field audit and laboratory work is done. All audits involve the gathering of information, analyzing facts, making judgements about the status of the site and reporting the results.

The activities performed for Phase I include, but are not limited to the following:

- Property ownership review;
- Royalties review;
- Lease/patents review;
- Compliance records review;
- Examination of available data;
- Facility personnel interviews;
- Exploration records review;
- Land use evaluation;
- Facility operation evaluation;
- Pollution control review;
- Solid and hazardous waste management evaluation;
- PCB record review;
- SPCC review;
- Hazardous/chemical waste disposal review; and
- QA/QC procedures review.

The purpose of Phase I is to perform a detailed review of existing available information pertaining to the site or property so the development of a focused, cost-effective and optimal field audit program is achieved.

The on-site, Phase II, activities are based on the findings of Phase I and include, but are not limited to:

- Process stream sampling;
- Process waste stream sampling;
- Groundwater investigation;
- Surface water investigation;
- Air sampling;
- Soil sampling;
- Hazardous waste sampling;
- Facility inspections; and
- Laboratory analyses.

As stated above, the Phase II activities should be directed towards identifying the property or site "hot spots" in order to develop worst case scenarios for evaluating the property or site.

Phase III activities include, but are not limited to:

- Laboratory data analysis;
- Report preparation; and
- Report submittal and review.

The purpose of the report is to inform the owner/operator and/or buyer of the property's or site's environmental risk. The report should also provide recommendations to the owner/operator/buyer whether the deal should be reconsidered or evaluated further.

Analytical Requirements

Although there are no specific analytical requirements or parameters that should be analyzed for in an environmental audit, the analytical portion of the audit should also be designed to target specific parameters. Each analytical program should be site specific and should be designed to minimize costs as the laboratory costs associated with any environmental audit can account for the majority of the costs.

Analytical parameters to be analyzed for include but are not limited to:

- Priority pollutant heavy metals;
- Cyanide including free cyanide, cyanide amenable to chlorination and total cyanide;
- Organic priority pollutants;
- PCB's; and
- Pesticides.

The above list is fairly all inclusive and not all the tests may be necessary. Again, the analytical program should be site specific.

Other tests to consider are the EP Toxicity test, bioassay tests and other site specific tests that are developed.

Costs

As stated earlier, the objective of using a phased approach is to minimize costs. The audit is a quick in and out effort while maximizing and optimizing the information needed to make an assessment of the site.

The costs to conduct an audit are directly related to site and facility location, the site size, mining and facility operational procedures, how detailed the an audit is performed and the environmental sensitivities assocaited with the site. The costs can vary from approximately $2,000 and up depending on the site and the level of effort desired by the owner/operator/buyer.

CONCLUSIONS

The potential financial liabilities associated with acquiring a mining property without performing an environmental audit could have an adverse impact on the environment and the financial viability of the new owner/operator. The environmental audit procedure presented has been used successfully at several mining properties in the western United States in identifying potential environmental risks. All potential owners considering acquiring new mining properties should be aware as it is becoming imperative that the environmental risks be identified prior to mine land transfers.

References

Haller, P., 1987, "Mining, EPA, and toxic waste: Will your mining company be liable?," Mining Engineering, February, pp. 109-112.

Chapter 34

GOLDEN APPLES OR SOUR GRAPES: MINING IN APPLE COUNTRY

Marie Garrett
Environmental Specialist
Asamera Minerals (U.S.) Inc.
Wenatchee, Washington

Timothy M. Dyhr
Permit and Environmental Coordinator
Asamera Minerals (U.S.) Inc.
Reno, Nevada

Abstract The Cannon Mine gold project has been perceived as unique for being developed in an area characterized by orchards and suburban residences. The Cannon Mine was permitted and developed successfully in this setting. However, the ability of companies to permit, develop and operate mines in populated areas is becoming more difficult as a result of opposition by residents and increasingly stringent regulation. The challenge to the mining operator in the future on otherwise economically attractive mineral deposits will be the ability to explore and develop these types of projects successfully and cost-effectively.

PROJECT DESCRIPTION

The Cannon Mine lies on the outskirts of the town of Wenatchee, Washington. Wenatchee is located on the Columbia River in the central part of the state, midway between Seattle and Spokane. The Cannon Mine project location is shown on Figure 1.

The relatively mild climate and abundance of water for irrigation has contributed to the development of the Wenatchee Valley as a major fruit-producing center. The town calls itself the "Apple Capital of the World". The area has a gold mining history that dates back to the 1880's.

Proven and probable ore reserves at the Cannon Mine are in excess of 3.4 Mt (3.8 million st) at an average gold grade of 8.72 g/t (0.279 oz per st). Present throughput averages 1360 tons (1500 st) of ore per day.

PROJECT BACKGROUND

The Cannon Mine began as a gold exploration project by Asamera Minerals with initial acquisition and leasing in 1981. With the core land package completed in mid-1982, exploration drilling was begun. Delineation of known or suspected areas of mineralization and, most importantly, the major discovery of the previously unknown mineralized areas, was achieved by March, 1983. At about the same time the decision to proceed with development was made. In September, 1983, the permitting process was officially commenced.

Environmental baseline studies, engineering studies and permit applications were initiated at that time. Approximately twenty-seven permits and approvals were required for construction of the project. The primary approval was the county zone change from suburban residential and general use zones to a Planned Unit Development. Prior to action on any of the permits, approval of the Environmental Impact Statement under Washington's State Environmental Policy Act (SEPA) was required. In July, 1984, approximately eleven months after commencing the permitting process, the EIS and zone change were approved by the Chelan County Board of Commissioners. Within weeks afterward, construction of the major above-ground facilities, the mill, tailings embankment and warehouse/office complex commenced.

PROJECT PERMITTING - THE STORY BEGINS

Washington Regulatory Framework

In the State of Washington, the principal environmental regulatory agency is the Department of Ecology. The Department was empowered with broad regulatory authority under the SEPA. Additional state programs created to comply with minimum federal standards (air quality, water quality and dangerous/hazardous wastes) are under the jurisdicition of the Department of Ecology.

Under SEPA, any activity by a project sponsor which requires a decision, usually a permit, by an agency, triggers the

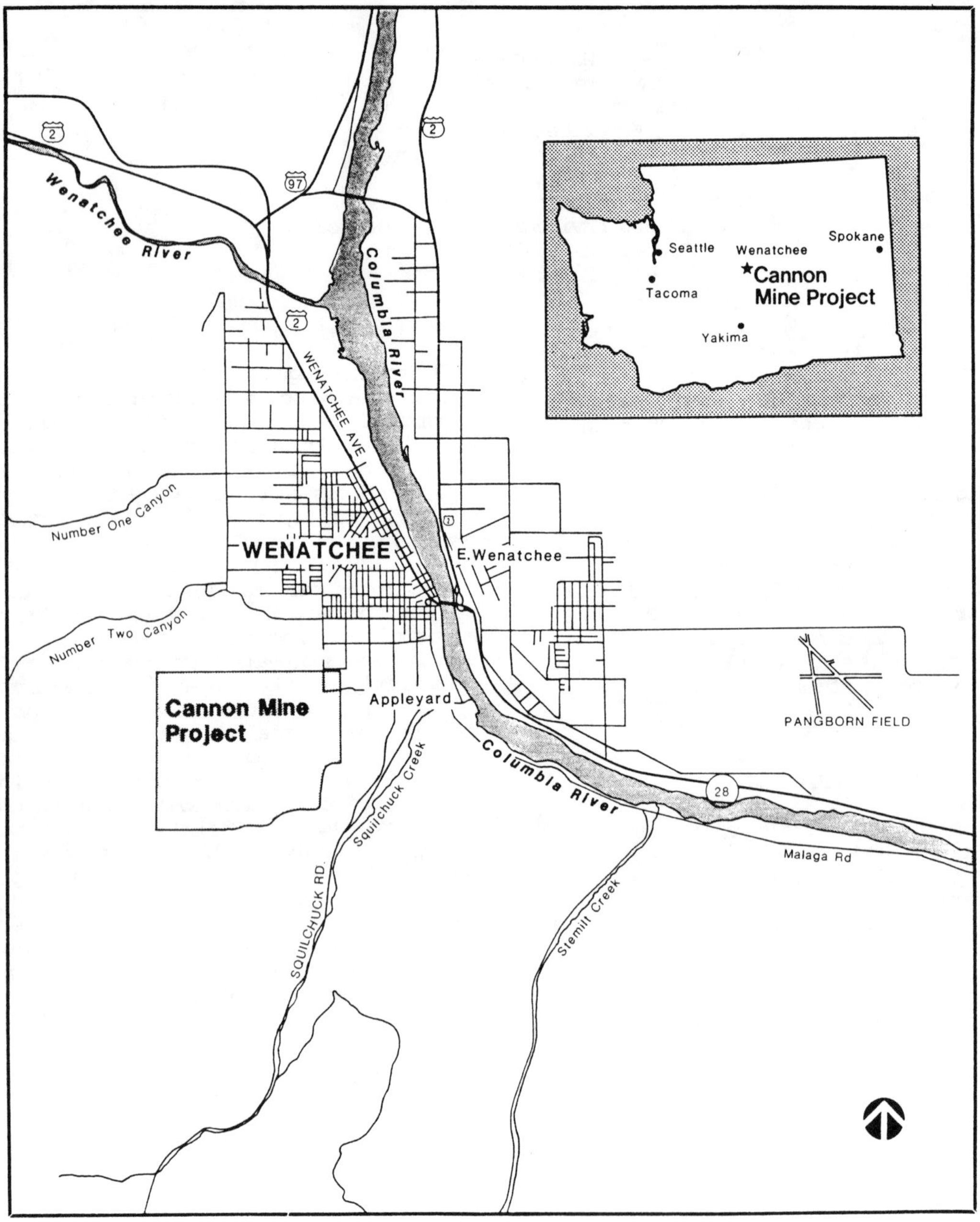

Figure 1. Cannon Mine Project Location

environmmental review process. There are numerous exemptions to SEPA, primarily agricultural, residential and minimum threshold activities. In many cases, a standard environmental checklist can be completed, resulting in a "determination of non-significance". In those cases the permit or approval can be issued without preparation of a formal Environmental Impact Statement.

If, in reviewing the checklist, an agency determines that the action or project could result in significant adverse environmental impacts, an Environmental Impact Statement is required.

Although the possibility of achieving a declaration of non-significance was examined at the onset of the Cannon Mine project, it was apparent it was highly unlikely for a project of this size.

Once it was apparent that an EIS was required, the procedures for completing the EIS needed to be identified. Also, under SEPA, as with other state and federal actions where a number of permits or approvals are required, it was necessary to designate a lead agency.

In the case of a mining project, for which the majority of approvals come under the jurisdiction Department of Ecology, it automatically becomes the lead agency under SEPA. However, with the proximity of the Cannon Mine project to residential neighborhoods, Chelan County strongly desired lead agency status. Ultimately, the County and Department of Ecology became co-lead agencies, with the County assuming the principal clearing house role as "nomimal lead". The SEPA process and agencies and permits are outlined in Figure 2.

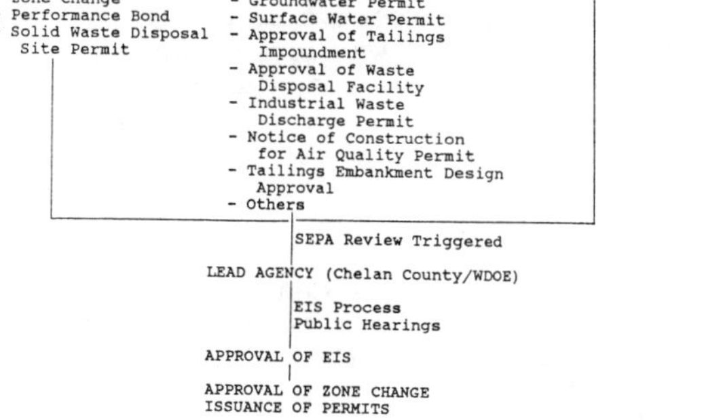

Figure 2. The SEPA Process

There were no federal permits required for the Cannon Mine, principally because Washington has federally approved state programs in most federally mandated areas. The potential requirement for a federal NPDES (National Pollutant Discharge Elimination System) permit was removed by designing the tailings impoundment and other facilities as "zero discharge" facilities.

Key Elements of the Permitting Process

Aside from the regulation rule-book, other aspects of the permitting maze were identified initially in the Cannon Mine permitting process. In addition to scoping out the regulatory and technical aspects of the envronmental review procees, it was necessary to identify the key issues and key decisions necessary to successfully permit the Cannon Mine.

The Agencies In early 1982, contacts were made with the two principal agencies affecting the Cannon Mine Project, the Department of Ecology and Chelan County. These early contacts greatly facilitated the permitting process. What was apparent in those early contacts and during initial permitting was that both agencies would act in a cooperative and facilitating role. This was in contrast to indifferent and perhaps adversarial roles that can be exhibited by regulatory agencies.

The level of cooperation was evidenced in the county's approval of shaft-sinking, concurrent with preparation of the EIS. Typically, this type of activity would not be allowed in an environmental review process. However, the company presented plausible arguments that access to the ore body was crucial toward development of the recovery process. In fact, access to the ore body provided important information pertinent to the permitting process. This approval was clearly conditioned with requirements that a full environmental assessment be done, a formal Environmental Impact Statement be prepared, and that a zone change for a Planned Unit Development was necessary. Public concern surfaced during the permitting process relative to this activity, and its lack of more formal review and approval. However, the shaft-sinking had relatively few environmental consequences and Asamera did implement procedures to mitigate the impacts, primarily noise and traffic. In the end, Asamera's ability to sink the shaft and decline were a boon to the development of the project on a very favorable time schedule.

During the review process there were almost no turf battles between agencies. In regard to typically socio-economic considerations, Chelan County played the lead role. In regard to technical considerations (air, water, dam safety, wildlife, reclamation) Chelan County deferred to the respective state agency with jurisdiction. There was, however, one case where there was some difficulty. This was in regard to the performance bond for project reclamation and closure. Toward the end of the review process, the performance bond issue remained unresolved. Neither the amount or jurisdiction had been settled. It appeared that both the Washington Department of Natural Resources and Chelan County would require a bond covering the same portions of the project. In the end, the Department of Natural Resources determined that its jurisdiction was over the reclamation of "surface mining" portions of the project and the primary bond covering all other aspects of the project was posted in favor of Chelan County. Ultimately, the critical factor in resolution of the performance bond issue was timing. Lack of resolution would have delayed approval of the EIS and zone change, delaying commencement of project construction activities which had been tentatively scheduled.

Finally, the level of cooperation amongst agencies was evidenced by the company's ability to convene meetings amongst them when critical decisions were required.

The Project Sponsor Some of the key factors in facilitating the permitting process were the various players working on behalf of Asamera. The key players were the project manager and project personnel, environmental consultant, engineering consultant and legal counsel.

From the onset, the project manager took an active, if not lead role in the permitting process. Not only did he monitor the permitting process, but had a thorough understanding of both the permitting aspects and the project development, construction and operation details. This gave the company great credibility with agency personnel, the press and the public. During public hearings there were only a few technical questions that the manager could not answer and these were answered by technical specialists in attendance at the meetings. During several extended, late night public hearings when the various decision-makers were wavering on a decision, this greatly improved the company's ability to obtain necessary approvals.

Both the environmental consultant and engineering consultant were active participants in the permitting process. Both had experience in the state of Washington and had thorough technical expertise. There was on-going dialogue amongst company representatives, agency personnel and the consultants. Our ability to prepare and present technically sound, clearly written documents, not only under excruciating time schedules, but also in compliance with regulatory requirements, virtually eliminated delays due to inadequate information. Perhaps most importantly, these documents were understandable to the general public and decision-makers.

Numerous company employees actively participated in the permitting process. This had two major effects: to facilitate preparation of the Environmental Impact Statement and permit applications and to communicate to the public a general understanding of the project.

Finally, legal counsel was kept involved during the permitting process to absolutely insure compliance with statutory and regulatory procedural requirements. Inadvertent procedural errors by the agencies and Asamera were prevented. These types of errors can often cancel out even the most sound technical base and present substantial opportunities for opponents to delay a project. There are numerous cases where errors of this type have delayed projects indefinitely.

The Public Inevitably, someone is against whatever you propose. The Cannon Mine project was not without public opposition. In fact, prior activity in the project area had generated a small, but real, undercurrent of concern by the public. Prior to Asamera's appearance on the scene, there had been much discussion of, and opposition to, mining, particularly open pit mining, on the site. However, early in Asamera's involvement, it was apparent that an open pit mine, although feasible and desirable from a strictly technical point of view, would run into a brick wall of public opposition. Therefore, Asamera made the decision to develop an underground mine.

Even with the decision to mine by underground methods, it was apparent that there was still significant potential for opposition. The project site was literally in suburban Wenatchee in a residential area. Major disturbances associated with the project would be readily visible from much of the town. In addition, the major discovery associated with the project had generated a flurry of exploration activity in the area. Residents had rigs drilling twenty-four hours a day, literally in their backyards. The local paper regularly featured mining and exploration activity occuring in Wenatchee, and how it was affecting people in this otherwise pastoral orchard community. A committee called Citizens Concerned About Mining was formed, which articulated concerns about the environmental review process, primarily espousing the need for more studies and more time to adequately evaluate environmental and social impacts of the project.

Ultimately, a strong technical basis, regular communication with regulatory personnel and an open and receptive attitude toward the public prevailed in obtaining necessary approvals to proceed with development of the mine.

ENVIRONMENTAL PERMIT COMPLIANCE: THE REST OF THE STORY

Obtaining the environmental permits and approvals necessary for start-up of a mining facility may require a major effort by the project proponent. Successful permitting is, however, only a first step in the regulatory process. The relationships developed between the mine operator and the regulatory agencies is one that continues through the life of the project, as each permit will contain conditions for compliance and there will be statutory requirements for reporting of non-compliance situations.

The importance of permit compliance cannot be stressed enough. When the permits required for project construction are obtained, there is a tendency to believe that the permitting phase is over. However, operations that are chronically out of compliance or that fail to report upset, non-compliance situations take a substantial risk. Regulators take an understandably jaundiced view of "problem" operations, and it can be very difficult to approach an agency for additional or modified permits once the operatory/agency relationship is damaged. An operation's public image is, if anything, even more fragile, and public relations efforts can be undone by a single unfortunate incident.

Regulatory agencies may in the past have taken a laissez-faire approach to verification of permit compliance. There is, however, a strong trend toward close scrutiny of permitted activities, particularly as individual states obtain primacy over federal environmental programs. Sponsors of mine projects now coming on line can attest to the increasing importance being placed on demonstrable compliance, and operators in the near future can expect requirements for environmental monitoring to become routine.

Environmental Monitoring Programs

Since few governmental agencies have the resources to conduct regular, frequent monitoring of environmental conditions at a given project site, the burden of monitoring will most often fall on the project proponent. Because permits are likely to contain monitoring and reporting requirements, development of a monitoring strategy is best initiated early and in concert with the rest of the permitting process.

In formulating a monitoring program, there are again several benefits to be gained by working closely with regulatory personnel. Agencies may have discretionary powers on items such as sampling parameters and frequency, and close contact will encourage discussion and help to insure that the project sponsor has input on these important issues. The agencies will in many cases publish detailed sampling procedures that can be incorporated directly into the monitoring program. The regulatory agencies represent a significant fund of technical knowledge, and it makes sense to take advantage of that expertise when possible.

A monitoring program need not be highly complex as long as it provides sufficient meaningful data; the final objective is to develop a monitoring plan that is both workable for the operator and acceptable to the agencies involved.

Ideally, the mine operator's aim in monitoring will be to validate impact predictions and to demonstrate that no unacceptable adverse environmental impacts result from his operation. In practice, monitoring will also identify unanticipated impacts and allow the modification of operating procedures in order to insure compliance with regulatory standards.

With these goals in mind, it should be obvious that a monitoring program will be largely meaningless unless environmental baseline conditions are defined prior to project start-up. Again, early initiation of environmental sampling and study will establish preexisting conditions and help to form a rational basis for discussion of potential impacts.

The various aspects of the environment that must be monitored will depend on the characteristics of the project and the project site. The specific interests of the permitting agencies overseeing the monitoring program are also likely to play a role. For any of the increasing number of mines located near population centers or other sensitive areas, a monitoring porgram may well include the full range of environmental elements.

Monitoring at the Cannon Mine Project

Environmental monitoring and baseline studies at the Cannon Mine site commenced in mid-1983, two years before the mine went into full production. The data obtained from the initial studies were compiled into a single document and served as technical support for the Environmental Impact Statement.

The initial studies addressed seven major areas:

- Air quality
- Meteorology
- Surface water quantity and quality
- Ground water quantity and quality'
- Noise levels
- Vegetation
- Wildlife

Monitoring and sampling in these areas have continued throughout the project construction and operations phases, and in some cases will continue after decommissioning of the mine. The monitoring schedule through the post-decommissioning phase is shown in Table 1.

	CONSTRUCTION	OPERATION OF MINE YEAR 1 YEAR 2 YEAR 3 YEAR 4 PAST YEAR 4	DECOMMISSIONING
AIR	24 Hours Every 3 Days →	→	Until Decommissioning Activities Cease
METEORLOGICAL	Continuous →	→	Until Decommissioning Activities Cease
GROUND H_2O QUALITY	Monthly 1 Year, Then Quarterly →	→ Semi-Annually →	Annually for 4 Years
GROUND H_2O QUANTITY	Monthly →	→ Quarterly →	Semi-Annually for 4 Years
SURFACE H_2O	Quarterly →	→ Semi-Annually →	Annually for 4 Years
SURFACE H_2O QUANTITY	Every 3 Days	Continuous →	
NOISE	Monthly	Quarterly →	
VEGETATION	Annually	Monitor All Revegetated Areas for 2 Years Following Initial Revegetation	Until Acceptable Standards of Reclamation are Achieved
WILDLIFE	Checked Monthly	Follow Schedule Established from First 6 Months	

Table 1. Monitoring Program Schedule

The impact of mining-related activities on air quality can be an important issue in even relatively unpopulated areas; at the Wenatchee mine site, fugitive and process-related dust is a primary concern to area residents and mine management. For that reason, ambient air quality measurements are obtained on a frequency of every three three days. Meteorological monitoring is conducted in concert with air sampling so that particulate concentrations can be correlated with local weather conditions.

Surface water conditions are monitored to gauge the effect of the project on surface water quality, and to support the calculations used in the design of the tailings impoundment. Similarly, monitoring of ground water conditions helps to track the effect, particularly of the tailings impoundment, on the local hydrogeologic regime.

Noise from mining, milling and surface drilling operations is an ongoing concern to local residents. In reality, both noise and dust resulting from mine operations tend not to cause far-reaching degradation in that they do not linger in the environment and cease when mining ceases. However, both these impacts can serve as major irritants to nearby residents and may therefore become major issues with the regulatory agencies.

At the Cannon Mine, several relatively simple measures have been successful in dealing with the noise issue. Mill facilities were located in such a way as to be as unobtrusive as possible, and the main concentrator building is insulated with sound-absorbing material. Diamond drilling conducted on the surface takes place behind strategically located plywood and foam noise barriers, and areas that are being surface-mined are furnished with concealing earthen berms.

Monitoring of impacts on wildlife is a minor aspect of the environmental program. The mine does, however, participate with the state Department of Game in a supplemental feeding program for deer and game birds.

Vegetation monitoring is used to assess the success of reclamation efforts on drill sites and other mine-related disturbances. This portion of the program will assume a greater importance at mine decommissioning when final reclamation of the mine site and tailings impoundment is accomplished. As shown in Table 1, monitoring of reclamation success will continue for some time after final mine closure.

The results of the monitoring program are compiled quarterly and are presented to regulatory personnel and a mine advisory group consisting of six local residents. Concerns and questions regarding compliance and general mine operations are aired and addressed at these meetings, which serve as an important communications link between the mine, the agencies and the public.

CONCLUSIONS

The permitting situation experienced at the Cannon Mine is representative of a growing number of mine projects where mining is competing with other land uses. Mining is often perceived as an activity that occurs in remote areas. However, more situations are occuring where ore bodies are found, and projects will be developed, literally in people's backyards. There is a contingent of public sentiment that has concluded that mining is categorically incompatible with other land uses.

This attitude presents tremendous challenges for the mining industry to successfully develop projects confronted with competing land uses. Although this attitude itself must be challenged and dealt with by mining interests, the industry must also demonstrate that with reasonable procedures, it can co-exist with other land uses and be a good neighbor.

Permitting of the Cannon Mine is one of several current examples of mine projects that have been developed successfully in a competitive land use situation. There are a number of others that have not succeeded in this context and others pending that are confronted with significant permitting problems or public opposition utilizing the permitting process to hinder or kill projects.

Permitting succeeded at the Cannon Mine because a thorough technical basis was developed for the project and was professionally prepared and presented. Real environmental impacts were distinguished from presumptions. Adverse impacts associated with the project were addressed in a direct fashion and procedures planned and implemented to eliminate or reduce these impacts. An ongoing and open dialogue was maintained with regulatory agencies, the public and the press.

Chapter 35

MEETING AIR QUALITY REQUIREMENTS IN CRUSHING OPERATIONS

by

Joe M. Bryan
Brown & Root U.S.A., Inc.
Environmental Scientist
Houston, Texas

George E. Rodger
Brown & Root U.S.A., Inc.
Senior Process Engineer
Houston, Texas

ABSTRACT

The development of small precious metal mines in the western states has been directed at heap leach operations. Crushing plants are an essential adjunct to heap leach operations. They must meet environmental air quality regulations which generally follow a similar pattern from state to state. Prior to mine construction a permit must be obtained and an operating permit granted during or shortly after the start-up period. The typical construction permit application consists primarily of drawings, specifications and an explanation sufficient to indicate that the mass of particulate matter entering the atmosphere is below minimum standards. The operating permit may be granted following demonstrations that air emission control equipment or methods comply with information submitted in the construction permit application.

INTRODUCTION

This paper on meeting air quality requirements covers two broad subject areas. The first area discusses in detail the requirements for application and for maintenance of permits. The second area briefly shows how emission reductions are determined and it also gives descriptions for four different types of dust suppression equipment. Finally the paper summarizes the main steps to follow for new permit applications.

PERMITS

General Permitting Structure

Meeting the regulations for air quality involves two related activities. Failure to comply with either of these activities can mean non-compliance, with associated penalties. These penalties are discussed later in this paper. One activity involves the filing of proper papers, documents and data with the appropriate governmental agency. The other activity is installing, using and maintaining the equipment necessary to control air emissions. Since a permit application must be filed before equipment is installed,the first topic of discussion will cover the types of applications required and the agencies with which they are filed.

Three major levels of authority may issue regulations regarding air emission permits. The U.S. Environmental Protection Agency (E.P.A.), acting under authority of the Clean Air Act, is authorized to promulgate regulations which encourage states to maintain ambient air quality standards. In certain cases, the Clean Air Act also requires direct EPA approval of air emission rates. The former is accomplished by state implementation plans and emission control technology guidelines.

The latter involves a mechanism called "Prevention of Significant Deterioration." Such a PSD review may be required for new major facilities in attainment areas. A nonattainment review is required in nonattainment areas. Attainment and nonattainment areas are designated geographic locations where airborne pollutants either meet or do not meet National Ambient Air Quality Standards. The same area can be attainment for one pollutant, such as total suspended particulates, and nonattainment for another, such as sulfur dioxide. Another special area designation, which may be important in the location of a plant, is called a "Class 1" area. A Class 1 designation normally covers national parks or designated wilderness areas where the maintenance of pristine air quality is desirable. Air emission requirements are more stringent for those sources which may impact a Class 1 area.

The majority of small mine operations will not have sufficient emissions to force a PSD review. The threshold for PSD requirements for precious metal mining is 250 tons per year of any regulated pollutant. Specific information regarding PSD requirements can be found in applicable state and federal regulations and will not be detailed here.

Construction Permit Regulation

Prior to the construction or significant modification of a crushing plant, an application for the construction of an air emission source should be filed with the proper state agency. In Nevada, for example, this would be the Department of Conservation and Natural Resources, Air Quality Office. Construction permits may also have different names in different states. In Arizona, it is an installation permit; in Nevada, it is a registration certificate for a permit to construct. Most states, however, require similar information on the application although administrative procedures may vary considerably. Similar information is required because most air emission permit programs are outgrowths of E.P.A. requirements. Thus states authorized to issue permits follow essentially the same pattern of requirements. Agencies and procedures differ somewhat because many of these agencies were established before EPA's regulations were promulgated. Thus, agencies retained many of their old administrative procedures.

Coordination with State Agencies

It is always good practice to confer with the appropriate state agency prior to preparation of a permit application for the construction or significant modification of an air emission source. This consultation should take place before equipment selection and may be helpful even before site selection. For example, a site in an area already in violation of ambient air quality standards, a nonattainment area, or near a Class 1 area, may prove expensive in terms of air pollution control equipment. Placement of the crushing plant and other equipment on the site is also an important consideration. Many states view potential violations as those which occur at the property line on the leeward side, opposite the direction from the prevailing wind. As distance from a source increases, pollutant concentration from that source decreases. Therefore, if the plant is located on the windward side of the property, higher potential emissions may be allowed. In one example of how this becomes a factor, a plant which went to tertiary crushing bought property on its leeward side in order to increase its allowable emission rate by increasing the distance from the plant to the property line downwind.

In addition, regulations change, boundaries of nonattainment and Class 1 areas change, and forms and procedures change. It is always helpful to be current on these items when preparing a permit application. Also, regulatory personnel are not fond of surprises. If they know that an application is being prepared and that it can be expected within a week or a month, it allows them to budget their time. This will ultimately speed the processing for all permit applications. A preapplication conference should be scheduled at least 120 days prior to the start of construction. An application for a construction permit should be submitted 75 to 90 days prior to construction. Although approval for small projects will usually require less time, early submission of an application is good practice in order to allow for contingencies in the approval process.

Construction Permit Application

Information required in construction permit applications typically includes the company identification, plus a process description and/or flow diagram, a map of the project location, major equipment specifications, a pollution control equipment description and projected emission rates. For surface mining operations, a description of the area to be disturbed is also required.

Equipment specified for emission reduction should at least meet "Best Available Control Technology" and may require "Lowest Achievable Emission Rate." The section on determination of emission reductions details these terms.

Detailed studies are required for air emission sources which discharge significant amounts of pollutants. In Nevada, this requirement is imposed on sources greater than 23 pounds per hour. These special studies are made in order to determine whether an air emission source will cause the violation of ambient air quality standards. These studies may involve the use of air quality models to simulate the aggregate impact of the proposed source and other sources in the area. Some states perform these studies for the applicant and use "screening" models while other states require the applicant to perform the studies. "Screening" models are used to determine the highest pollutant concentration and plant distance from the highest concentration using presumed meteorological conditions.

Applications for registration or construction permits are generally subject to public participation. The appropriate state agency will publish notices of intent to grant air emission construction permits and public as well as other agency comments are received. At the state air quality agency's discretion, a public hearing regarding the permit application may be held if sufficient negative comments are received. However, such a hearing for a small mining and crushing operation would be an unusual occurrence. Generally, public hearings are not held except for large emission sources or sources which would have significant impact on local residences or on other sensitive areas.

Operating Permits

Operating permits and, in many cases, compliance testing are required after start-up but before or during the beginning of permanent operation. They are designed to assure regulatory personnel that the air pollution control equipment specified in the construction permit application was installed properly and that it will operate as specified. Inspections by regulatory personnel are typically made when an operating permit is submitted and air pollution control equipment may be tested during these inspections. Fees are often charged for operating permit applications and vary in amounts from state to state. Typically, applications for operating permits or notifications to the air pollution control agency of the commencement of operations are submitted after 60 days of operation but no later than 180 days after start-up. These time frames vary and the state agency will generally notify the applicant, after a construction permit is received, of these time limits.

Monitoring

A program for monitoring emissions may be required during mining and crushing operations. The applicant will be appraised of any monitoring requirements by the appropriate state agency. Monitoring reports may be required by the state agency on a regular basis and monitoring records should be kept for at least three years.

PERMIT MAINTENANCE

Notifications and Updates

Notification should be given to the appropriate state air pollution control agency at the start of construction, 30 days prior to start-up and at actual start-up. These notification requirements are detailed in some state regulations, such as Colorado, but should be followed even if they are not. Thirty days advance notice should also be given to the state agency in the case of operational or physical changes which may increase emission rates. Small modifications are usually handled as variances to existing permits. Large modifications, such as significant increases in production rates, require a reapplication for a construction permit.

Operating permits are renewable about every five years and renewal fees, in many states, are to be paid each time the permit is renewed. Operating permits are also transferable to another company, provided the new owner does not change the operations or equipment in such a way as to increase emissions. In some states, such as Nevada, a fee is associated with a permit transfer and in others, such as Texas, ownership transfers are exempt from fees.

Maintenance

It is essential that the pollution control equipment which has been installed be maintained and operated as specified. Any equipment not in working order should be repaired or replaced in a timely fashion. Should an inspector from the state air pollution control agency or from the E.P.A. appear, he will expect the pollution control equipment to be in good operating order or require a reasonable explanation of why it is not working. Ordinarily, inspections are announced at least one day in advance. However, surprise inspections may also occur, particularly in cases where there is reason to believe that air emission limitations are not being met.

Noncompliance

Generally, it is the object of state air pollution control agencies to allow permittees every opportunity to achieve compliance with the conditions of their permits. Agency engineers and inspectors know that there will be times when deviations from the permitted emission will occur. These excursions, which could occur during start-up or when equipment malfunctions, are usually overlooked as long as a reasonable pace of progress is maintained to correct the problem. If such violations continue to occur, a schedule of compliance will be established whereby the operator will be required to insure that pollution control equipment is repaired or installed in a certain time frame. The compliance schedule is usually negotiated between the state agency staff and the permittee.

Several items can trigger the need for punitive enforcement action by the state agency. These include failure to install required equipment, failure to maintain and operate air pollution control equipment, failure to apply for or obtain required permits and failure to meet a compliance schedule. In addition, most states have provisions for levying fines in the case of minor violations, usually not associated with a permitted activity. These include open burning, fugitive dust, odors, and other nuisance conditions.

Continued violation of permit conditions without reasonable explanation, in spite of various warnings, is viewed as a serious offense. Some states handle serious violations in a different manner than others, although the disposition of each case is highly dependent upon the circumstances of the violation. The issuance of fines is the more common punitive action and is usually levied on a daily basis for each day the operator is out of compliance. Shutting and padlocking the plant until the proper equipment has been installed is another effective method used by the state agency in achieving compliance. Occasionally, the matter will be referred to the state attorney-general for civil or criminal prosecution. After that occurs, room for negotiation will be severely limited. It is recommended that every effort be made to avoid that stage.

DETERMINATION OF EMISSION REDUCTIONS

The determination of emission reductions begins once the operating criteria are set for a project. For emission reduction purposes,the key items are: the hourly operating rate, the operating hours per day and per year and the crushing circuit itself. The emission reduction requirements are a direct function of the circuit complexity. In other words, the required emission reduction factors will increase as the crushed product becomes finer. Initially, a decision will be made to use "Best Available Control Technology" (B.A.C.T.) or "Lowest Achievable Emission Rate" (L.A.E.R.) rules. Generally, L.A.E.R. rules will be applied.

On the basis of L.A.E.R. rules, the estimated emissions are then calculated. This will be based on E.P.A. report AP-42 entitled "Compilation of Air Pollutant Emission Factors." Section 8.23 covers metallic minerals processing and Table 8.23-1 gives the "Uncontrolled Particulate Emission Factors for Metallic Mineral Processes." This is reproduced as Table I. Heap leach operations will generally be concerned with the factors for crushing, material handling and transfer for low moisture (less than four percent) ores. No model exists for Run-of-Mine dump hoppers and open ore stockpiles. Since these points are the worst offenders for generation of uncontrolled particulate emissions, anything that can be done to suppress these emissions will have a positive effect in supporting an air quality application.

Figure 1, a hypothetical 500 S.T.P.D. (50 T.P.H.) crushing flowsheet, has been set up to show how Estimated Emissions (EE) and Allowable Emissions (AE) are calculated. Each source of emissions and the applicable emission factors from Table 8.23-1 are identified. The formula for Allowable Emissions is based on Nevada Administrative Code Chapter 445 Article 732. The required emission reduction for this circuit works out to 74.9%. Since water sprays only give a 70% reduction, it is necessary to go to the next best suppression method. This would be with wet scrubbers or a dry fog system which are rated as being 95% efficient. At the 95.0% level of efficiency the emissions would be 8.9 lbs/hr or 1.12 grams per second. The 1.12 g/s figure will be used in an environmental department model to determine (in micrograms per cubic meters) a 24-hour Impact and an Annual Impact at some point from the source. Generally, this point will be at the property boundary in the direction of prevailing winds or at the closest public access. Both the 24-hour and Annual Impact figures will be corrected for the actual hours per day and days per year of operation. For each pollutant, an assumed background level will be added to the figure. If the sum of these results falls under the 24-hour and annual ambient standards, a permit (generally with restrictions) will be granted. The restrictions will cover the following areas:

- Production limitations (by rate and hours)
- Notification and reporting requirements
- A note of permit authority limitations
- Notice that the L.A.E.R. must be approved by an environmental commission
- Emission offset requirements if the project is in a nonattainment area
- The type of emission control method to be used for the crushing plant

In reviewing Table 8.23-1 and the hypothetical flowsheet, note that the emissions for crushing are based on weight of material entering the primary crusher. Therefore, the quantity of uncontrolled emissions can be substantially reduced by scalping the undersize ahead of the crushing operation.

TABLE I

TABLE 8.23-1 UNCONTROLLED PARTICULATE EMISSION FACTORS FOR METALLIC MINERAL PROCESSES

Process	Low Moisture Ore		High Moisture Ore	
	Emissions kg/Mg (lb/ton)	Particulate Emissions 10 um kg/Mg (lb/ton)	Emissions kg/Mg (lb/ton)	Particulate Emissions 10 um kg/Mg (lb/ton)
Crushing[1]				
Primary	0.2 (0.5)	0.02 (0.05)	0.01 (0.02)	0.004 (0.009)
Secondary	0.6 (1.2)	NA	0.03 (0.05)	0.012 (0.02)
Tertiary	1.4 (2.7)	0.08 (0.16)	0.03 (0.06)	0.001 (0.02)
Wet Grinding	Negligible	-	Negligible	-
Dry Grinding				
With air conveying and/or air classification	14.4 (28.8)	13.0 (26.0)		
Without air conveying or air classification	1.2 (2.4)	0.16 (0.31)		
Drying				
All minerals but titanium/ zirconium sands	9.8 (19.7)	5.9 (12.0)		
Titanium/zirconium with cyclones	0.3 (0.5)	NA		
Material Handling and Transfer[2]				
All minerals but bauxite	0.06 (0.12)	0.03 (0.06)	0.005 (0.01)	0.002 (0.0006)
Bauxite/alumina	0.6 (1.1)	NA	NA	NA

1 Based on weight of material entering primary crusher.
2 Based on weight of material transferred. Applies to each loading or unloading operation and to each conveyor belt transfer point.

From E.P.A. Report AP-42

FIGURE 1

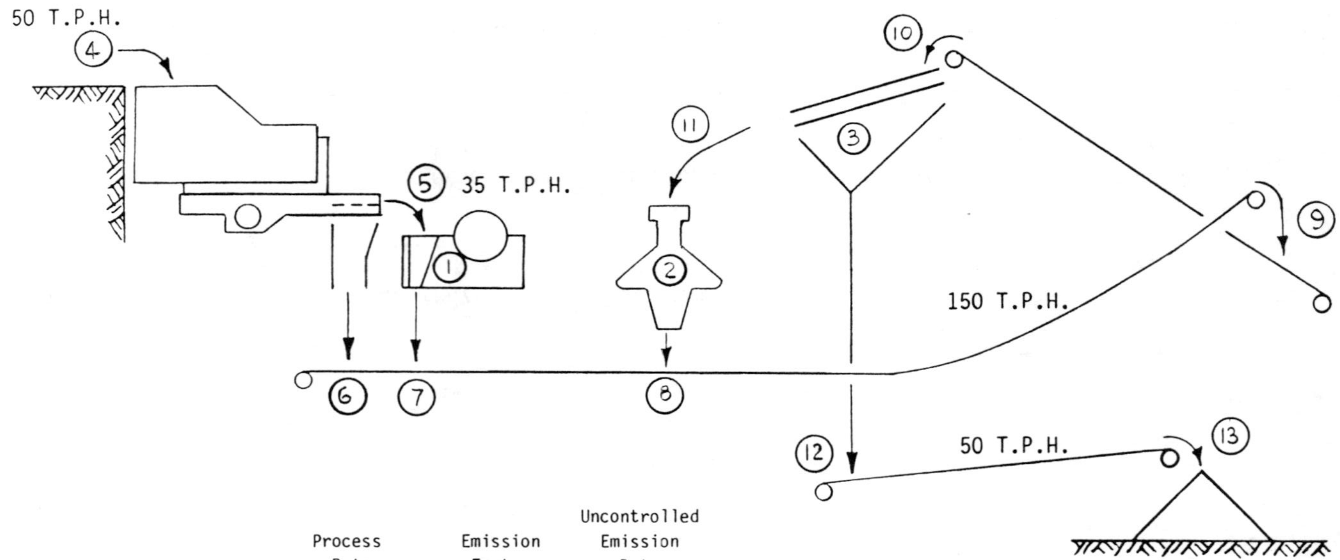

Dust Source	Station	Process Rate (S.T.P.H.)	Emission Factor (lb/S. Ton)	Uncontrolled Emission Rate (lbs/hr)
Primary Crusher	1	35	0.5	17.5
Secondary Crusher	2	35	1.2	42.0
Screen	3	150	0.12	18.0
New Feed	4	50	0.12	6.0
Transfer	5	35	0.12	4.2
"	6	15	0.12	1.8
"	7	35	0.12	4.2
"	8	150	0.12	18.0
"	9	150	0.12	18.0
"	10	150	0.12	18.0
"	11	150	0.12	18.0
"	12	50	0.12	6.0
"	13	50	0.12	6.0

Total Uncontrolled Emissions = 177.7 lb/hr

Allowable Emissions = $55(50)^{0.11} - 40 = 44.6$ lb/hr.
(NAC 445.732)

Reduction Required = $\left(1 - \frac{44.6}{177.7}\right) \times 100$

= 74.9%

With a scrubber or dry fog sprays at 95% efficiency, the estimated emissions will be 177.7 x (1.00 - 0.95) = 8.9 lb/hr. or 1.12 grams/sec.

DUST SUPPRESSION EQUIPMENT DESCRIPTIONS

The equipment descriptions for dust suppression systems have been limited to the four types commonly used by small mines. These are: 1. hydraulic (water) sprays, 2. dry fog nozzles, 3. wet scrubbers and 4. bag-type dust collectors. The characteristics and costs of each are compared in Table II.

Water Sprays

Water sprays provide the simplest and most economical dust suppression system available, but they are also the least efficient. Use of this type of dust suppression would be limited to simple low capacity crushing circuits where emission reductions under 70% would meet the requirements. Typically, such a circuit could handle up to 40 S.T.P.H. with two stage crushing and one stage screening that runs for approximately ten hours per day, 250 days per year. In this type of system, pressurized liquid is forced through an orifice and liquid is sheared into droplets. The droplets vary in size from 100 to 200 microns with the 100 micron droplets formed around 100 P.S.I. Spray velocity varies between 2000 and 4000 F.P.M. Minimum flow rates per nozzle are in the range of 0.5 gal/hr. and there is no limit on the maximum.

The hardware for these systems consists of water pipes, line filters, pressure regulators, gauges, valves and spray nozzles. In colder climates these systems can be heat traced and insulated to protect the hardware. However, operation below 32°F is not practical.

Minor problems are experienced with the hardware for these systems. These are due to plugged line filters or external damage to nozzles and valves. The major problem area has to do with the amount of water added. Generally, an excess is added which results in a clean up problem under the transfer points and under conveyor belts. Water sprays are not effective in bringing down dust in cone crushers.

Dry Fog Nozzles

Dry fog nozzles are slightly more complex than water sprays, but they can provide up to 95% emission reductions. In the two stage crushing-one stage screening circuit mentioned previously, up to 350 S.T.P.H. could be processed through the circuit and still meet the emission reduction requirements.

In this type of system, 80% of the dust suppression comes from having a well sealed dust enclosure around the dust generating point. The other 20% comes from fine droplets of water that are in the same size range as the dust particles. Therefore, the probability of collision between the two is high.

The fine water droplets are produced with air and water. The air passes through a venturi where it is accelerated beyond the speed of sound. Water is introduced at the end of the venturi and both air and water are carried forward in a shock wave. A portion of this shock wave is reflected back by a resonating chamber. The resulting collision of these two waves renders the water into very fine droplets. Air which bypasses the resonating chamber serves to carry these fine droplets forward into the dust enclosure. The highest air pressure produces the finest sized droplets. On the coarse end droplet diameters can vary from 200 to 600 microns and on the fine end from 1 to 10 microns. Pressure variations are from 50 to 65 P.S.I.G. for air and 5 to 25 P.S.I.G for water. The volume variations per nozzle are 3 to 9 S.C.F.M. for air and 2 to 8 G.P.H. for water. Spray velocities are in the range of 1000 to 4000 F.P.M. The fine water droplets suppress the dust by agglomeration in one of two modes. In a "dry" mode an optimum volume of dry fog will precipitate out the dust generated within the enclosure and the product will not be wetted to any significant degree. In this mode there is no carry-over effect from one dust suppression point to the next. In a "wet" mode a greater volume of dry fog will be added to permit some carry-over of moisture on the product. This is done in cases where it is necessary to serve remote transfer points that are not enclosed, and which are at the end of a processing line where some level of additional water can be tolerated.

The hardware for these systems is the same as that for the water spray systems but includes an air line system and fog nozzles instead of water spray nozzles. Freeze proofing this hardware with heat tracing and insulation allows operation below 35°F, the lower limit being around 0°F. This can be done because there is a critical mass below which water will not freeze, the critical mass size being in the 20 micron size range. Operation under these conditions also requires a post purge system for drainage on shutdown. Care must be taken in the design of the enclosure, because if the fog is allowed to coalesce on a cold surface it will freeze, even though the fog itself does not freeze.

TABLE II

COMPARISON OF DUST SUPPRESSION SYSTEMS

Characteristic	Water Sprays	Fog Nozzles	Wet Scrubbers	Bag Type Collectors
Capital Costs	Low	Moderate	High	High
Operating Costs	Low	Moderate	Low	High
% Collection Efficiency	70	95	95	99
Operating Characteristics				
- Energy Consumption	Low	Moderate	High	High
- Operator Attention	Minimal	Required	Minimal	Minimal
- Maintenance Required	Some	Constant	Some	Constant
- Fines Disposal	Not required	Not required	A Problem	A Problem
- Minus 32°F Operation	Not practical	Possible (with heat tracing & insulation)	Possible (with heat tracing & insulation)	Possible
- Tanner Gas	Not required	Required	Not required	Required
Dust Suppression-In/On				
- Hood Enclosures	Feasible	Feasible	Feasible	Feasible
- Dump Hoppers	Feasible	Feasible	Feasible but not practical	Feasible but not practical
- Stockpiles	Feasible	Feasible	Not feasible	Not feasible
Hardware Required				
- Fans	Not required	Not required	Required	Required
- Ductwork	Not required	Not required	Required	Required
- Dust Enclosures	Optional	Required	Required	Required
- Piping for 100 psi Air	Not required	Required	Not required	Required
- Piping for 40 psi Water	Required	Required	Required	Not Required
- Fines Disposal Equipment	Not Required	Not required	Required	Required

1416m

Minor problems are experienced with the hardware for these systems. The most frequent problems are due to solids in the water lines which plug the filters and with nozzles if the filters are improperly placed or the pore size is too big. Damage can occur to the fog nozzle resonators from abrasive wear or by impact of rocks. In freezing weather care must be taken not to damage the heat tape or insulation and to adequately drain the system on shutdown.

Wet Scrubbers and Bag-Type Dust Collectors

Both wet scrubbers and bag-type dust collectors use the same hardware for the collection of dusts. The differences in the two methods occur in the filtering of the air and in disposal of the dusts.

The collection of dust starts with dust tight enclosures at each dust generation point. Each enclosure is connected by branch ducts to main duct(s). Main ducts go to the filter. A fan is connected to the filter outlet which produces a negative pressure within the dust collection system. The lowest negative pressure occurs near the entrance to the fan. This negative pressure decreases at the extremities of the duct/enclosure system. The dust generated within the enclosures follows the increasingly negative pressure towards the filter. This is a simple description for these dust gathering systems but a more technical description is beyond the scope of this paper.

In both of these systems, it is the filter that determines the dust collection efficiency.

Wet Scrubbers. In wet scrubber systems, the dust is accelerated in either a centrifugal motion or through a venturi and it is impacted on a wetted surface. Venturi type wet scrubbers are more efficient than centrifugal types but it is generally accepted that the collection efficiency of the latter is around 95%. With this type of system, the dust is collected as a slurry. In heap leach operations this slurry can be placed on the crushed ore conveyors going to a stockpile or to the leach pad. In milling operations this slurry would be added to the ball mill feed or at some point where it would be included in the metallurgical accounting for the plant.

The most common problems associated with wet scrubbers are due to abrasive wear. In the duct system this problem is usually found in the elbows or points where the streams impinge at an angle greater than 30°. In the scrubbers themselves, wear takes place on the impingement plates in wet centrifugal scrubbers and in the first turn after the venturi. With some ores "mudding" problems can occur in the sludge drain system.

If operations below 32°F are anticipated, extensive freeze proofing with heat tape and insulation is a requirement on all wetted surfaces. In heap leach operations, disposal of the slurry could be a problem under these conditions.

Bag-Type Dust Collectors. Bag-type collectors can provide up to 99% emission reductions for particles in the 5 to 10 micron size range. Therefore in the two stage crushing-one stage screening circuit mentioned previously, a bag-type collector would allow processing of up to 2500 S.T.P.H. without exceeding the emission guidelines.

In bag-type collectors dusty air is passed through a fabric at a low velocity. The fabric can be arranged in a envelope or tubular shape. Removal of dust is by a straining action through the fabric. On start-up the filtering action is not efficient until a mat of dust builds up on the dirty side of the fabric. The straining action through this mat of material is very effective in removing particulates even in the submicron size range. As the dust collects on the fabric, the resistance to air flow increases. Eventually, the fabric must be reconditioned by shaking or vibrating to remove the bulk of the mat. The fabric must retain some of the mat to maintain maximum efficiency. Rates of flow through the collector vary with the collector type, application and dust concentration but generally pressure drops across the filter will not exceed 6 in. W.G. (water gage).

With bag-type collectors, the dust is collected in a dry powder form. Dust disposal can be a problem but the most practical solution is to pug mill the fines at a high solids content with water and lime before reintroducing them back into the circuit.

The most common problems associated with bag-type dust collectors are abrasive wear on the equipment and bags. Bag wear is reduced by insuring there is an even distribution of dusty air flow through the dirty side of the collector. High humidity is always a problem because it can completely blind the fabric. Where it occurs, provision should be made to heat the air to the collector or consider high voltage precipitators which are even more expensive than bag type collectors.

Cold weather operations below 32°F are generally not a problem, except in cases where the fines are disposed of by wet methods. In those cases freeze protection methods are mandatory.

SUMMARY

When contemplating the installation of a mining and/or crushing operation, early planning for air pollution control is essential in order to avoid later problems with the regulatory authorities. The following steps should be made a part of the engineering and installation of such a facility:

° Become familiar with applicable state air pollution codes. Arrange a meeting with personnel at the state agency in order to explain plans for the proposed facility. Bring information regarding the size and location of the planned facility to the meeting.

° Complete a set of registration or construction permit applications, following instructions in the regulations or on the appropriate forms. The application should be submitted 75 to 90 days prior to start of construction in order to allow time for full agency review and public participation activities.

° Keep the state agency informed of construction starts, design changes during construction and compliance testing. For the latter at least 5 to 30 days notice should be given prior to the start of compliance testing, if any. Also, notification should be given prior to plant start-up.

° Between 60 and 180 days after start-up an operating permit application should be filed.

° Good equipment maintenance and record keeping of monitoring activities during operation are essential in satisfying air pollution control requirements.

In general, the majority of air pollution problems can be avoided by knowing and following the regulations, and by working with the state agency. Of course installing, using and maintaining the proper air pollution control equipment is also required.

7

Late Papers

Chapter 36

DILEMMA OF GOLD RECOVERY PROCESS
MAKING THE RIGHT SELECTION

BY
ROBERT MARTINEZ
MINE GENERAL MANAGER

AMSELCO MINERALS INC.
ALLIGATOR RIDGE GOLD MINE
ELY, NEVADA

ABSTRACT

A proven high grade reserve of gold mineralization located in Mexico is evaluated to determine which gold recovery process is best suited for this deposit. The processes evaluated are; Heap leaching, Vat leaching and a CIP circuit. All qualify as viable options, however each one presents a different economic picture which is impacted substantially by the financial constraints of the project. Taking this into consideration the most economic process was selected.

INTRODUCTION

A gold deposit located in Durango County, Mexico with a proven reserve of 287,582 mt (317,000 st) of ore containing an average grade of 112.5 gm/mt (3.28 oz per st) can be processed three different ways; Vat leaching, Heap leaching and Carbon in Pulp (CIP). Selection of the appropriate method taking into consideration the financial position of the owner and project location is discussed.

HEAP LEACHING

Metallurgical testwork indicated that crushing to 38.1 mm (1.5 inches) would produce a recovery of 32 percent in 35 days of leaching. The owner has sufficient funds to finance a plant of 45.36 MTPD (50 STPD) without any outside financing. It is the owners intent to leach this material and save the tails for Carbon in Pulp (CIP) processing at a later date when appropriate cash flows are available.

The Heap leaching flow sheet consists primary and secondary crushing, fine ore stockpile leach pads and ponds and a processing plant consisting of carbon adsorption, desorption, electrowinning and smelting. It is anticipated that the mine equipment would be used to construct the heaps.

Of the three methods, Heap leaching requires the smallest amount of capital to start production, no outside financing is required and it has the lowest operating cost of the three options 9.92 per mt ($9.00 per st ore). At a design rate of 45.36 MTPD (50 STPD), cash flows of $21,000 per day are generated thus producing the capital required for constructing a CIP plant within a period of less than six months.

VAT LEACHING

Metallurgical testwork indicated that reduction of the material down to minus 2 mm (10 mesh) would yield 62.6 percent recovery in only three days. The operating cash costs per mt of ore are $9.85 ($10.86 pst) at a design rate of 45.36 MTPD (50 STPD). As with the Heap leaching option the tails would also be saved for future processing through a CIP plant as soon as the cash flows provide the necessary funds. The incremental capital required for this option is approximately three times that of the Heap leaching scenario thus 20 percent outside financing is required for this option.

The Vat leaching flowsheet will consist of primary, secondary and tertiary crushing, fine ore stockpile and reclaim, a grinding circuit, thickener, Vat leach cells and tailings storage area. The process plant would be identical to the Heap leach scenario and it is anticipated that mine equipment would be used to remove the tailings from the cells.

As stated the incremental capital for this option is approximately three times that of the Heap leaching however the cash flows are not only double but they are obtained at a quicker rate. Cash flows on a daily basis are $41,000 and at this rate a period less than three months is required to provide the necessary funding to construct a CIP plant.

CIP PLANT

The testwork indicated that a plant designed to reduce the ore down to 90 percent passing .053 mm (270 mesh) would yield a recovery of 97.5 percent. The plant design rate would be set at

90.7 MTPD (100 STPD) with a operating cost of $13.28 per mt ($12.05 per st) and outside financing of 49 percent would be required for construction. The flowsheet for the CIP plant would be identical to the Vat leach scenario up to the grinding section where a second ball mill would be installed in addition to a full CIP circuit. A tailings dam complete with reclaim circuit would also be needed along with a slightly larger process plant as used in the Heap and Vat leach scenarios.

It is calculated that capital costs would be four times as high as Heap leaching and two times high as Vat leaching, however the daily cash flows are expected to be $127,000.

SUMMARY

All the options produce positive cash flows at a rapid rate. The Heap leaching option requries no outside financing, the least amount of capital but produces the lowest cash flow in a given period of time. The Vat leaching option requires 20% outside finacing at twice the capital of Heap leach but the cash flows are doubled. In addition, except for the Vat leach cells, all the remaining capital costs can be applied to CIL plant. Due to the quick payoff (less than four months) a large tailings storage area is not required and a CIL plant can operate for two months on tailings during which time crusher modifications can take place. The CIL plant produces the highest cash flows, three times higher than Vat leaching and 6 times higher than the Heap option but it does requires 49 percent outside financing. See Figure 1. On an annual basis the cash flows produced by the CIL plant are the largest by far even after paying the initial capital investment however, the combination of the Vat leaching option and the CIP plant produces the highest revenues for the owner. When one compares to the outside financing cost between the Vat and the CIP plant operation there is a 6 fold difference between the two. Since both cases exhibit a very high DCF (investment recovered in less then 3 months) an analysis was completed to determine which option produced the highest rate of return from the outside financing. On this basis the CIP option exhibited a 9.3:1 return on outside financing while the Vat leach option produced 18.5:1 return in the initial three month period. In addition since approximately one half of the capital costs required for the CIP Plant are paid for by Vat leaching, processing of the tails from Vat leaching during final construction of the CIP Plant will more than pay for the additional capital required. The Heap leach option was not considered in this comparison since the option requires 6 to 9 months to generate the cash flow required for CIP financing.

FIGURE 1
Financial Summary ($)

Design Rate	45.6 MTPD (50 STPD)	45.6 MTPD (50 STPD)	90.7 MTPD (100 STPD)
Op. Cost/Day	492.50	450.00	1,205
Cash Flow/Day @ $400	41,064	20,928	127,920
Profit Margin/Day	40,572	20,478	126,715
Annual Profit	14,200,000	7,167,300	44,350,250

Thus it is recommended that the Vat leaching option followed by CIP processing be selected for this project. The return on outside financing is higher, there is less impact on the environment, the owner has the flexibility to buyout the outside inventor, the owner having complete flexibility from a production standpoint. At a design CIP rate of 90.7 MTPD (100 STPD) the project will have a life of eight years with excellent cash flows. These cash flows are of the amount that capital expenditures requried for production increases can be financed quite readily.

Chapter 37

INNOVATIVE FINANCING FOR SMALL GOLD MINING PROJECTS

Dwane K. Johnson

Vice President
First Interstate Bank, Ltd.
Los Angeles, California

INTRODUCTION

The small mining company faces the dilemma of how to finance the development of its properties because it doesn't have the financial resources to pay for the development costs from its own funds and doesn't have the financial strength which will enable it to borrow the money needed for development. However, with a little imagination and the right set of circumstances, a financing can be arranged for the small mining company.

This paper is an attempt to describe the fundamental steps on how such financing can be obtained --in other words--what is required to provide adequate capital at a low cost for developing and bringing a mining property into production. Innovative financing follows from this.

The basic keys for building and maintaining a successful mining company are "the four M's": Management, Market, Mine, and Money. The writer believes their order of importance is as presented and each is a building block required in order to obtain the type of financing that best suits the borrower and lender.

MANAGEMENT

Management must be experienced, highly organized, realistic, able to take advantage of opportunities and fulfill promises.

Mine financing requires a variety of skills and concentration on the part of management. Management should have a thorough understanding derived from experience in geology, construction, mining and metallurgy and environment, marketing, and financial matters. Management must realistically visualize what will be produced at what cost and truly evaluate the risks associated so the proper security can be correctly designed.

Since risk cannot be eliminated, the objective of management is to identify the risk which will be present in a given venture and assess the level of that risk which will be acceptable to the firm. Generally the risk which is present is subject to little or no modification. After management identifies and measures the acceptable risk the firm will proceed in a way that it will be least exposed.

Management must determine its long-term objectives and strategies within the context of a constantly changing world. This question must be addressed before examining the sorts of risk which affect the development and operation of a specific property. The compilation and interpretation of "hard data" by competent people is good to a point but the experience and instinct of seasoned individuals are the important factors in management choosing a long-term direction for the corporation.

The attitude towards risk greatly affects the goals of the corporation.

Major risks in a mining project can be placed in four categories:

1. Market - Price, demand, substitution.
2. Costs - Capital, operating, financial.
3. Regulation.
4. Taxation.

The feasibility study for a project is the fundamental tool used in the management of risk. Management should employ an experienced team with an established and well-understood set of ground rules for the preparation and assembly of a feasibility study. The necessary degree of realism is built into all levels of the study and if the project is technically feasible, the company's hurdle rate is then used to discount the base-case project cash flow in order to determine its financial viability.

Understanding the geological significance of the mineral deposit is critical. There should be strong interplay among the pure regional geologist, the detailed mine geologist, the mining engineer, and the metallurgical engineer. The collaboration among these different disciplines will yield a feasibility study that is more reliable.

The essence of this work must be appreciated and understood by the executives planning the financing. It is vital that they thoroughly understand the risks and make the proper risk assessments to cope with our rapidly changing market environment.

Projecting future revenue values is paramount and is subject to much estimation. One way to mitigate fluctuations and the risk of falling prices is by selling production for future delivery.

MARKET

Since mining is a worldwide industry it is mandatory that bankers be aware of what is occurring in the marketplace, both foreign and domestic.

The marketing of the production is a very important facet when considering financing for the small mining company. Historically, lenders have not properly understood market risk. Currently, knowledgeable lenders are devoting considerable time and effort to better understand the particular intricacies of the markets, e.g., metal movements, physical markets, commodity exchange functions, substitution effects, and price determination methods and therefore are structuring better loan agreements which result in a better working relationship between the borrower and lender.

As an example of a metal that is readily salable, consider gold. We know that gold can be sold almost anywhere in the world any hour of the day. The question which concerns us is what price will be realized? Therefore, gold does not really pose a problem as to salability.

Gold, however, is unique. For other metals and minerals, bankers are considering many different items as to marketability. Some of these are:

1. Is the market expanding or contracting?
2. What is the percentage of scrap to primary produced metal?
3. Is the commodity traded on an established exchange?
4. Are there several producers?
5. Are there several purchasers?
6. Is there a cartel or an intracountry production agreement?
7. Is there a supported floor or ceiling price?
8. What is the effect of substitution?
9. What is the mineral's price volatility both short- and long-term?
10. Is the price of the mineral or metal market or cost driven?
11. What effect does transportation play on the price of the mineral?
12. What are the environmental regulations pertaining to the mineral?
13. In what other countries is the product produced and does their government influence the price?
14. Is the current market in a "backwardization" or "contango"?
15. What is worldwide capacity and what percentage of that capacity is being utilized?
16. What are the various producers' cost of production?
17. What are import/export regulations?
18. What is the government stockpile quota?
19. What is the general economic climate, both foreign and domestic?

It is important to understand the effects of the marketplace on the mine and to employ all possible techniques available to management in order to reduce its risk and ensure the cash flow that is needed to either repay debt, expand the operation, or pay dividends to the stockholders.

MINE

There should be no mystery surrounding the bank's mine evaluation process. The banks will look at the same information as the borrower. These are:

1. Reserves (all the background needed to evaluate reserves).
2. Process network.
3. Mine plan.
4. Metallurgical processing.
5. Project sequencing and optimization planning.
6. Capital cost.
7. Operation cost.
8. Market study.
9. Cash flow projections.

Banks seldom use only net present value (NPV), discounted cash flow rate of return (DCFROR), or payback period (PbP) in a minerals finance evaluation. Sensitivity analyses are readily used with considerable time spent in understanding the fixed and variable costs of production. We have been and are spending more time using Option Pricing Theory in valuing properties; we lean heavily on the Cash Flow Debt Coverage Ratio (an undiscounted number) for structuring our loans.

Cash Flow Debt Coverage Ratio

The ratio is computed by the following:

$$CFDCR = \frac{NPAT_i + NCC_i + I_i - CAPEX_i + CWC_i - CO_i}{P_i + I_i}$$

where i is the fiscal year ended, NPAT is the net profit after tax, NCC is the noncash charges, I is the total interest payments on debt, P is the principal payment on debt, CAPEX is capital expenditures, CWC is change in working capital, and CO is cash outflow.

Depending on commodity, economic cycle, and the realistic value of revenue and expenses used the CFDCR ratio we prefer is normally between 1.3 or 2.

Risk Sensitivity

Typical mine risks we identify and assess are:

1. Management.
2. Operator experience.
3. Reserves - quantity and quality.
4. Production level.
5. Price of product.
6. Operating costs.
7. Completion risks.
8. Technological risks.
9. Environmental permits and risks.
10. Insurance coverage.
11. Capital expenditures.
12. Cost overrun risks.
13. Operating costs.
14. Availability of energy and supplies.
15. Transportation infrastructure requirements and costs.
16. Interest rate.

In order to evaluate the effects on cash flow

projections, each risk is analyzed with its impact on the cash flow derived from the mine. Sensitivities are run in order to meet our minimum credit standards. Examples of sensitivity runs are:

1. Reserve sensitivity - proven reserves.
2. Price sensitivity - expanding or contracting market.
3. Production sensitivity - 80-95% capacity.
4. Operating senstivity - time and cost.
5. Interest rate sensitivity.
6. Capital expenditure sensitivity.
7. Cost overrun sensitivity - delays and changes.
8. Environmental sensitivity - delays and reclamation costs.

We have been using and are trying to better understand the Option Pricing Model. As we all know, due to the price volatility over the past few years, the evaluation of a mine is made particularly difficult by the high degree of uncertainty in product prices. As the Option Pricing Model accounts for volatility, it is thus another tool that is assisting us in structuring credits. In historical periods of rather stable prices, the discounted cash flow technique was ample but in today's environment, one may be over- or understating asset value. This is but another tool that may be used to value projects, and to determine the optimal policies for developing, managing, and abandoning a mining property.

MONEY

The degree of how innovative one can be in providing financing for a mining company will depend heavily upon the experience and sophistication of management, the condition and type of market, and the quality of the mine. For example:

1. An excellent mine will carry weak management and a weak market for a considerable time.

2. A strong market will carry weak management and a marginal mine for some time but one can observe that in a mine's lifetime, strong markets don't last long.

3. Discerning management will be able to get the most out of a marginal property (efficiency); they will have credibility with buyers in the marketplace, and be able to profitably operate a property which would in other circumstances be unprofitable or closed down.

4. If there is good management, a good market, and a good mine, one can become more innovative in financing and better maximize the earnings of the company.

The degree of innovation in financing is directly proportional to quality of management, market, and mine. Creativity will arise if the basics are all sound and proven.

Knowledge of how the banker evaluates a credit will aid the firm in providing the necessary particulars which will greatly enhance the chances of the credit request being approved.

<u>All</u> financial institutions will evaluate a financial request by considering the age-old, tried and proven "FIVE C's" of credit:

1. Character of management.
2. Financial capacity of the project.
3. Financial condition of the project.
4. Condition of the mining industry.
5. Collateral for the loan.

One of the more important aspects to stress is that financing should be tailored to meet the mining company's needs and not the other way around. There are many different types of financing vehicles available. The primary concern is to ascertain the risks and then structure the vehicle around these risks so that both borrower and lender are protected.

In today's investment climate, the cheapest source of funds for the small gold producer is obtained from the equity market. One readily observes price earnings ratios (PE's) in the 30 to 60 range which translates to a cost of money between 2 to 3½%. I believe that (PE's) will become more realistic when the profit margins decline, the inventory of mining stocks becomes larger, and the investment funds' demand for gold plays stabilizes.

Various techniques that are available to mining companies in raising money and in some cases hedge the risk associated with their capital investment decisions, are as follows:

Project Finance

This form of limited recourse financing reduces the risk taken by the common shareholders. The loan to develop the property is obtained with the property itself as the predominant collateral. The lender is repaid from the cash flow of the property financed. This type of financing is well developed and understood by both borrowers and lenders.

Debt Instruments with Coupons Linked to Commodity Prices

The linking of interest rates on debt instruments to an index of prices of commodities produced is a useful approach. The formula for calculating interest is a form of call on the metal by the lender when the metal price reaches a predetermined strike price. The advantage to the mining company is that it pays only a minimum level of interest in bad times and as a tradeoff also gives away some of the upside benefits in good times.

Debt Instruments with Principal Linked to Commodity Purchased

An alternative approach to the aforementioned is to link the redemption of the principal of a redeemable debt instrument to a quantity of a commodity produced. This type of financing gives the investor a direct metal exposure, and if issued at a time when metal prices are depressed will yield a lower coupon rate.

Precious Metal Loans

The following precious metal loans are examples of transactions for which we have provided funds. There are essentially two principal risks that must be addressed; the production risk and the price risk. The production risk to the lender is that, due to operating or other conditions, the mine is not able to produce. The price risk is the risk of what revenues will be realized when the metal is ready for sale. This can be at least partially offset by an active hedging program.

Example of a Gold Loan: A gold loan will not usually exceed 50% of the annual mine's production. The loan involves borrowing gold to be repaid from future production. Interest rates are minimized and are often in the 3 to 5% per annum range with a term of 3 to 5 years.

The concept is fairly simple and is generally envisioned as a fixed interest rate variable principal loan. If the producer knows his production costs, the principal payment could then be considered fixed.

The bank implements the loan in the following manner:

1. The mining company (the "Buyer") enters into a metal purchase agreement with the bank.
2. The bank delivers the metal to the Buyer, and charges the cost of the metal to the Buyer's loan account.
3. The Buyer immediately sells the metal to the bank and the proceeds are credited to the Buyer's bank account.
4. The Buyer pays interest to the bank on a monthly basis at the agreed rate.

When the Buyer repays the loan:

5. The Buyer may repay the loan either by delivering the same quantity of metal as borrowed, or
6. The Buyer may make payment to the bank in an amount equal to the quantity of metal taken down in the transaction multiplied by the effective price at the time of repayment.

Precious Metals Loan as a Lease: The following illustrates the mechanics of a loan structured as a lease where a company had a term loan and a working capital line:

1. The mining company leases metal from the bank.
2. The mining company locks in its revenue for the next 18 months by selling forward in the commodity exhange.
3. The price realized with the hedge is higher than the metal price when the metal was leased from the bank.
4. The company liquidates its forward sales contract for a profit.
5. The company reduces its interest cost from approximately 11% floating to 3% fixed.
6. The company makes its metal payments to the bank rather than delivering it to the commodity exchange warehouse.

An option is also provided whereby the borrower could go in and out of the transaction in cash or metal on a revolving basis, thereby enhancing liquidity.

Debt-Equity Swaps

This newly created mechanism is a result of the loans made to countries which are currently unable to perform as agreed. The lender receives approval from the central bank to take a portion of its outstanding debt, converting the same into the local currency which in turn is invested in a local going concern or a start-up company. One of the primary targets for this type of arrangement occurs with natural resource related companies. This method of investment helps the bank in moving the loan from a nonperforming to performing status. It also assists the company in obtaining a cheaper source of local currency. This is also another way to encourage investment in an otherwise disinvestment environment.

Flow-Through Shares

Flow-through shares is a relatively new method of financing made available to individuals or small Canadian mining companies which are conducting exploration or development of mining properties, and are not now or expected to be in a taxpaying situation for the next several years.

The use of flow-through shares helps to finance the cost of exploration or development programs without creating any additional liabilities for the company.

Individuals or mining companies using flow-through shares receive more money than they would from an issue of regular common shares. They also acquire the funds at a significantly lower after-tax cost.

The premium allows them to obtain additional government incentive grants because the company is able to spend more money than otherwise would have been available.

The advantages to the investor are that he is entitled to a deduction from income for the amount invested, he will own shares of the capital stock of the company which can be sold after the exploration or development costs are incurred, or he may obtain capital appreciation by holding the shares as an investment.

Other forms of gold financings are common shares with gold warrants attached to purchase gold at a predetermined price within a specified time period, preferred shares with gold warrants, and debentures with gold warrants.

Hedging of Metals, Foreign Exchange, and Interest Rates

Additional financial risks pertaining to metal price movements, foreign exchange exposure, and interest rate fluctuations can be alleviated through forward selling or proper utilization of the current options markets. Proper hedging allows one to sleep better when one knows some of the future risks have been reduced.

Interest Rate and Currency Swaps

Interest swaps effectively convert a floating rate loan to a fixed rate loan or vice versa. Currency swaps convert the loan or the cash flow from one currency to another. We have been very active in these markets and have seen the positive results from companies utilizing these products at the appropriate time.

CONCLUSION

There is always money available for good projects, but many good projects are never developed as they lack the preparation and presentation which allows the financial community an opportunity to analyze their potential for financing. Two major categories of risk are in developing and operating a mine and cost estimates and product price forecasts, Many of the risks associated with a mining venture can be reduced with careful preparation while others have to be accepted as part of a mining operation. The best overall procedure for qualifying and reducing risks is a meticulous evaluation by qualified and experienced management. Also, the proper timing of construction and mine operation within the economic cycles will help enhance property profitability.

It must be stressed that the most sophisticated financing techniques will not make a marginal property good.

Chapter 38

WHAT FINANCIAL INSTITUTIONS LOOK FOR IN SMALL MINE DEVELOPMENT

Stanley Dempsey, Chairman

Denver Mining Finance Company
Denver, Colorado

INTRODUCTION

The mining industry has been undergoing a significant restructuring for the past several years. Some firms have been liquidated. Some have been merged with other firms. A number of mining divisions of larger firms have been spun off, in some cases in leveraged buyouts by management. Recently, there have been a number of spin outs of gold subsidiaries. There are also a number of start-up situations where new firms are emerging as significant gold producers through mine development or acquisition.

This restructuring has created an attractive market for specialized mining finance firms, and a number of firms have recently entered that field in the United States. These firms provide financial advisory and valuation services to mining firms and project developers, and access to venture capital. In some cases, they invest for clients or for their own account.

I have been asked to describe how our firm works, and what kind of information we expect to receive when mine developers seek our assistance in financing a project.

HOW WE OPERATE

Denver Mining Finance works for both mine developers and investors. We also provide expert advice on mining issues to governments and to firms who are in litigation over mining issues. A lot of our work involves valuing mineral interests for various purposes. Our principal expertise is in the technical, commercial, and financial area. We help put mining deals into shape commercially so that they can be financed, and we help investors get into mining deals that are commercially sound. We work on projects, but we also advise mining firms on the structure of their business and how to make the most out of one or several mines that they might own. In addition to our work for clients, we are also prepared to invest in mining projects for our own account, and to recommend gold properties for acquisition by our parent company, Royal Gold, Inc.

We are, to some extent, a venture capital firm, in that we sometimes work with projects that are not "bankable," and are prepared to invest in attractive early-stage situations in mining. However, I want to emphasize that our principal focus is on development of commercially attractive mining projects that can be brought into production in the best possible financial terms. If we can make a project qualify for a project loan from a commercial bank, we will not advise the developer to use high cost venture capital.

For the purpose of this paper, I will focus primarily upon our functions with respect to development of precious metal projects. In a typical case, a firm with an advanced-staged project approaches us for help with financing the final feasibility study, and construction and operation of a mine. In some cases, the developer has already developed a financing plan and wants only our access to investors. If we agree that the project is sound and that the financing scheme will be attractive to investors, we will help the developer reach investors on a commission basis. In such cases, we are particularly insistent that the financing plan and our participation meet all regulatory requirements, particularly compliance with state and federal securities laws.

In most cases, however, the developer asks us for advise both on how to structure the financing, and how it can be implemented. Here we work for fees based on hourly rates for advisory work, and on commission or success fees for implementation. These assignments typically involve a review of both project fundamentals, and the structure and financial condition of the development firm. With information coming out of this review, we can determine whether to recommend debt or equity financing, and how such financing can be accomplished. In many cases, we recommend that the developer secure development funds by farming out part of the project in a joint venture. Where a firm wants to keep control and can handle the debt, we work with them on project financing approaches, including the use of gold loans. In some cases, we can help a developer raise funds through private placements of shares.

We generally find we can be most effective when we can have the freedom to use accountants, lawyers, and engineers to come up with an optimum commercial and financial program. This often includes development of economic models at both the project and entity level. I will say more about entity models later in this paper. It is our view that the cost of our work with the developer in putting together models and developing an integrated project construction and operating commercial/financial strategy will be saved when the actual financing takes place. In that case, the developer is in a good position to bargain with commercial and investment banks over the terms of a financing, and we are in a good position to help the developer keep the transaction costs of a project financing or an underwriting under control.

We are very selective in the work that we do, and work hard at screening projects. Our biggest source of income is success fees, and we will not take on advisory clients where we feel a project is marginal.

We receive a lot of inquiries about our services, and several submittals of projects come in each day. We log each inquiry into a docketing system that helps us keep track of every inquiry and submittal. A new matter form prepared for each matter received and a file is prepared. Each of the firm's principals receive either a copy of the new matter form or a weekly summary of new matters that have come into the office. Our files are set up on a geographic basis, by country and state or province. Mining projects are often known by several names, and we find geographic filing works best. We also send an acknowl edgement of receipt, usually the same day.

We are small enough that all of our principals can be involved in the review of inquiries, but in the usual case one person quickly reviews the inquiry to see what the inquirer wants, and in the case of project submittals, whether the firm or any of our existing clients will be interested. Properties which clearly do not meet our interest criteria are rejected promptly and submittal materials are returned.

INFORMATION REQUIRED IN SUBMITTALS

We are looking for two principal types of information in initial project submittals: (1) project data, and (2) entity data. We like to receive a two or three page summary of the project (and the deal if one is proposed) up front, with some reasonable backup material attached. We do not like to receive massive data packages with the first inquiry. We pay strict attention to caring for data entrusted to us and do not want to be responsible for lots of data and rare maps when we can tell from a summary that we have no interest. At later stages of review, we expect to receive access to all data, and often expect to travel to a mine site to see it.

FACTUAL PROJECT DATA

We expect to see a project summary that describes such features as land ownership, ore reserves, mining and recovery methods, infrastructure, environmental compliance, marketing assumptions, and development and operating costs. We also expect to see a description of the commercial aspects of the project such as royalty burdens and regulatory requirements (environmental, safety) that will effect capital or operating costs. A sophisticated document will also point out project risks and how they are to be dealt with.

We recognize that in most cases where venture capital is sought, some element of a so-called "bankable" feasibility study will be missing, but even venture capitalists want to see projections of the economic performance of the project on a stand alone basis.

Property Ownership

A project description should set forth how the developer holds the property. Is it owned outright or leased? What are the royalty and work obligations? A property map is helpful.

Ore Reserves

Ore reserves are a primary element of a mining finance firm's initial "sniff test." Reserves which have been developed to meet Security and Exchange Commission requirements will receive the most favorable attention, particularly if their existence is confirmed by a well-known, independent engineering firm. If you do not have reserve data of this quality, the next best thing is a report estimating reserves based upon your own engineer's work, with a clear statement of how he did the work and what assumptions he used. Finally, if you have done enough work to show that reserves may be developed, but not enough to call any of the mineralized material ore, we expect to see a report from a geologist or an engineer that describes the hard data available, and his or her estimate of the mineral inventory and potential, showing the assumptions used.

A mining finance firm will look at a property before the proven and probable reserve stage, and may help you find funding for reserve development, but you will not find prudent mining investors jumping past this step. Ore reserves are the critical factor in any mining venture, and must be adequately proven before any project can be safely undertaken.

Mining and Extraction

Mining methods and metallurgy come next. We expect to receive conceptual mine plans and flow sheets, and in the case of advanced-staged projects, fully engineered mine plans and flow sheets supported by bench, and in some cases, pilot scale metallurgical testing. Audits of mine plans and metallurgical tests by independent engineering firms add credibility. Comparables are okay for conceptual cost estimates, but engineered estimates for the actual mine and plant are eventually required.

As with ore reserves, the venture capitalists in a mining finance firm may help fund mine planning and metallurgical work, but will not short cut these elements to put a mine into production.

Infrastructure

We expect to see a description of access, water supply, housing, power availability and other infrastructure matters. We are particularly interested in hard information about the cost of securing power and water, and whether permitting delays might effect their availability.

Environmental Complaince

We are paying more and more attention to environmental and health and safety matters that might impact upon a mining project. We want to see a discussion of permitting issues, and how the project proponent expects to meet permit requirements. As with performance, we want to see some evidence of the feasibility of various compliance measures.

Commerical Considerations

A project description should include a summary of all of the various burdens on the property, including landowner royalties, overriding interests, local taxes, and the costs associated with any long-term contracts that are a liability of the project. Many properties are subject to royalties or net profits interests. These must be clearly described so they can be taken into account in projecting project economics. Likewise, if the property is burdened with a long-term power contract or an equipment lease, these facts need to be disclosed up front.

Project Management

We are particularly interested in receiving information concerning the individuals and firms involved in the project to date, and your plans for building and running the mine. Obviously, "name brand" firms and people clear project review hurdles first, but we do go beyond firm name, and like to know who in a big firm actually did the work.

Risk Analysis

We also appreciate receiving your risk analysis. It saves us time when we prepare our own, and shows us that you are realistically trying to deal with risk issues. A mining finance professional recognizes that every mine involves some risk, and is willing to accept some risk if it has been reduced and is taken into account when project appraisal discount rates and rates of return for investors are set.

Project Economics

We expect to receive the most detailed cost and project economic numbers available. We recognize that these may be "back of the envelope" in an early stage project, but we always need some estimate of project economics. We particularly urge you to spell out your assumptions. Are your mining costs based on comparable mines, or have you made a detailed estimate? Are equipment costs comparables, or are they based on actual quotes? How did you arrive at a discount rate for the DCF/ROI and NPV calculations.

What are your assumptions on the timing of first production? Of the time of receipt of first cash from a refiner?

ENTITY DATA

Let's turn now to what I call "entity data." When I talk about "entity," I mean the business organization that the project proponent intends to use to develop a mining project. Such an entity is usually a partnership, joint venture, or corporation.

We want to know in the first submittal all that we can about the technical, managerial and financial capability of the entity now holding the property. If the mine developer contemplates forming a new entity when an investor comes in, we want to know how that entity would be operated. In any financing, the investor will eventually focus on the questions of how the funds invested will be used, and how the investor will "exit," or get his money back out of the project. Entity data is essential to answering these questions.

To illustrate the point, consider a situation where a small corporation seeks funds to put a mine into production. An investor puts in money in return for a private placement of shares. The investor receives a minority position. In Case A, the corporation uses most of the money to pay off old debts and to try to start new projects. They invest little of the investor's money in the mine development project, and the firm ultimately runs out of money before the mine is completed. Here we have a disappointed investor. In Case B, same investment, but this time the corporation spends the money properly, develops a successful mine, but uses income from that mine to pay high wages to management and to buy a company airplane, with no dividend to the investor. Absent share price appreciation, and an opportunity to sell his shares, we again have a disappointed investor. In both cases, it would have been possible to protect the investor using a variety of legal mechanisms. Development of these mechanisms requires the kind of information about the entity which I will now describe.

First, we need a description of the business organization. If you have tied up a property with your own funds and have not formed a firm, just say so. You are acting as a sole proprietor. Where a partnership or corporation is involved, we want to know when the firm was organized, and in what state.

Business History

You should include in a submittal a few paragraphs detailing the history of the entity. Indicate major milestones in the firm's existence. A typical history might read:

> Joe Brown and Bill Smith formed a Mine Partners partnership, to mine the XYZ uranium mine in San Juan County, Utah in August 1956. The mine operated successfully from 1956 until 1972. In 1973, the firm was incorporated as MINECO, and entered into the contract drilling business through its wholly-owned subsidiary DRILLCO. DRILLCO has been consistently profitable. It finances short term working capital requirements with a $1 million line of credit from the First National Bank. Earnings from DRILLCO permitted the firm to weather the down turn in the uranium business, and to seek new opportunities in other commodities. In 1984, MINECO took an option on a gold property in Lander County, Nevada. Exploration financed by internally generated funds (DRILLCO earnings) has demonstrated the existence of a small gold deposit which appears to be amenable to heap leach extraction. The firm is currently seeking financing for development of the gold property, and expects to place the mine into production in late 1988. DRILLCO opened a branch office in Reno in August, 1987, and expects to benefit from the current gold boom in Nevada.

Business of the Future

Include a statement of your firm's plans for the future. If you have a formal business plan, attach it. If not, at least set out the principal goals of the firm. Do you plan to develop into a broad ranging mining company? Will you limit yourself to precious metals? What are your longer term goals and policies with respect to funding growth of the company?

If you have an entity model, attach it too. By entity model I mean a projection of your firm's revenues and costs over future periods, including all projects such as the one you propose to finance, and all entity expenditures such as G&A. Such a model also includes financing assumptions. These are usually set up as computer-based spreadsheets, and should give you projectional future cash flows. In the case of a corporation, it should be able to generate share price based upon assumed p/e ratios.

Structure

Include an organization chart or a description of how you are organized.

Management

We are particularly interested in knowing who is involved in your firm. You should include biographical information for directors and officers, and for key employees. We are particularly interested in information about the ability of the directors, officers and key management people to carry out the project proposed for financing, and to undertake overall management of the entity and all of its projects.

We are also interested in knowing whether you use long-term employment agreements and stock options. We do not need details of remuneration in an initial proposal, but we will eventually need that kind of information if we are seriously considering a financing plan for your firm.

Compliance

You should include a statement concerning the condition of your filings with government agencies. If your firm is a corporation, we want to know whether all of your state filings are current. If you are a pubic company, we want to know the condition of your filings with the SEC. You should include copies of recent filings.

Other Projects

If your firm is involved in other projects and other lines of business, provide a brief discussion of these matters. We are trying to assess where the project to be financed fits into your firm and how much attention it will receive.

Litigation

You should describe any litigation or threats of litigation that involve your firm. The mining finance firm will find out about these eventually through its due diligence efforts, and it is better to get these matters out into the fresh air at the earliest possible time.

Conflicts of Interest

We are particularly interested in knowing about any real or apparent conflicts of interest in your firm. Such conflicts are not necessarily evidence of wrongdoing, but must be disclosed. It is very typical for small firms to employ family members and for principals to rent a building to the firm. In start up situations, it is also not unusual for a firm to employ people on a part time basis. The important thing is to have these relationships disclosed up front.

Consultants, Lawyers, etc.

We also like to see who your firm uses as consultants. Mining is a small industry, and "good names" make a difference. We also like to know the name of your bank.

Financials

A statement of your firm's financial condition should be included. If you feel you want to hold this back in the case of a private firm, at least make a general statement concerning condition. We receive a lot of requests for financing from firms who are teetering on the edge of bankruptcy. In a rare few cases, we might be able and willing to help, but we need to know what we are up against up front.

We also want to know whether your financials are audited. Audited financials are often required in a financing, particularly a public financing, and can take a long time to accomplish.

Financing Data

Discussion of the more detailed data required to put together a financing plan for a property is beyond the scope of this paper, but I will say that even in the initial submittal, we want some idea of how much money you think you need, when you need it, what you will do with it, and how the investor will get it back.

A. Use of Proceeds

We want to know, in detail where an investor's money will be spent. Stating that the money will be used for "working capital" is not enough. If you need $200,000 to fund a study by Bateman, or Behre Dolbear, and another $20,000 to keep your office open while they do the work, say so.

B. Exit

The most critical concept in finance is "exit," or how does the investor get his money and hoped for profit back out of the deal. Generally, a mining company must either dividend the money out, or provide a way for the investor to sell his share of the firm. Your view of how the investor will exit should be expressed in the first proposal for financing.

CONCLUSION

Summing up, a mining finance firm wants to know about both your project and your firm, and appreciates receiving well-organized and complete initial data packages. Thorough preparation of such packages forces you to think through the fundamentals of your project and your business, and makes it easier for the financier to see the merit of your proposal. And a well-developed data package is easy to supplement when more detailed information is required at more advanced stages of the financing process.

I hope that you will find this listing of submittal data requirements useful, and thank you for your attention.

Chapter 39

PRECIOUS METALS SLAG TREATMENT USING AN ELECTROSTATIC SEPARATOR

by

Mr. Ted D. Maki
FMC Paradise Peak Corporation
Refinery Supervisor
Hawthorne, Nevada

Mr. Joseph B. Taylor
Carpco, Inc.
Technical Manager
Jacksonville, Florida

INTRODUCTION

FMC's Paradise Peak mine, located 13 km (8 miles) south of Gabbs, Nevada, became operational in April of 1986 (Figure 1). It was designed and built by Davy McKee, who was instrumental in equipment design and selection. It is the 7th largest gold deposit in the United States with defined reserves of 10.9 mt (12 million st) containing 34 million gms (1.1 million tr oz) of recoverable gold and 933 million gms (30 million tr oz) of recoverable silver.

The mine extracts ore by the open-pit method, taking advantage of a 1.5:1 stripping ratio. The mill operates at a 3600 mt/day (4000 st/d) capacity. Crushing is done in three stages to achieve an ore size of minus 0.635 cm (0.25 inches). Grinding further reduces the ore to 85% minus 100 mesh. The ground ore is treated with cyanide in agitated leach tanks and then washed in the counter-current decantation (CCD) thickeners. Zinc is added to the clarified, deaerated pregnant solution to precipitate the precious metals. The precipitate is acid digested to eliminate excess zinc, filtered and retorted to drive off contained mercury. The retorted precipitate is then fluxed, melted and poured through cascade molds. Dore bars are cleaned for shipment and the slag is sent to an in-house slag treatment system.

ELECTROSTATIC SEPARATION

The dry electrostatic slag treatment system at Paradise Peak is the first installation of its kind. Electrostatic separation has been widely used in mineral processing since the early 1950's. A brief discussion of the theory behind the process is helpful to those not familiar to electrostatic separation. Charging and separating slags, dry minerals or other materials by ion bombardment is the most common form of electrostatic separation. Millions of tonnes of minerals are processed each year by this method.

In an ion bombardment separation, granular material is fed onto a grounded metal cylinder (or roll) and charged by a corona-producing electrode placed above the roll's surface (Figure 2). While both conductors and nonconductors become charged, only the conductor is able to lose its charge. The charged nonconductor, as it rests on the roll's surface, "sees" an oppositely charged image of itself in the metal surface. It is attracted to the image charge, becomes electrostatically pinned to, and moves with the roll's surface. The conductive particle also sees an image and is attracted to it. But upon touching the roll's surface, it discharges rapidly to the grounded surface and is thrown free from the roll's surface with a projectile motion.

In the case of precious metal slags, the conductive particles would be metallic prills of dore metal; and the nonconductive particles would be the slag, free of metal. Middling grains, generally, are in the form of a metallic prill encased in or incompletely liberated from slag.

LABORATORY TESTING

A number of precious metal slag samples have been tested in the laboratory. It has been found that the composition of slags vary widely from one refinery operation to another. Typical laboratory procedure is to crush and size the slag followed by two-stage lab-scale electrostatic separation. The conductor fractions from the first and second pass are then combined as a prill concentrate; middlings fraction and clean slag tailings from the second pass are held separate. Selected results from laboratory testing are shown in Table 1. It is important to note in Table 1 that the assays represent overall silver and gold, not metallic values.

From experience, the laboratory results are typically lower in grade and recovery than the industrial installations. This is largely due to the hydroscopic nature of precious metal slags coupled with the high local humidity in Carpco's Jacksonville, Florida, location. Relative humidity in Jacksonville can range form 60-95%, while most mining locations in the

Western United States are typically 15-20% relative humidity. Heating laboratory feed material tends to alleviate the problem.

One should note that not all precious metal slags are amenable to electrostatic separation. From laboratory tests, it has been found that remelted slags with low head assays are typically free of metallic prills. An electrostatic separation will not concentrate chemically entrained metal but will only separate physical metal from the slag.

OPERATION AND MAINTENANCE OF SLAG PROCESSING SYSTEM

The purpose of slag handling at Paradise Peak is to recover the precious metal prills from the slag for subsequent remelt. The cleaned slag is then recycled to the ball mill in the grinding circuit. This allows for higher recoveries of metal values in the refinery and eliminates the costs of shipping and processing the clean slag at a smelter. Flowsheet is shown in Figure 3.

Slag from the refinery furnaces is poured directly into cast iron slag pots. Upon cooling, the resulting slag "buttons" are manually loaded into a 907 kg (2000 lb) portable hopper. This hopper is then transported by a crane to the 1814 kg (4000 lb) capacity slag bin atop the processing unit. An electromagnetic vibrating feeder supplies slag to a 178 mm x 305 mm (7 in x 12 in) laboratory-type jaw crusher. The minus 127 mm (1/2 in) jaw-crusher product is carried via bucket elevator to a 305 mm (12 in) face width by 457 mm (18 in) diameter double-roll crusher. The roll-crusher discharge drops onto a triple- deck circular vibrating screen. The plus 6.35 mm (1/4 in) material is flattened metal prills and some platy slag which is sent to remelt. The minus 6.35 mm (1/4 in) plus 0.85 mm (20 mesh) product is recirculated back to the roll crusher by means of a second bucket elevator. The minus 0.85 x 0.15 mm (20 x 100 mesh) portion contains the majority of the prills and is the feed to the two-stage, high-tension 1219 mm x 1346 mm x 3073 mm (48 in x 53 in x 121 in) electrostatic separator. The minus 100 mesh is virtually metal free and reports to the clean slag hopper (Figure 4).

Feed to the separator is heated, as even a slight increase in humidity will decrease efficiency. The two-stage Carpco electrostatic separator (Figure 5) utilizes the difference in electrical conductivities to separate the clean slag from the metal prills. The clean slag reports to the clean slag hopper and the prills to a small drum for future remelt. The middlings generated are recycled through the electrostatic separator.

Due to the dryness and fineness of crushing, a 2400 acfm capacity baghouse is required with several collection points. Baghouse dust is collected in a hopper and recycled to the ball mill as clean slag. The entire system processes 136-159 kg/hr (300-350 lbs/hr). Typically, one day's production of slag can be processed the following day. The whole package requires only one operator on a part-time basis for changing and weighing hoppers.

Maintenance is limited to that normally associated with crushers and bucket elevators. The electrostatic separator is self-cleaning and virtually maintenance free.

METALLURGICAL DATA

Table 2 contains typical data obtained from processing a batch of slag in May, 1987. The recoveries (Table 3) are, at best, an approximation due to the difficulty in sampling and assaying material containing metallics. Calculations were made on the assumption that silver and gold in the baghouse dust are chemically entrained. Microscopic observations have confirmed this.

CONCLUSION

The slag processing system at FMC Paradise Peak has been operational since April 1986. With the exception of minor problems at start-up, the system has performed without major difficulty. The advantages of the dry electrostatic route are:

- Dry processing -- no water is required and moisture-related hazards are consequently reduced in the refinery.
- Process requires minimal operator attention.
- No slag remelting is required.
- Slags are processed on a daily basis, thus reducing precious metals inventory.

Figure 1. FMC Paradise Peak Mine, Gabbs, Nevada.

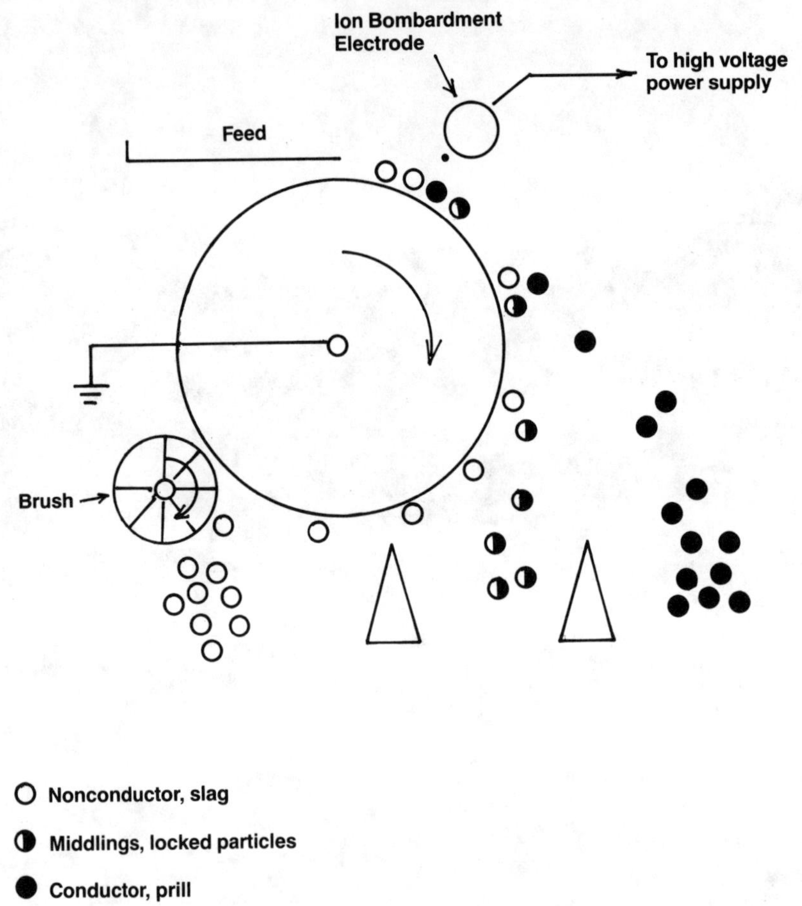

Figure 2. Diagram of an electrostatic separator.

Figure 3. Slag processing system, FMC Paradise Peak.

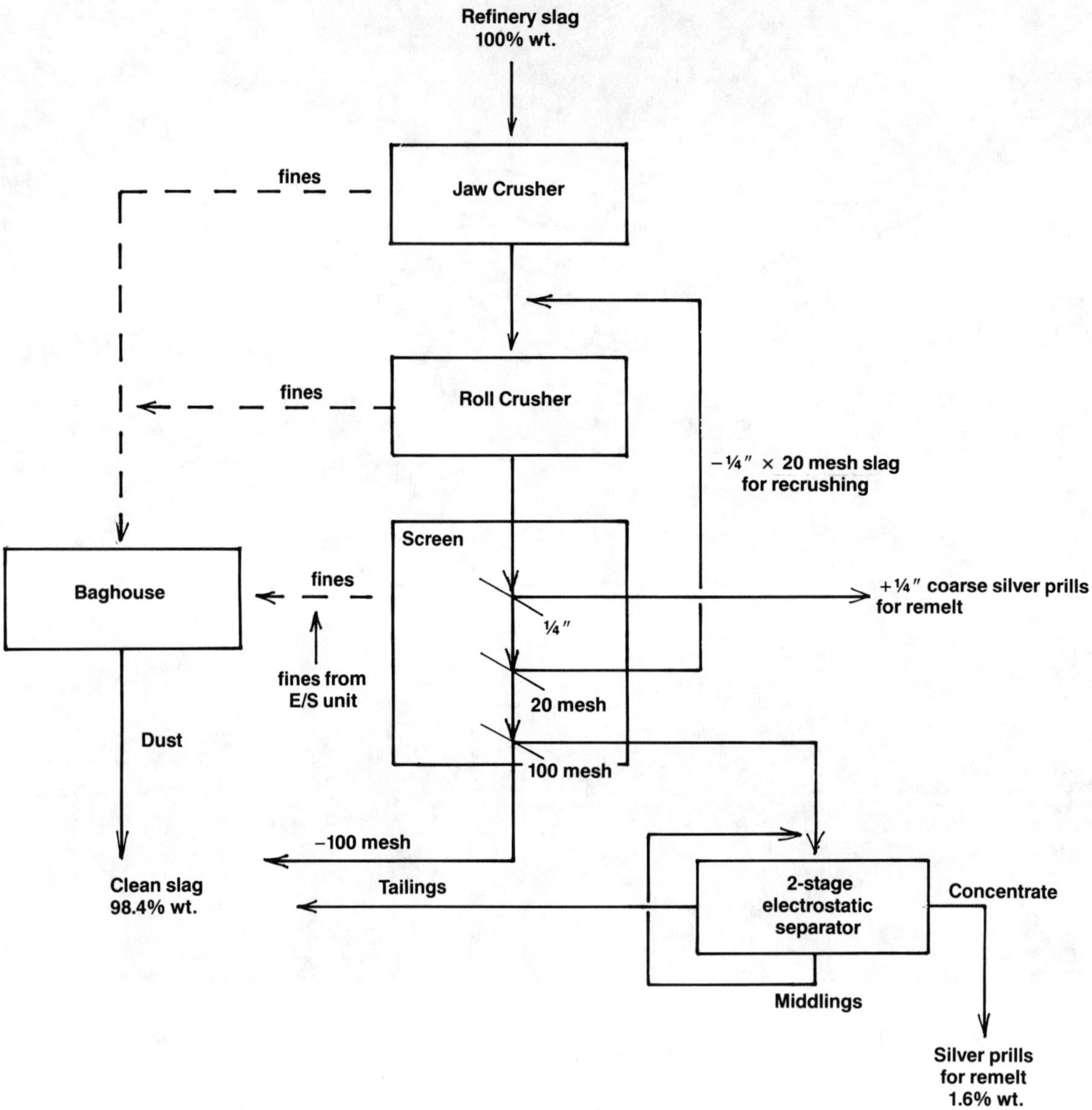

Figure 4. Slag treatment flowsheet, FMC Paradise Peak.

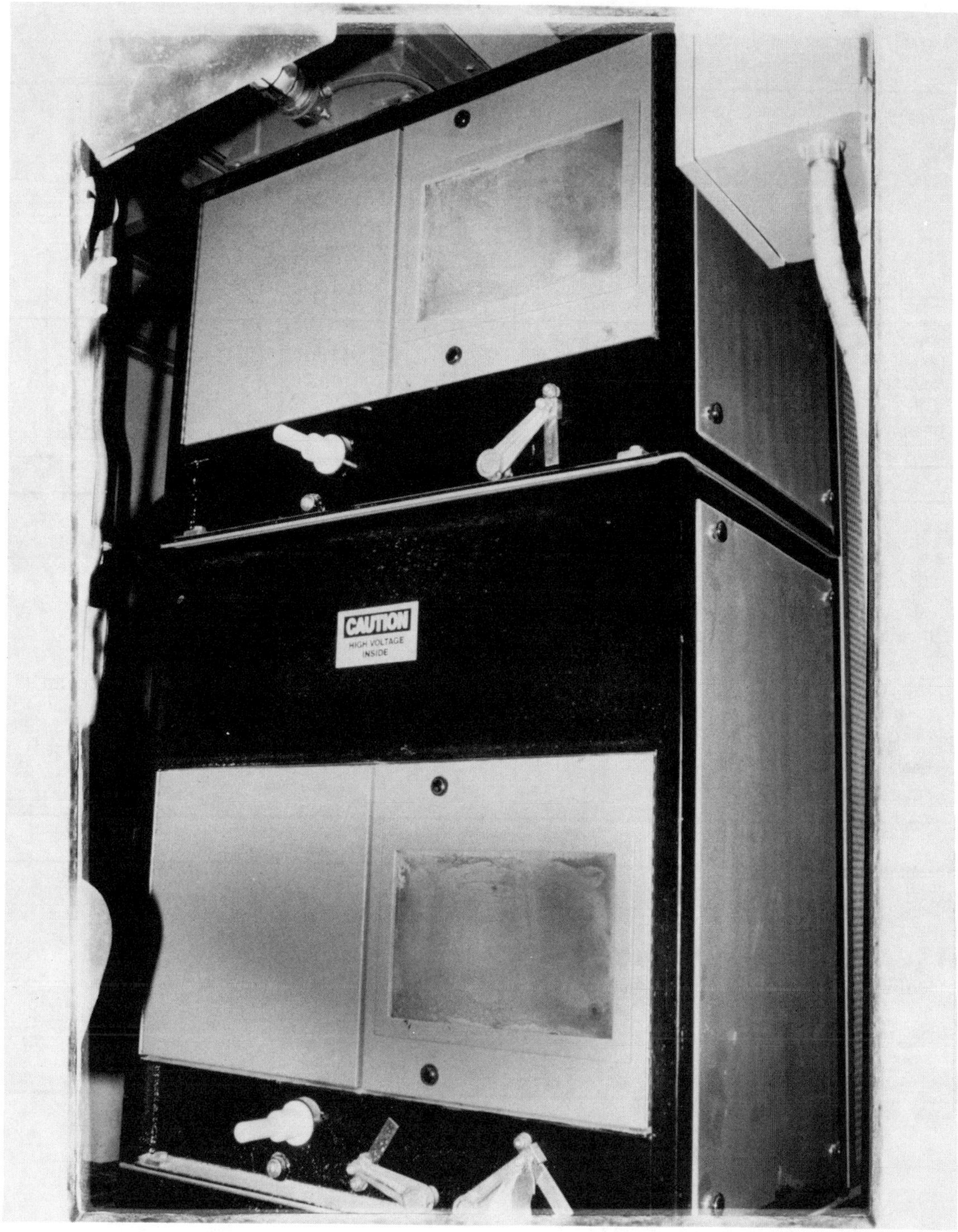

Photograph courtesy of Valerie Parks.

Figure 5. Two-stage Carpco Electrostatic Separator, Model HTE (36) 121-46.

Table 1. Selected Laboratory Electrostatic Slag Processing Results.

Location	Material	Feed Grade (oz./ton) Gold (Au)	Feed Grade (oz./ton) Silver (Ag)	Prill Concentrate Grade (oz./ton) Gold (Au)	Prill Concentrate Grade (oz./ton) Silver (Ag)	Overall Recovery % Gold (Au)	Overall Recovery % Silver (Ag)
Nevada	Gold Slag	55	—	359	—	84	—
Utah	Silver Slag	35	870	200	4470	79	76
Nevada	Gold Slag	25	—	81	—	90	—
Utah	Refractory	50	1340	118	2663	94	85
Nevada	Gold Slag	180	—	1700	—	95	—

Table 2. Typical Batch Slag Processing Results.

Product	Weight kg. (lbs.)	Weight (%)	Assay Gold (Au) g/mt (oz./ton)	Assay Silver (Ag) g/mt (oz./ton)
Head Feed	3,160 (6,970)	100.0	155.1* (4.52)	2,070* (60.3)
Cleaned Slag	2,300 (5,070)	72.8	8.3 (0.24)	566 (16.5)
Prill Concentrate	50 (110)	1.6	9,049.8 (264.0)	89,300 (2,600.0)
Baghouse Dust	810 (1,790)	25.6	24.4 (0.71)	939 (27.4)

*Calculated

Table 3. Gold And Silver Recovery To Prill Concentrate.

Element	Recovery (%) Overall	Recovery (%) Metallics
Gold (Au)	92.1	95.9
Silver (Ag)	68.2	77.4

Chapter 40

EXPLORATION PERMITTING: THE PROSPECTOR'S FIRST BRUSH WITH THE LAW

by

Collon C. Kennedy* **
Gold Fields Mining Corporation
Land Manager & Assistant General Counsel-Exploration
Lakewood, Colorado

Frances C. Hartogh
Gold Fields Mining Corporation
Senior Attorney
Lakewood, Colorado

C. Joel Casburn
Gold Fields Mining Corporation
Regional Land Manager
Sparks, Nevada

INTRODUCTION

During the nineteenth century, the federal mining law was enacted with the objective of promoting the development of the western United States by providing unrestricted access to vast areas of federal land for mineral exploration. Prior to the advent of comprehensive federal, state and local land use plans, there was little, if any, regulation or restriction of mineral exploration activities. Today, the laissez-faire policies of the past that encouraged mineral exploration have gradually given way to the public's legitimate concerns regarding environmental protection, multiple use and planned land development. As a result, the contemporary prospector confronts a maze of statutes, ordinances and regulations with which he must comply to obtain the requisite approvals and permits before commencing mineral exploration on federal, state or private lands.

This presentation will provide a broad overview of the exploration permitting process. The first section of this paper recommends certain general management practices to follow when attempting to obtain exploration permits. The second section presents two hypothetical cases, one involving a drilling program on private property in California and the other involving a drilling program and the removal of a bulk sample for metallurgical tests from federal land in Nevada. The cases are used to discuss jurisdictional issues and application of certain federal, state and county laws and regulations that pertain to mineral exploration. The final section contains a brief analysis of three important recent United States Supreme Court decisions. These decisions address the scope of the federal pre-emption doctrine as it applies to activities conducted on unpatented mining claims and the constitutional limits of permissible state and local land use and zoning regulations.

PERMITTING PRACTICES

The environmental impacts associated with mineral exploration generally result from the construction of drill sites, roads or trenches. Consequently, after the regulatory authority has considered the issues of permissible land use, the primary emphasis of the permitting process is to address problems related to pre-construction site identification and clearance, and post-construction site reclamation. The approaches taken to address these and other related issues vary with the scope of proposed work, the site, the appli-

* This presentation is not intended to be an exhaustive legal analysis of the permitting process. Before commencement of any permitting effort, the permittee should consult with an attorney to ensure full compliance with all applicable laws. The states of California and Nevada are used as examples only; each state will have its own applicable state and local laws, rules and regulations.

** The authors wish to express their gratitude to Pauline Wilber, Carolyn Griffin, Sandy Hetsko and Linda Bates, all of whom work in the Land and Legal Departments for Gold Fields Mining Corporation, for their research and contribution to this paper, and to Debra S. Glader, Graphic Services Coordinator for Gold Fields Mining Corporation, for her coordination and development of the slide presentation.

cant and the reasonable concerns of the permitting authority. An applicant, however, should consider the following recommended permitting management practices to insure that permits and approvals containing fair and reasonable conditions are timely obtained.

Formulate Project Description

Prepare a written description of the exploration project. The project description should contain two elements: a geologic or technical description and a land use/regulatory description. The geologic or technical description should include, among other things, a scope of work, access routes, equipment and material requirements, site identification, extent of surface disturbance, and water availability and needs.

The land use/regulatory element of the project description should include relevant information relating to property ownership both within and immediately adjacent to the project area, pertinent restrictions in existing acquisition documents, and any new land use developments or environmentally sensitive areas located within or near the proposed exploration project. Finally, the name and telephone number of the contact person and any relevant information regarding the regulatory authority primarily responsible for processing the permit application should be included in the project description.

Identify and Analyze All Applicable Statutes and Regulations

Prepare a list and a summary of all applicable statutes, ordinances and regulations, including all local land use maps, plans and zoning requirements. Include in the summary a description of the procedural process and time frames in which the regulatory authority must review, comment on and process the permit application.

Prepare Complete Permit Application

It is imperative that the permit application be complete when filed. All legal issues and environmental, technical and, if warranted, social impacts resulting from the exploration project should be addressed. The permit application and supporting documentation form the basis upon which a regulatory agency will decide whether a permit should be issued. As a general rule, no regulatory action will be taken on a pending permit application until the application is determined to be complete.

Strict Notice and Procedural Compliance

Whenever possible (after taking into account all political, technical and non-legal considerations), insist on strict statutory notice and procedural compliance by the regulatory authority. Remember, however, timely compliance is a mutual responsibility. Prepare for your own reference a chronology of due dates. Do not depend on the regulatory authority to have the correct information. Establish an active and cooperative working relationship with the regulatory authority. If the initial exploration program is successful, the cooperation, good will and positive working relationship established during the first phase of permitting should contribute to subsequent permitting efforts.

Documentation

An exploration permit is the final product of a negotiated agreement between the permittee and the permitting agency that describes, among other things, the approved scope of work and any reporting and bonding requirements. Many of the specific details, however, regarding the implementation of permit conditions are routinely not incorporated in the permit. Consequently, documentation which elaborates on the details of the parties' understanding is an essential and necessary supplement to the permit. Thorough documentation of all matters pertaining to the permit both before and after issuance facilitates permit compliance, promotes a positive and professional working relationship with the regulatory agency, and helps prevent misunderstandings after work has commenced on the property.

CASE STUDY ANALYSIS OF APPLICABLE LAW

At times, determining which governmental agencies have jurisdiction over a proposed exploration project can be a monumental, if not impossible, task, as witnessed by the recent United States Supreme Court decision in <u>California Coastal Commission et al. v. Granite Rock Company</u>. Permit requirements vary with each exploration project. The following two hypothetical exploration projects will be referred to for the purpose of discussing the issues of jurisdiction and analyzing applicable federal, state and local laws and regulations. Case 1, the "Green Gold Project," involves unpatented lode mining claims located on federal land in Nevada. Case 2, the "American River Project," involves private fee land in California.

Case I - Green Gold Project

The Green Gold Project consists of seventeen (17) unpatented lode mining claims ("Mining Claims") located in northeastern Nevada on federal land managed by the Bureau of Land Management ("BLM"). Though no mining has been conducted on the Mining Claims, outcrops containing significant gold mineralization on the property, and assay results from extensive diamond drilling conducted by several major mining companies, indicate the possible presence of a small high grade gold deposit. Initial examination of the drill core indicates there may be metallurgical problems that would prohibit the use of a heap leach recovery process. An initial work program is formulated that would involve geochemical and geophysical surveys, a sage sampling program, a ten (10) hole diamond drilling program to confirm the results of the previous drilling and the removal of a 50 ton bulk sample from the surface of the Mining Claims by trenching across the mineralized outcrops. The bulk sample will be tested offsite. Water for drilling is required. A building will be constructed for the purpose of storing drill core and housing sample preparation facilities. Cattle bearing the Bar None brand have been seen on the Mining Claims.

Public Lands. The United States owns 740 million acres of land or roughly one-third of the nation's land mass. Approximately 648,000,000 acres make up what is commonly classified as the public domain,[2] a portion of which is open to mineral entry under the General Mining Law of 1872 ("General Mining Law").[3] Excluding Alaska, more than 90% of the public domain is located in the 11 western states of Arizona, California, Colorado, Idaho, Montana, Nevada, New Mexico, Oregon, Utah, Washington and Wyoming.[4] Public domain land comprises 42% of the land in California and 86% of the land in Nevada.[5] Congress has the power to dispose of, and to promulgate rules and regulations to administer, the federal lands pursuant to the Property Clause in Article IV[6] of the United States Constitution.

Federal Law. Hardrock mineral exploration and development on federal land is authorized pursuant to the provisions of the General Mining Law[7] and the Mining and Mineral Policy Act of 1970.[8] Unlike state or local laws and ordinances, the federal law contains an orderly scheme for the exploration and development of all natural resources discovered on federal lands within the context of a policy that promotes multiple use.

Surface Management Regulations. In 1955 Congress passed the Surface Resources and Multiple Use Act,[9] which provided that the federal government would retain and manage the surface resources of subsequently located unpatented mining claims. Congress has delegated to the Secretary of Agriculture the authority to regulate the use and occupancy of national forests.[10]

In 1976 the Federal Land Policy and Management Act (FLPMA)[11] and the National Forest Management Act (NFMA)[12] were enacted. These acts required the development of federal land use management plans affecting unpatented mining claims on federal lands. Under FLPMA, the BLM is responsible for managing mineral resources on federal lands. NFMA provides that the United States Forest Service (Forest Service) under the Secretary of Agriculture is responsible for management of surface impacts of mining on national forest lands.

Today the BLM administers most of the public domain land open to mineral entry under the General Mining Law, which consists of approximately 341,000,000 acres.[13] The Forest Service administers approximately 146,000,000 acres of the public domain.[14]

Both the BLM and Forest Service regulate access to mining claims through surface management regulations. The regulations recognize that operators are entitled to access consistent with provisions of the mining laws, but require regulatory compliance to prevent unnecessary degradation of surface resources.

The principal objectives of the federal surface management regulations are to:

(a) Provide for access to lands open for mineral entry in a manner that will prevent unnecessary or undue degradation and provide for protection of nonmineral resources;

(b) Provide for reclamation of disturbed areas; and

(c) Coordinate with the appropriate state agencies to prevent unnecessary or undue degradation with respect to mineral operations.[15]

The main difference between BLM and Forest Service regulations is that the BLM employs an acreage threshold to apply surface management regulations. The acreage categories are (1) casual use, (2) disturbance of five acres or less, and (3) disturbance which exceeds five acres.[16] The BLM District Manager for the area in which the mining claims are located is responsible for the review and approval of all permit applications.

By contrast, the Forest Service criteria distinguish between operations which "might" cause disturbance and those which are likely to cause "no significant" disturbance to national forest surface resources. The District Ranger having jurisdiction over the forest in which the operations will be conducted determines whether the proposed operations will likely cause significant disturbance of surface resources.[17]

Whether on federal land administered by the BLM or the Forest Service, a mining claimant or operator is entitled to access to his operations provided that measures are taken to prevent unnecessary or undue degradation of the federal lands. If a notice or plan of operations is required, the location of access routes must be identified.

Recently, the Interior Board of Land Appeals ("IBLA") ruled that FLPMA required a right-of-way application to be approved prior to transporting water across public land for any mining purposes.[18] The administrative court elaborated by stating that "the implied right of access to mining claims never embraced the right to convey water from outside the claim for use on the claim."[19]

Finally, areas of disturbance within the Green Gold Project may require fencing or other protection if a danger is presented to cattle or other authorized multiple uses. Watering holes, however, may not be fenced.

The State of Nevada and its counties generally do not regulate exploration for non-energy minerals. Because of the large percentage of federal land in Nevada, the state and the county governments rely on the federal surface manager to regulate disturbance on and under federal lands. Consequently, the potential financial burden imposed by the overlap of federal, state, and local regulations which can plague similar projects in other states is not present in the case of the Green Gold Project.

Case II - American River Project

The American River Project consists of approximately 250 acres of fee land located on the scenic American River in northern California. The county has zoned the property R-10, which permits mineral exploration drilling upon obtaining a Special Use Permit issued by the County Planning Department. Single family residences and ranches are located within one mile of the property. Within the boundaries of the property is located the Golden Dream Mine, an abandoned underground gold mine. Access to the property is provided by an all-weather road maintained by the county. Based on a review of existing technical data, old mine maps and a field inspection, the property appears to be a good prospect for a possible open pit low grade gold deposit. A proposed exploration program is formulated that would consist of geochemical and geophysical surveys followed by a fifteen (15) hole reverse circulation drilling program. The program would disturb less than 1000 cubic yards of earth. Water for the drilling program would be obtained from a "glory hole" located on the property.

State and County Regulations. Mineral developers on private land in California are exposed to the full range of state and local regulation. California sets minimum standards for private and non-federal land; counties may adopt additional standards.

Assessment of Environmental Effects. California has enacted legislation, similar to the federal National Environmental Policy Act, which requires any local public agency (such as the county planning department), prior to making a decision on a project which affects the environment, to comply with the California Environmental Quality Act ("CEQA").[20] CEQA requires the county to consider the possible environmental effects of the proposed project. If the county determines that the project will not significantly affect the environment, a negative declaration is issued. If it is determined that the project may have a significant effect on the environment, the county must perform an initial study to determine if an Environmental Impact Report ("EIR") is required. If the study concludes that the project will significantly affect the environment in a manner that cannot be adequately mitigated, an EIR must be prepared.[21] Preparation of an EIR will usually require a minimum of several months to one year.

The effect of county and other local regulation on the proposed exploration program cannot be underestimated. Mineral exploration, including exploratory drilling, is a conditional use in most California counties. Since local permitting procedures may take months and require extensive public hearings, necessary land use permits must be applied for well in advance of the anticipated start-up date. For any project where there is residential development within or near the project area, obtaining the requisite local permits and approvals may take as long as 6 to 12 months. Application to conduct an exploration program is normally made to the county planning department.

Surface Reclamation. In addition to complying with the requirements of CEQA, a permittee must also comply with the California Surface Mining and Reclamation Act of 1975 ("SMRA"). SMRA exempts "prospecting for or the extraction of minerals for commercial purposes and the removal of overburden in total amounts of less than 1,000 cubic yards in any one location of one acre or less".[22] With respect to the American River project the initial drilling program should be exempt from SMRA. Many counties, however, have adopted ordinances requiring a special use permit and/or rezoning as a prerequisite to exploration drilling.

Miscellaneous Drilling Requirements.

DRILL SITE ABANDONMENT REQUIREMENTS: California has no surface abandonment regulations for drilling or other exempt exploration activities. The county may regulate such activities through conditional use permits or by adopting ordinances, rules, and/or guidelines.

DOWN HOLE ABANDONMENT REQUIREMENTS: The legal liability for improper abandonment is not limited in duration or amount. Therefore, it is essential to specifically have the state and county approve the drill hole abandonment procedures. By proceeding in this manner, it is possible that the operator's liability may terminate when bond release by the state or county is approved.[23]

DRILLER REQUIREMENTS: California has a licensing program for water well drillers.[24] Because a licensed well driller's signature may be necessary for the abandonment report, the driller should be state licensed.

RECORDING AND REPORTING: Filing of a Water Well Driller's Report for abandonment of the drill holes may be required.[25] County regulations should also be reviewed in this regard. Several counties require posting of a Notice of Abandonment and subsequent filing of a complete well installation and abandonment report.

Water

All exploration, mining and milling operations require water. Consequently, it is important to develop a water acquisition and permitting strategy during the early stages of exploration.[26] An effective permitting strategy will take into consideration both federal and state water laws and regulations that affect acquisition of water rights and water quality protection.[27]

Acquisition of Water Rights. The basic legal framework for acquiring water rights has been developed under two doctrines, the appropriation doctrine and the riparian rights doctrine. Each state has developed laws and regulations controlling water rights.[28] In addition, the western states have taken the lead in developing comprehensive groundwater statutes.[29] Excluding California and Texas, "it is generally true that the 17 western states follow appropriation law and the remaining states are thought of as riparian states."[30]

APPROPRIATION RIGHTS The concept of prior appropriation of surface water is that priority of legal right to water is based on actual beneficial use. The three essential elements of this doctrine are:

(i) Priority of right is based on priority in time.

(ii) Beneficial use is the measure of the right.

(iii) Ownership of riparian land is irrelevant to the right.[31]

To determine the relative rank of appropriated water rights, each state has implemented preferences for certain uses in the state constitution and/or by statute. Colorado, for instance, pioneered the judicial method of determining priorities as early as 1879.[32] Most states, however, rely upon an administrative permit system whereby an official (usually the state engineer) "is responsible for compiling and administering existing appropriations in the order of their priorities."[33]

RIPARIAN RIGHTS "Unlike appropriated water rights which have relative priorities or may be subject to constitutional or statutory preferences," riparian rights "are an incident of land ownership rather than the result of some act, such as diversion or use."[34] Because under the riparian rights doctrine, the right to use water in reasonable amounts is based on the continued ownership of property adjacent to the water source, riparian rights are not lost by nonuse. A riparian right "is a right of use, arising from the ownership of land."[35]

GROUND WATER RIGHTS Because of early misconceptions that surface and underground water were or could be physically distinct, "the law of subsurface water rights has developed independently from the law of surface water rights."[36] At least ten western states have declared that they own all of the ground water.[37]

Several states, including Nevada, have applied the appropriation doctrine for surface water "to one or more classes of groundwater as well, usually in a modified form under which (the state engineer) may control pumping of wells."[38] In contrast, California groundwater law follows the doctrine of "correlative rights," similar to the riparian doctrine of reasonable sharing, wherein the emphasis in determining relative groundwater rights is not so much on who used the water first but rather on reasonable quantity of use by each owner. Priority is given to owners of land overlying the groundwater basin."[39]

Water Quality. Potential impacts to surface and groundwater quality have been under increased scrutiny over the years. This concern for water quality has resulted in the development of an extensive set of federal and state water quality laws and regulations which (although more applicable to the development and operations stages of mining) are briefly mentined below. In general, surface and ground water quality are subject to separate regulatory programs.

The principal objective of the federal Clean Water Act,[40] is to maintain and restore the chemical, physical and biological integrity of the nation's waters. In order to help achieve this objective, Section 301 of the Act requires an operator to obtain a National Pollutant Discharge Elimination System (NPDES) discharge permit, prior to any point source discharge of pollutants into surface waters of the United States (which includes dewatering of workings). The term "pollutants" is broadly defined under the Clean Water Act by including all types of waste and, under certain circumstances, rock and sand.[41] NPDES permits are issued and administered by the EPA; however, the states may assume responsibility for the permitting authority if they develop a discharge program consistent with the federal requirements.[42] Ground water protection is specifically addressed by the federal Safe Drinking Water Act and may also be governed by regulations relating to state mined land reclamation and state or local solid waste disposal.[43]

RECENT SUPREME COURT DECISIONS

Three recent United States Supreme Court decisions will have future impacts on the mining industry in the area of land use and environmental permitting. These cases address two fundamental elements of the U.S. Constitution regarding property rights. The Federal Preemption doctrine, based on the Supremacy Clause of the Constitution, provides that the laws of the United States are the supreme law of the land and that contrary state or local laws are invalid. Over the years, however, the courts have recognized that state and local law supplementing federal law are permissible. The Just Compensation Clause of the 5th and 14th Amendments provides that neither the federal, state, nor local government may take private property without making "just compensation" to the owner. Laws that impose reasonable conditions on the use of private land are permissible; however, as witnessed by a long line of U.S. Supreme Court cases, the distinction between permissible regulation of use and a taking of private property requiring compensation is difficult to make.

A landmark decision recognizing state and local jurisdiction over mining activities conducted on federal lands was issued this spring. In the case of California Coastal Commission, et al., Appellants v. Granite Rock Company,[44] the Supreme Court in a 5-4 decision overturned a Ninth Circuit Court of Appeals ruling and held that the California Coastal Commission could require the Granite Rock Company to obtain a coastal development permit for work on unpatented mining claims located on federal land, even though Granite Rock has been conducting active mining operations since 1981 under the terms of a plan of operations issued by the U.S. Forest Service. It is important to note that the Supreme Court's decision is narrow in that Justice O'Connor, in the majority opinion, stressed that the court was not giving the state a green light to block all mining on federal land.[45] The holding in this case follows a growing trend of state supreme court decisions that delegate to state and local governments ever increasing permitting authority over activities conducted on federal land.

The Supreme Court held in the case Nollan v. California Coastal Commission[46] that the Fifth Amendment required the California Coastal Commission to pay just compensation to the Nollans for requiring them to grant a public easement across their property. The California Coastal Commission issued a permit to the Nollans authorizing them to replace a bungalow on their beachfront lot with a larger house. The permit contained a condition requiring the Nollans to grant the public an easement to cross their property, located between two public beaches. The Supreme Court, in a 5-4 decision, overturned the California Court of Appeals' decision upholding the permit condition. The Coastal Commission attemped to justify the easement requirement by arguing that it would serve the legitimate public purpose of access to public beachfront by alleviating the "psychological barrier" to use the public beachfront caused by private development.[47] The Supreme Court agreed that the condition would not constitute an unconstitutional taking of private property if it "substantially" advanced legitimate state interests and did not deny the Nollans economically viable use of their lands.[48] The Supreme Court stated, however, that where "the actual conveyance of property is made a condition to the lifting of a land use restriction, the Supreme Court will be "particularly careful" in verifying that the condition substantially advances legitimate state interests because "...[I]n that context there is a heightened risk that the purpose is avoidance of the compensation requirement, rather than the stated police power objective."[49] Analyzing the permit condition in this manner, the Supreme Court found it "impossible to understand" how the condition substantially advanced any legitimate state interest.[50] The Supreme Court concluded by suggesting that California is free to advance the Coastal Commission's program to provide a continuous strip of publicly accessible coastline by using its eminent domain power to purchase easements. However, "the Nollans (and other coastal residents) alone cannot be compelled to contribute to [the program's] realization without awarding just compensation."[51]

At the same time, the Supreme Court overturned the California Appellate Court in an inverse condemnation action.[52] A Lutheran church sought to recover damages for the temporary loss of all use of their property resulting from a county interim ordinance prohibiting all construction in a temporary flood protection area. Although the Supreme Court remanded for trial the question of whether the interim ordinance resulted in an actual taking of the church's property, the church's prayers for relief were somewhat answered by the Chief Justice's landmark decision. The Supreme Court held that land use restrictions may constitute a taking, even for only a temporary period of time, and that the remedy for the taking is monetary compensation based on the Just Compensation Clause of the Fifth and Fourteenth Amendments.[53]

One of the Supreme Court's most significant findings was that temporary regulatory takings, which deny a landowner all use of his property, are no different in kind from permanent takings for which the Constitution clearly requires compensation.[54] The fact that the confiscatory regulation is eventually invalidated, whether by a court or by the regulatory agency, does not amount to adequate compensation to the landowner.[55] The Supreme Court concludes its opinion as follows:

> Here we must assume that the Los Angeles County ordinances have denied appellant all use of its property for a considerable period of years, and we hold that invalidation of the ordinance without payment of fair value for the use of the property during this period of time would be a constitutionally insufficient remedy.[56]

The Supreme Court's recent decisions may be a signal that the Court is looking for a fair balance in the land use game. Although the individual owner still bears a heavy burden of proving an unconstitutional taking, if a taking is established the government must provide a meaningful remedy.

CONCLUSION

In the United States, prospectors and junior mining companies make a significant contribution to the discovery and development of precious metal deposits. Their continued success depends, in part, on the ongoing development of a systematic approach to understanding the permitting and regulatory process and acquiring environmental and land use permits, an essential first step in conducting mineral exploration. By adhering to standard permitting practices and staying informed on changes and recent developments in the law, the prospector should be successful in traversing the exploration permitting maze.

FOOTNOTES

1. California Coastal Commission v. Granite Rock Company, 590 F. Supp. 1361 (W.D. Cal. 1984), rev'd. 768 F.2d 1077, rev'd and rem'd 55 U.S. Law Week 4366 (3/24/87).

2. 1 AM. L. OF MINING §6.02 (2d ed. 1984).

3. 30 U.S.C. §21-42; 17 Stat 91 (1872).

4. Public Land Law Review Commission, One Third of the Nation's Land 22 (1970) (G.P.O.: 1970-0-399-696).

5. Id.

6. Art.IV §3 cl.2.

7. 30 U.S.C. §21-42; 17 Stat. 91 (1872).

8. 30 U.S.C. §21a; 81 Stat. 1876 (1970).

9. 30 U.S.C. §601 et seq.

10. 16 U.S.C. §551.

11. 43 U.S.C. §1701 et seq. (1982 ed. and Supp. III).

12. 16 U.S.C. §1600-1614 (1982 ed. and Supp. III).

13. Id. Note 2 at §6.02.

14. Id.

15. 43 C.F.R. 3809; 36 C.F.R. 228.

16. 43 C.F.R. 3809. 1-2.

17. 36 C.F.R. 228.4(a).

18. Desert Survivors, 96 IBLA 193 (1987)

19. Id.

20. CAL. PUB. RES. CODE §21000-21176 (Deering's 1976).

21. Id. §21100.

22. CAL. PUB. RES. CODE §§2710-2793 (Deering's 1976).

23. CAL. WATER CODE §§13700-13806 (Deering's 1976).

24. CAL. WATER CODE §13750.5 (Supp. 1987).

25. Id. §13751 (Supp. 1987).

26. 5 AM. L. OF MINING §169.01 (2ed. 1984).

27. 4 AM. L. OF MINING §113.01[2] (2ed. 1984).

28. CAL. WATER CODE §100-192 (Deering's 1987); 7c UTAH CODE ANN. §73-3-1 - 73-3-29 (1943); 15 ARIZ. REV. STAT. ANN. §45-141 - 145-154 (1980); 9 MONT. CODE ANN. §85-2-302 (1983); 22 NEV. REV. STAT. §533-030 (1985).

29. The Law of Ground Water - Does Modern Buried Treasure Create a New Breed of Pirates? Moses, 11 ROCKY MTN. MIN. LAW INST. 279 (1974).

30. 4 AM. L. OF MINING §113.02(2)(a)(2ed. 1984).

31. 4 AM. L. OF MINING §113.02(1)(a)(2ed. 1984).

32. Id. and Colo. Sess. Laws 94-108 (Act of Feb. 19, 1879).

33. 19 MONT. CODE ANN. §85-1-204 (1986); 22 NEV. REV. STAT. §533.090; 12 N.M. STAT. ANN. §72-2-1. The Colorado State Engineer has statutory jurisdiction to administer decreed water rights, COLO. REV. STAT. §37-92-501-504 (1973).

34. 4 AM. L. OF MINING 113.02(2)(b)(2ed. 1984).

35. James B. Lee, The Mining Industry as a Water Use and Water Polluter - Legal Aspects, 16 ROCKY MTN. MIN. L. INST. 110 (1971)

36. 4 AM. L. OF MINING 113.02 (4)(a) (2ed. 1984).

37. 15 ARIZ. REV. STAT. ANN. §45-101(a) (1980); 15 COLO. REV. STAT. §37-90-102(1)(1969); 8 IDAHO CODE ANN. §42-226 (1953); KAN. GEN. STAT. ANN. §82a-702 (Supp. 1961); 22 NEV. REV. STAT. §534.020(1)(1931); 12 N.M. STAT. ANN. §72-12-1 (1978); N.D. CENT. CODE §61-01-01 (1963); 7c UTAH CODE ANN. §73-1-1 (1943); WASH. REV. CODE ANN. §90.44.040 (Supp. 1963); MONT. CONST. Article IX, §3.

38. 4 AM. L. OF MINING 113.02(4)(a) (2ed. 1984); 22 NEV. REV. STAT. §534.010-060 (1985).

39. Frank J. Trelease, Water Law, pp. 12-13, 475 et seq. (3ed. 1979).

40. 33 U.S.C. §1251-1376 (1982).

41. 33 U.S.C. §1362(12),(13) & (19) (1902).

42. 33 U.S.C. §1313 (1982); 8 MONT. CODE ANN. §75-5-605(a) (1985); 36 UTAH CODE ANN. §26-11-9; 12 N.M. CODE ANN. §74-6-1 - 74-6-4, §74-6-6 - 74-6-13 (1978); CAL. WATER CODE §13379 (Deering's 1976); 11 COLO. REV. STAT. §25-8-101-25-8-703 (1973); NEV. REV. STAT. §4.45.221 (1973).

43. Safe Drinking Water Act, 42 U.S.C. 300f et seq, as amended by Pub.L. 99-339, 99th Cong., 2nd Sess. (June 19, 1986). See generally 5 AM. LAW MIN. 169 (2ed. 1984); Emerging Policy and Stategy Choices for Protection of the Groundwater Resource, R. Braun and The Clean Water Act, Water Quality; and Water Use, B. Ray. Both papers may be found in the manual for the Univ. of Colo. Natural Resources Law Center, June

1-3, 1987 Program: Water as a Public Resource: Emerging Rights and Obligations.

44. Id. at note 1.

45. "The question presented is merely whether the state can regulate uses rather than prohibit them. Put another way, the state is not seeking to determine basic uses of federal land; rather it is seeking to regulate a given mining use so that it is carried out in a more environmentally and resourceful protective fashion." Id. at Granite Rock, 4370.

46. James Patrick Nollan, et ux., Appellant v. California Coastal Commission, S.Ct. No. 86-133 (June 26, 1987); See also Keystone Bituminous Coal Association v. DeBenedictus 107 S. Ct. 1232 (1987), involved an action by a coal mining association challenging state statutes which prohibited mining of approximately 50% of the coal under houses and publicly-used buildings. The mining companies argued that the state laws acted to confiscate almost 27 million tons of coal. The Supreme Court held that the state statutes did not result in an unconstitutional taking of private property, for two reasons:

1. The statutes substantially advanced the state's legitimate public interest in protecting public safety, and

2. The statutes did not prevent the coal companies from engaging profitably in their business, and did not unduly interfere with their "investment-back expectations." In this regard, the Court found it significant that the 27 million tons of coal represented "only" 1% to 9.4% of each mine's total and noted that "petitioners failed to prove that the statutes made any of their mines unprofitable."

47. Id. Nollan at 8.

48. Id. at 8.

49. Id. at 15.

50. Id. at 12-13.

51. Id. at 4333.

52. First English Evangelical Lutheran Church of Glendale v. County of Los Angeles, California, No. 85-1199, 47 C.C.H. Sup. Ct. Bull. (June 9, 1987).

53. Id. at B3252.

54. Id. at 3251.

55. Id. at B3252.

56. Id. at 3254.

Authors' Index